Monitoring the Comprehensive Nuclear-Test-Ban Treaty: Regional Wave Propagation and Crustal Structure

Edited by
Howard J. Patton
Brian J. Mitchell

Springer Basel AG

Reprint from Pure and Applied Geophysics
(PAGEOPH), Volume 158 (2001), No. 7

Editors:

Howard J. Patton
Los Alamos National Laboratory
P.O.Box 1663, MS C335
Los Alamos, NM 87545
USA
e-mail: patton@lanl.gov

Brian J. Mitchell
Dep. of Earth & Atmospheric Sciences
Saint Louis University
3507 Laclede Avenue
St. Louis, MO 63103
USA
e-mail: mitchbj@eas.slu.edu

A CIP catalogue record for this book is available from the Library of Congress, Washington D.C., USA

Deutsche Bibliothek Cataloging-in-Publication Data

Monitoring the comprehensive nuclear test ban treaty. - Basel ; Boston ; Berlin : Birkhäuser
(Pageoph topical volumes)

Regional wave propagation and crustal structure / ed. by Howard J. Patton ;
Brian J. Mitchell. - 2001
ISBN 978-3-7643-6550-9 ISBN 978-3-0348-8262-0 (eBook)
DOI 10.1007/978-3-0348-8262-0

Originally published by Birkhäuser Verlag, Basel - Boston - Berlin in 2001
Printed on acid-free paper produced from chlorine-free pulp

9 8 7 6 5 4 3 2 1

Contents

Pure appl. geophys. 158 (2001) 1117–1118
0033–4553/01/071117–02 $ 1.50 + 0.20/0

Pure and Applied Geophysics

Monitoring the Comprehensive Nuclear-Test-Ban Treaty

Preface

The first nuclear bomb was detonated in 1945, thus ushering in the nuclear age. A few political leaders quickly saw a need to limit nuclear weapons through international cooperation and the first proposals to do so were made later in that same year. The issue of nuclear testing, however, was not formally addressed until 1958 when the United States, the United Kingdom, and the Soviet Union, initiated talks intended to establish a total ban on that testing (a Comprehensive Test-Ban Treaty or CTBT). Those talks ended unsuccessfully, ostensibly because the participants could not agree on the issue of on-site verification.

Less comprehensive treaties did, however, place constraints on nuclear testing. The United States, the United Kingdom, and the Soviet Union, in 1963, negotiated the Limited Test-Ban Treaty (LTBT) which prohibited nuclear explosions in the atmosphere, outer space and under water. The Threshold Test-Ban Treaty (TTBT), signed by the United States and the Soviet Union in 1974, limited the size, or yield, of explosions permitted in nuclear tests to 150 kilotons.

Seismological observations played an important role in monitoring compliance with those treaties. Many of the world's seismologists set aside other research projects and contributed to that effort. They devised new techniques and made important discoveries about the Earth's properties that enhance our ability to detect nuclear events, to determine their yield, and to distinguish them from earthquakes. Seismologists are rightfully proud of their success in developing methods for monitoring compliance with the LTBT and TTBT.

Although seismologists have also worked for many years on research related to CTBT monitoring, events of recent years have caused them to redouble their efforts in that area. Between 1992 and 1996 Russia, France and the United States all placed moratoria on their nuclear testing, though France did carry out a few tests at the end of this period. In addition, the United States decided to use means other than testing to ensure the safety and reliability of its nuclear arsenal, and all three countries, as well as the United Kingdom, agreed to continue moratoria as long as no other country tested. Those developments, as well as diplomatic efforts by many nations, led to the renewal of multilateral talks on a CTBT that began in January 1994. The talks led to the Comprehensive Nuclear-Test-Ban Treaty. It was adopted by the United Nations General Assembly on 10 September, 1996 and has since been signed by 160 nations. Entry of the treaty into force, however, is still uncertain since it requires ratification by all 44 nations that have some nuclear capability and, as of 30 April 2001, only 31 of those nations have done so.

Although entry of the CTBT into force is still uncertain, seismologists and scientists in related fields, such as radionuclides, have proceeded with new research on issues relevant to monitoring compliance with it. Results of much of that research may be used by the International Monitoring System, headquartered in Vienna, and by several national centers and individual institutions, to monitor compliance with the CTBT. New issues associated with CTBT monitoring in the 21st century have presented scientists with many new challenges. They must now be able to effectively monitor compliance by several countries that have not previously been nuclear powers. Effective monitoring requires that we be able to detect and locate much smaller nuclear events than ever before and to distinguish them from small earthquakes and other types of explosions. We must have those capabilities in regions that are seismically active and geologically complex, and where seismic waves might not propagate efficiently.

Major research issues that have emerged for monitoring a CTBT are the precise location of events, and discrimination between nuclear explosions, earthquakes, and chemical explosions, even when those events are relatively small. These issues further require that we understand how seismic waves propagate in the solid Earth, the oceans and atmosphere, especially in regions that are structurally complex, where waves undergo scattering and, perhaps, a high degree of absorption. In addition, we must understand how processes occurring at the sources of explosions and earthquakes manifest themselves in recordings of ground motion.

Monitoring a CTBT has required, and will continue to require, the best efforts of some of the world's best seismologists. They, with few exceptions, believe that methods and facilities that are currently in place will provide an effective means for monitoring a CTBT. Moreover, they expect that continuing improvements in those methods and facilities will make verification even more effective in the future. This topical series on several aspects of CTBT monitoring is intended to inform readers of the breadth of the CTBT research program, and of the significant progress that has been made toward effectively monitoring compliance with the CTBT.

The following set of papers, edited by Drs. Howard Patton and Brian Mitchell presents research results on regional wave propagation and crustal structure; areas that will be important in monitoring a CTBT. It is the third of eight topics addressed by this important series on Monitoring the Comprehensive Nuclear-Test-Ban Treaty. Previously published topics are Source Location and Hydroacoustics. The topics to appear in later issues are Surface Waves, Source Processes and Explosion Yield Estimation, Infrasound, Data Processing and Source Discrimination.

Brian J. Mitchell
Saint Louis University
Series Editor

Pure appl. geophys. 158 (2001) 1119–1120
0033–4553/01/071119–02 $ 1.50 + 0.20/0

Pure and Applied Geophysics

Introduction

HOWARD J. PATTON[1] and BRIAN J. MITCHELL[2]

Regional phases will be considerably more important in monitoring compliance with a CTBT than they had been for monitoring compliance with earlier Limited Test Ban and Threshold Test Ban Treaties. Since the CTBT is intended to prevent all testing, smaller events than ever before must be detected, located and identified with high confidence. Adding to the challenge of monitoring a CTBT is the Earth's physical complexity, since regional phases will be affected by lateral variations in Earth structure for areas of monitoring concern.

Seismic phases in complex media can be strongly scattered by regional variations in elastic properties. In addition, amplitudes of regional phases attenuate rapidly with distance if they traverse regions that are highly absorptive. Attenuation varies regionally depending on the depth distributions of elastic and anelastic properties along different paths, and the severity of those variations is likely to be proportional to the length of the propagation paths over which the phases are recorded. Thus, optimizing our ability to discriminate between small nuclear explosions and earthquakes will require well-recorded seismic phases recorded over relatively short regional paths, especially in regions with complex and/or highly absorptive crustal structures.

The papers in this volume address various factors that contribute to regional variations in seismic wave travel-times and amplitudes. The first four papers are studies of wave propagation in regions of Eurasia important for CTBT monitoring. *Sandvol et al.* used a large data set to map zones of blockage and inefficient Lg, Pg, and Sn propagation throughout the Middle East. They also develop tomographic techniques to determine regions of lithospheric attenuation. *McNamara and Walter* map crustal heterogeneity throughout the Middle East, the Mediterranean, and continental regions, both north and south of the Mediterranean, using the Lg phase. They image crustal structure anomalies that significantly reduce Lg amplitudes. Both studies show clear relationships of seismic velocity variations to tectonics and crustal structure. *Phillips et al.* applied tomographic methods of Pg/Lg amplitude-ratio data for western China which resolve combinations of relative attenuation, source-

[1] Los Alamos National Laboratory, P.O. Box 1663, MS D408, Los Alamos, NM 87545, U.S.A. E-mail: patton@lanl.gov

[2] Department of Earth and Atmospheric Sciences, Saint Louis University, 3507 Laclede Avenue, St. Louis, MO 63103, U.S.A. E-mail: mitchell@eas.slu.edu

generation, site, and spreading effects. Their maps of relative Pg/Lg attenuation correlate well with previous tomographic results for Lg, Lg-coda and local S-coda data sets, and estimates of relative geometrical spreading agree with numerical wave propagation simulations. They point out that residuals from the inversion show coherent geographical variations that follow structural boundaries. This information seems to reveal small-scale zones of high attenuation or phase blockages, and could be used to reparameterize future tomographic models for this region.

Baumgardt draws on observations of wave propagation in three regions of Eurasia to develop a "basin capture" model for Lg blockage. The observations show that blockages are often associated with propagation paths through large sedimentary basins. Crustal S waves that normally constitute the Lg wave train are trapped, delayed and attenuated by the low-velocity sediments that replace "granite" layers in the upper crust of basin structures. Nevertheless, early-arriving Lg phases are observed in Sn codas, and this is believed to be caused by Sn waves propagating in the mantle far below the basin and later scattered into crustal S waves by lateral heterogeneities on the Moho interface.

Fu and Wu developed a hybrid numerical method for modeling regional wave propagation in which wave fields in regions of severe surface topography are modeled with one method whereas fields in regions of weak heterogeneity are modeled by a less time-consuming method. They apply their methodology to two real crustal waveguides in Tibet, one with, and one without, Lg blockage. They find that strong scattering is the most characteristic effect of irregular topography.

The final two papers describe research that utilizes results of path calibration studies to improve on event discrimination and retrieval of source parameters. *Jenkins et al.* conduct a study that improves our ability to discriminate between explosions and earthquakes by correcting the commonly used S/P discriminant for regional variations in S and P attenuation. They apply their method to more than 1000 Pn, Pg, Sn, and Lg phases from nuclear explosions, chemical explosions, and earthquakes in India and Pakistan. *Saikia et al.* review research that develops methods for improved hypocentral location estimates, with special focus on source depth. Precise depth determination is a powerful discriminant. In addition, the authors demonstrate that depth is a key for better origin time and epicentral location estimates. Various strategies for retreiving focal depth are presented using applications of teleseismic, regional and combined data sets.

This group of papers illustrates the important role that regional phases will play in monitoring a CTBT. Theoretical and numerical studies, in combination with observational results, have improved our knowledge of regional phases, giving them a prominent role in monitoring a CTBT. It should be noted that although regional phases have already contributed to the effectiveness of monitoring a CTBT, they will be even more useful in the future as new stations of the International Monitoring System (IMS) go on line and improve regional coverage in key regions.

Pure appl. geophys. 158 (2001) 1121–1163
0033–4553/01/071121–43 $ 1.50 + 0.20/0

Pure and Applied Geophysics

Tomographic Imaging of *Lg* and *Sn* Propagation in the Middle East

ERIC SANDVOL,[1] KHALED AL-DAMEGH,[1] ALEXANDER CALVERT,[1] DOGAN SEBER,[1] MUAWIA BARAZANGI,[1] RANDA MOHAMAD,[2] RENGIN GÖK,[3] NIYAZI TÜRKELLI[3] and CEMIL GÜRBÜZ[3]

Abstract—Observations based on relatively limited data recorded by sparsely distributed stations have indicated that regional seismic phase propagation (*Lg* and *Sn*) is very complex in the Middle East. Accurate characterization of regional seismic wave propagation in this region necessitates the use of a large number of seismic stations. We have compiled a large data set of regional and local seismograms recorded in the Middle East. This data set comprises approximately four years of data from national short-period networks in Turkey and Syria, data from temporary broadband arrays in Saudi Arabia and the Caspian Sea region, and data from GSN, MEDNET, and GEOFON stations in the Middle East. We have used this data set to decipher the character and pattern of regional seismic wave propagation. We have mapped zones of blockage as well as inefficient and efficient propagation for *Lg*, *Pg*, and *Sn* throughout the Middle East. Two tomographic techniques have been developed in order to objectively determine regions of lithospheric attenuation in the Middle East.

We observe evidence of major increase in *Lg* attenuation, relative to *Pg*, across the Bitlis suture and the Zagros fold and thrust belt, corresponding to the boundary between the Arabian and Eurasian plates. We also observe a zone of inefficient *Sn* propagation along the Dead Sea fault system which coincides with low *Pn* velocities along most of the Dead Sea fault system and with previous observations of poor *Sn* propagation in western Jordan. Our observations indicate that in the northern portion of the Arabian plate (south of the Bitlis suture) there is also a zone of inefficient *Sn* propagation that would not have been predicted from prior measurements of relatively low *Pn* velocities. Mapped high attenuation of *Sn* correlates well with regions of Cenozoic and Holocene basaltic volcanism. These regions of uppermost mantle shear-wave attenuation most probably have anomously hot and possibly thin lithosphere.

Key words: *Lg*, *Sn*, wave propagation, Middle East, tectonics.

[1] Institute for the Study of the Continents, Snee Hall, Cornell University, Ithaca, New York 14853, USA.

[2] Syrian National Seismological Center, General Establishment of Geology and Mineral Resources, Damascus, Syria.

[3] Department of Geophysics, Kandilli Observatory and Earthquake Research Institute, Bogaziçi University, Istanbul, Turkey.

Corresponding Author: e-mail: sandvol@geology.cornell.edu

Introduction

Regional seismograms are dominated by the phases *Pn*, *Pg*, *Sn*, and *Lg*. More often *Sn* and *Lg* are used to infer the attenuation structure of the lithosphere. The seismic phase *Sn* is a high-frequency shear-wave (typically from 1 to 4 Hz and occasionally higher) that travels in the lithospheric mantle above the negative velocity gradient which usually marks the lithosphere-asthenosphere boundary. *Sn* has been reported out to distances of 35° (e.g., MOLNAR and OLIVER, 1969; HUESTIS *et al.*, 1973). *Sn* arrives as a high-frequency wavetrain lasting tens of seconds and up to 1 to 2 minutes. *Sn* velocities are typically 4.7 km/s in stable continental and oceanic lithosphere (HUESTIS *et al.*, 1973) and as low as 4.3 km/s (KADINSKY-CADE *et al.*, 1981) in more tectonically active regions. *Lg* is a complex short period guided wave consisting of high-frequency *P* and *S* energy which travels primarily in the earth's crust at frequencies typically between 0.5 and 5 Hz. It has been modeled as higher-mode Love and Rayleigh waves as well as a sequence of multiply reflected post-critical *S* waves trapped in a crustal guide (BOUCHON, 1982; KENNETT, 1986; BOSTOCK and KENNETT, 1990). *Lg* has been observed not to propagate in oceanic or very thin continental crust (PRESS and EWING, 1952; SEARLE, 1975; ZHANG and LAY, 1995). Both observational and theoretical studies have demonstrated that *Lg* propagation characteristics are primarily a function of the shape of the crustal wave guide (KENNETT, 1986; BOSTOCK and KENNETT, 1990; BAUMGARDT and DER, 1997; RODGERS *et al.*, 1997). There is strong evidence to suggest that rapid changes in crustal thickness will block or weaken *Lg* propagation (BOSTOCK and KENNETT, 1990; FAN and LAY, 1998; MITCHELL *et al.*, 1997). Furthermore, it has been observed that both deep sedimentary basins and intrinsic attenuation can also severely weaken *Lg* propagation (e.g., NUTTLI, 1980; BAUMGARDT and DER, 1997). Recently there have been a number of studies which use *Lg*/*Pg* ratio behavior in terms of crustal waveguide structure (e.g., BAUMGARDT, 1996; FAN and LAY, 1998; HARTSE *et al.*, 1998). This work is critical to isolating path effects from source effects on *Lg*/*Pg* ratios.

Studies of regional wave propagation have yielded important constraints on crustal and uppermost mantle structure and rheology. Studies of *Sn* propagation have helped constrain the rheology of the uppermost mantle in regions such as the Tibetan Plateau where NI and BARAZANGI (1983) and later MCNAMARA *et al.* (1995) documented a region of *Sn* blockage in northern and central Tibet, indicating that this region is probably underlain by thin, hot lithosphere. Similarly, WEIR (1982) found a region in southeastern China where *Sn* was extremely attenuated. Studies of *Lg* propagation have provided valuable insight into crustal structure in the Afar (SEARLE, 1975), Tibet and central Asia (NI and BARAZANGI, 1983; KENNETT, 1986; MITCHELL *et al.*, 1997) and in the Arabian Peninsula (MELLORS *et al.*, 1999).

In the Middle East there are a number of studies of regional wave propagation and its correlation with regional tectonics. Several studies have found that *Lg* coda *Q* values were low (200–500) in tectonically active regions and very low within most of the Anatolian and Iranian plateaus (MITCHELL *et al.*, 1997; CONG and MITCHELL, 1998). KADINSKY-CADE *et al.* (1982) found a continuous zone of *Sn* attenuation extending from the northern Iranian plateau to central and western Turkey as well as strong evidence of *Lg* blockage across the Zagros fold and thrust belt (Fig. 1a). RODGERS *et al.* (1997) extended this original study using several broadband stations in the region as well as data from the Iranian Long Period Array (ILPA) to extend the regions of *Sn* and *Lg* blockage originally proposed by KADINSKY-CADE *et al.* (1982). RODGERS *et al.* (1997) found a relatively good correlation between regions of inefficient *Sn* propagation and low *Pn* velocities as determined by HEARN and NI (1994). Both MELLORS *et al.* (1999) and RODGERS *et al.* (1997) found evidence for *Sn* attenuation in the Gulf of Aqaba. MELLORS *et al.* (1999) identified evidence for efficient *Lg* propagation across the shield although several other studies had found evidence for low *Lg* coda *Q* in the region (e.g., GHALIB, 1987; SEBER and MITCHELL,

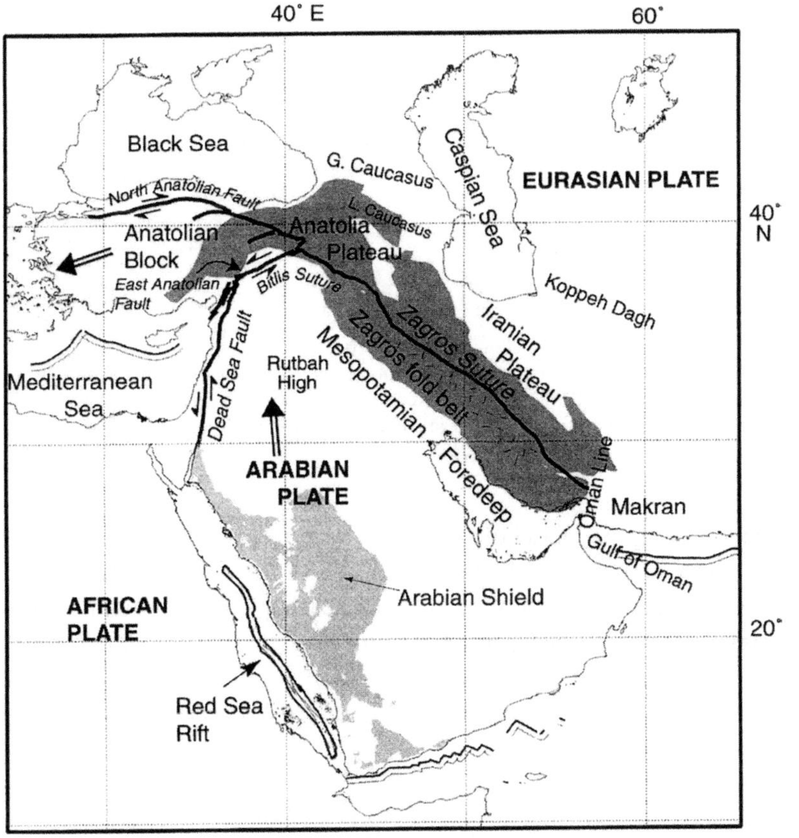

Figure 1a
A simplified tectonic map of the Middle East (after SEBER *et al.*, 1997). Also shown are the major topographic and geographic features. The regions of dark gray correspond to plateau elevations greater than 1500 m. The region of lighter gray corresponds to the Arabian shield.

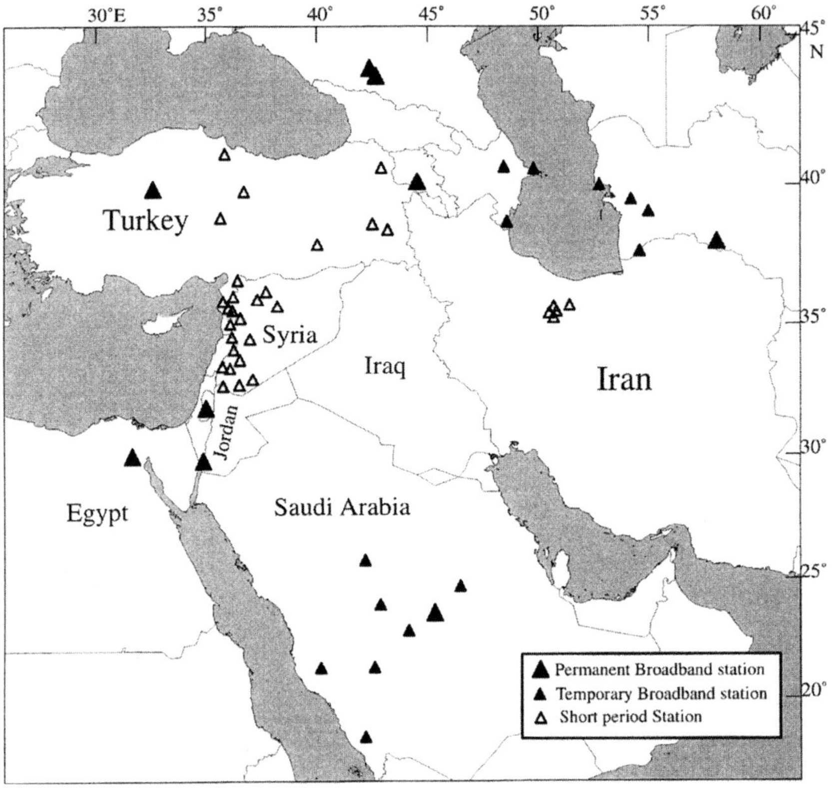

Figure 1b
A map showing all stations and networks used in this study. We have used 65 stations from 6 different networks throughout the Middle East. Open triangles correspond to short-period stations while solid triangles indicate broadband stations. The large solid triangles indicate permanent broadband GSN and GEOFON stations.

1992.). The tomographic model of Cong and Mitchell (1998) has *Lg* coda *Q* values of approximately 450 in the Arabian shield. All of these studies in the Middle East have utilized a relatively sparse station coverage; this leads to potential ambiguity in the interpretation of *Sn* and *Lg* propagation efficiency since most observations were made using very long paths (~1000 km). Because of the relatively dense station coverage used in this study, we have greatly improved on models of regional wave attenuation in the Middle East.

Tectonic Setting

The Middle East is tectonically and seismically active (Fig. 1a). The major topographic features in the region are the Anatolian–Iranian plateau and Zagros mountains which formed as a result of the collision between the Arabian plate and Eurasia. This collision is relatively young, having occurred approximately 12 Ma (DEWEY *et al.*, 1986). The Bitlis suture, the Zagros fold and thrust belt, and the East Anatolian left-lateral strike-slip fault mark a distributed, irregular, and young continental collision zone (e.g., SENGOR and KIDD, 1979). Prior to the continental collision, convergence was accommodated by northward subduction of the Neo-Tethys oceanic lithosphere beneath the southern Eurasian margin (DERCOURT *et al.*, 1986). The Anatolian block is escaping to the west (Fig. 1a), as evidenced by the right-lateral strike-slip movement along the North Anatolian fault system (BARKA and REILINGER, 1997) and with left-lateral movement along the East Anatolian fault system (MCKENZIE, 1972, 1976; JACKSON and MCKENZIE, 1988). Within the Anatolian plateau there is extensive Neogene basaltic volcanism with ages younger than 8 Ma (e.g., INNOCENTI *et al.*, 1982; PEARCE *et al.*, 1990) and the majority younger than 3 Ma. Based on their geochemical signature, all of these basaltic volcanics are inferred to be derived from the lower portion of the lithospheric mantle. PEARCE *et al.* (1990) suggests that these basalts were created through heating and localized delamination of the thickened Anatolian lithosphere and subsequent migration of the magma to the surface. PEARCE *et al.* (1990) also suggests a similar type of convective heating of the lithosphere south of the Bitlis suture where similar young basalts are also abundant. Northeast of the Anatolian plateau lies the Lesser and Greater Caucasus which formed, for the most part, synchronously with the collision of the Arabian and Eurasian plates during the middle to late Pliocene (e.g., PHILLIP *et al.*, 1989). Total shortening in the Caucasus has been estimated to be 200 km (DOTDUYEV, 1986). Current shortening rates have been estimated to be approximately 10 mm/yr (REILINGER *et al.*, 1997).

The only active oceanic lithosphere subduction within the Middle East is occurring in the Makran region of southern Iran. The oceanic lithosphere of the Gulf of Oman is being subducted beneath the Eurasian plate. An active volcanic belt associated with this subduction is located approximately 200 km north of the Iranian coast. The Makran is separated from the Zagros collisional belt by the Oman Line (BERBERIAN and KING, 1981; KADINSKY-CADE and BARAZANGI, 1982), which continues northward along the eastern portion of the Lut block in the central Iranian plateau. Northwest of the Makran and west of the Zagros fold and thrust belt is the Mesopotamian foredeep, consisting of thick sediments which increase in thickness eastward (SEBER *et al.*, 1997).

The western edge of the Arabian plate is bounded by the Dead Sea fault in the north and the Red Sea spreading center in the south (Fig. 1a). The Dead Sea fault system, often referred to as the Dead Sea rift, is a leaky transform and has experienced a total slip of about 105 km since the Miocene (e.g., QUENNELL, 1959; FREUND *et al.*, 1970). The majority of this deformation has occurred in the southern portion of the Dead Sea fault system. However, several portions of the fault system are leaky and have produced pull-apart basins such as the Dead Sea trough and the

Gulf of Aqaba (GARFUNKEL, 1981). GARFUNKEL (1981) suggests that spreading centers located in the middle of the pull-apart basins probably extend into the lithospheric mantle and are indicative of considerable lithospheric thinning.

The Arabian shield comprises a series of accreted Proterozoic island arc terranes which were possibly formed during several subduction episodes (GREENWOOD *et al.*, 1980). The northeastern portion of the Arabian shield is the Afif terrane which is composed of granites (640–580 Ma), volcanic rocks, and meta-sediments (660–600 Ma) overlying a crystalline basement (HUSSEINI, 1988). The volcanic fields in particular are very young; 21 eruptions are known to have occurred in the last 1500 years (CAMP *et al.*, 1987). Along the western margin of Saudi Arabia, there is a young active spreading centered along the axis of the Red Sea. To the east the Shield is bounded by the Phanerozoic sedimentary rocks of the Arabian platform which dip gently eastward and overlap the Shield unconformably (POWERS *et al.*, 1966).

Data

In order to understand regional wave propagation behavior in the Middle East we have collected data from local stations in Turkey and Syria, and combined these with broadband stations in the region as well as data from temporary arrays to form a data set of over 5000 seismograms from 65 stations (Fig. 1b) and 2200 events. This large data set has allowed us to achieve resolution in the Middle East that was not available in earlier studies. We used approximately two years of data from 21 short-period stations produced by the newly installed Syrian National Seismic Network (15 vertical component 1 Hz seismometers and six three component instruments); examples of these data are shown in Figure 2. We also used data from short-period single component (vertical) stations in eastern Turkey that are part of the Kandilli Observatory and Earthquake Research Institute network (Fig. 3), these stations are in a region within the Anatolian plateau that previously had no substantial recordings of seismic waveforms. Data from all available permanent three-component broadband stations from the Global Seismographic Network (GSN), Mediterranean Seismographic Network (MEDNET), and German Global Seismographic Network (GEOFON) are also included in our study. Additionally, data collected from single and three-component short-period data from the Iranian Long Period Array (ILPA) as well as data from temporary broadband field deployments in the Caspian Sea and Saudi Arabia were included in our database of seismograms. The

▶

Figure 2
Examples of seismograms produced by the Syrian National Seismic Network. All seismograms are bandpass filtered with corner frequencies at 1 and 4 Hz. These seismograms demonstrate the inefficient *Sn* propagation in the western Arabian plate as opposed to the efficient *Sn* propagation we observe in the Mediterranean. All three of the recordings are taken from single component short-period seismometers. The gray lines are faults, the solid triangles are seismic stations, and the solid circles are earthquakes.

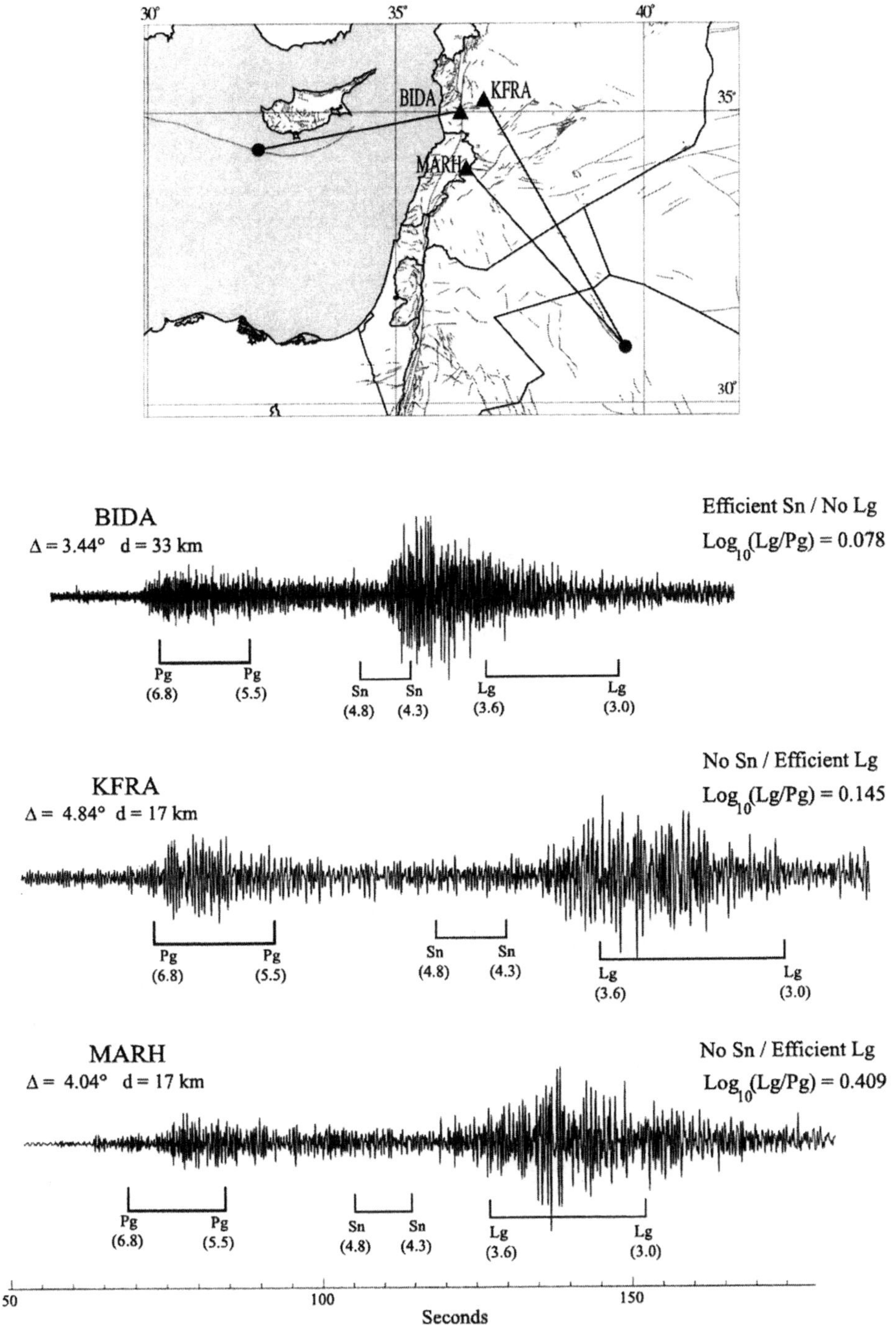
30°
35°
40°
35°
30°
BIDA
KFRA
MARH
BIDA
Δ = 3.44° d = 33 km
Efficient Sn / No Lg
Log10(Lg/Pg) = 0.078
Pg (6.8)
Pg (5.5)
Sn (4.8)
Sn (4.3)
Lg (3.6)
Lg (3.0)
KFRA
Δ = 4.84° d = 17 km
No Sn / Efficient Lg
Log10(Lg/Pg) = 0.145
Pg (6.8)
Pg (5.5)
Sn (4.8)
Sn (4.3)
Lg (3.6)
Lg (3.0)
MARH
Δ = 4.04° d = 17 km
No Sn / Efficient Lg
Log10(Lg/Pg) = 0.409
Pg (6.8)
Pg (5.5)
Sn (4.8)
Sn (4.3)
Lg (3.6)
Lg (3.0)
50
100
150
Seconds

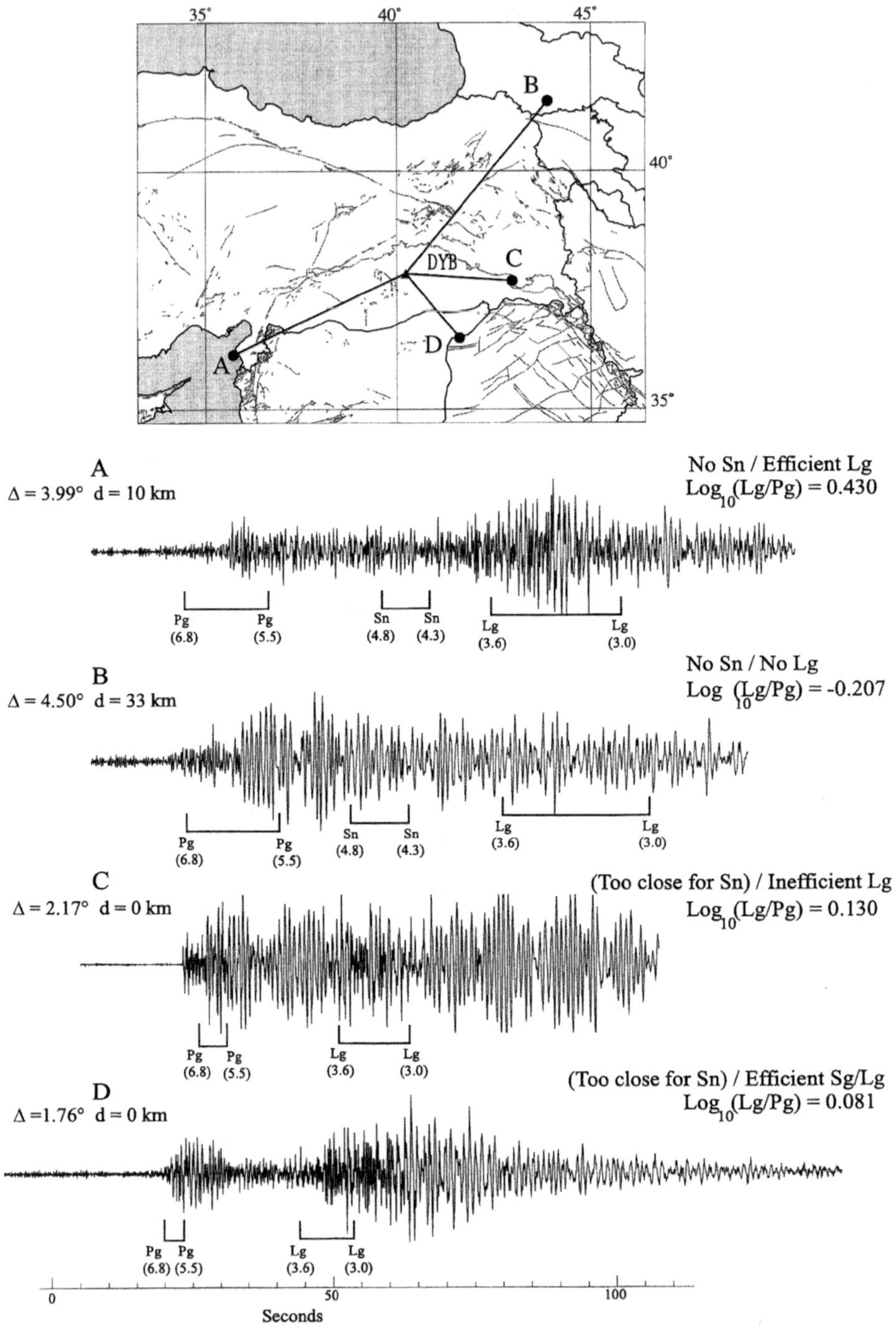
35°
40°
45°
40°
35°
B
C
DYB
D
A
A
Δ = 3.99° d = 10 km
No Sn / Efficient Lg
Log10(Lg/Pg) = 0.430
Pg (6.8)
Pg (5.5)
Sn (4.8)
Sn (4.3)
Lg (3.6)
Lg (3.0)
B
Δ = 4.50° d = 33 km
No Sn / No Lg
Log10(Lg/Pg) = -0.207
Pg (6.8)
Pg (5.5)
Sn (4.8)
Sn (4.3)
Lg (3.6)
Lg (3.0)
C
Δ = 2.17° d = 0 km
(Too close for Sn) / Inefficient Lg
Log10(Lg/Pg) = 0.130
Pg (6.8)
Pg (5.5)
Lg (3.6)
Lg (3.0)
D
Δ =1.76° d = 0 km
(Too close for Sn) / Efficient Sg/Lg
Log10(Lg/Pg) = 0.081
Pg (6.8)
Pg (5.5)
Lg (3.6)
Lg (3.0)
0
50
100
Seconds

Saudi Arabian broadband experiment recorded extremely high-quality broadband regional seismograms starting in November 1995 and terminating in January 1997 (Fig. 4); we collected over 1200 seismograms from this experiment alone. The Caspian Sea Experiment included eight three-component broadband stations both to the east and west of the Caspian Sea (MANGINO and PRIESTLEY, 1998) and recorded approximately eighty seismic events well enough to be included in our study (Fig. 5).

All 5000 seismograms in our regional waveform database have been reviewed by applying a range of bandpass filters in order to identify the relevant regional phases. Those seismograms that were saturated, had low signal-to-noise, or included large amounts of electronic noise were not included in our database. Seismograms with sampling rates of less than 10 sps were also discarded. Distances between 175 km to 2000 km were used for our *Lg* observations and a variety of bandpass filters were applied to test the frequency dependence of this phase. We discarded any *Lg* seismograms for events with hypocentral depths larger than 40 km. In order to identify *Sn* we applied bandpass filters between 1 to 4 Hz and 2 to 5 Hz. For *Sn* we included all seismograms between 330 and 2000 km. We avoided potential confusion with direct *S* arrival by applying filters in travel-time windows that should only include *Sn* rather than *S* at distances past 1500 km. We did identify direct *S*-wave arrivals for many regional seismograms but noticed a distinct difference in the waveform shape and frequency content; therefore, we were able to distinguish between the different arrivals.

Lg Propagation in the Middle East

Lg/*Pg* ratios are generally used in discriminating earthquakes from explosions. In order to determine the path effects on *Lg*/*Pg* amplitude ratios and to characterize *Lg* propagation (relative to *Pg*) we have computed approximately 4400 *Lg*/*Pg* amplitude ratios. We have tomographically mapped the variations in *Lg*/*Pg* ratios throughout the Middle East. For this study, we defined *Lg* as the energy within a group velocity window of 3.0 km/s to 3.6 km/s. For *Pg* we used a group velocity window of 5.5 to 6.8 km/s. We calculated the group velocity window for *Lg* and *Pg* group velocity window relative to the first arrival for all seismograms. This technique has the advantage of eliminating errors caused by network timing errors as well as minimizing effects from possible source mislocations. The amplitude ratios were

◄

Figure 3
Examples of vertical component seismograms recorded by station DYB. All seismograms shown are bandpass filtered with corner frequencies at 1 and 4 Hz. These seismograms show the difference in *Lg* propagation in the Anatolian plateau and in the northern Arabian plate. For events A and B we do not observe any *Sn* signal on any of the three components seismograms. The gray lines are faults, the solid triangles are seismic stations, and the solid circles are earthquakes.

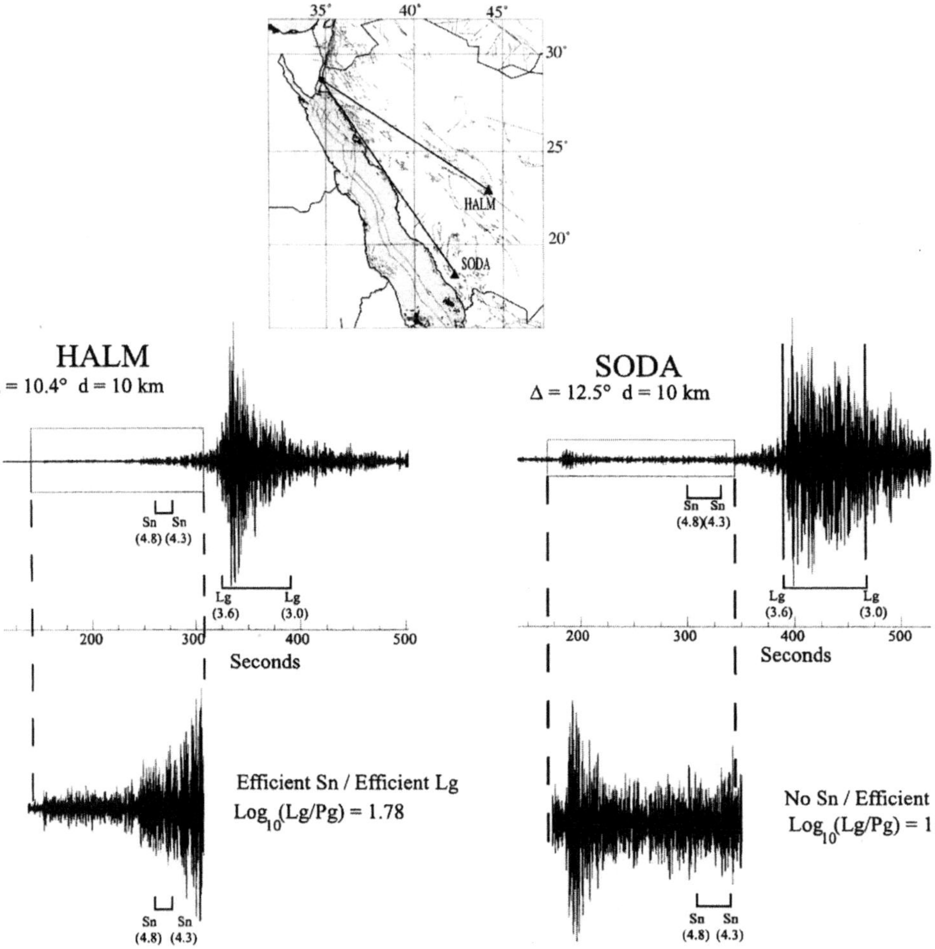

Figure 4
Two examples of vertical broadband seismograms taken from temporary stations in the Arabian shield. The seismograms were filtered with a bandpass filter with corner frequencies at 1 and 4 Hz. These records clearly indicate that *Lg* is an efficient phase across this portion of the Arabian shield. Furthermore, a very clear difference in *Sn* efficiency can be seen for the Red Sea coastal path (station SODA) and the interior Arabian shield path. The gray lines are faults, the solid triangles are seismic stations, and the solid circles are earthquakes.

calculated by taking the logarithm of the ratio of the root-mean-square (RMS) of the amplitudes in the group velocity window in the time domain (*Lg* and *Pg*):

$$D = \text{Log}_{10}\{\text{RMS } [Lg]/\text{RMS } [Pg]\} \ . \tag{1}$$

Maximum absolute amplitude ratios were also used although we found them to be less stable than the RMS amplitude ratios. We tested the sensitivity of our *Lg*/*Pg*

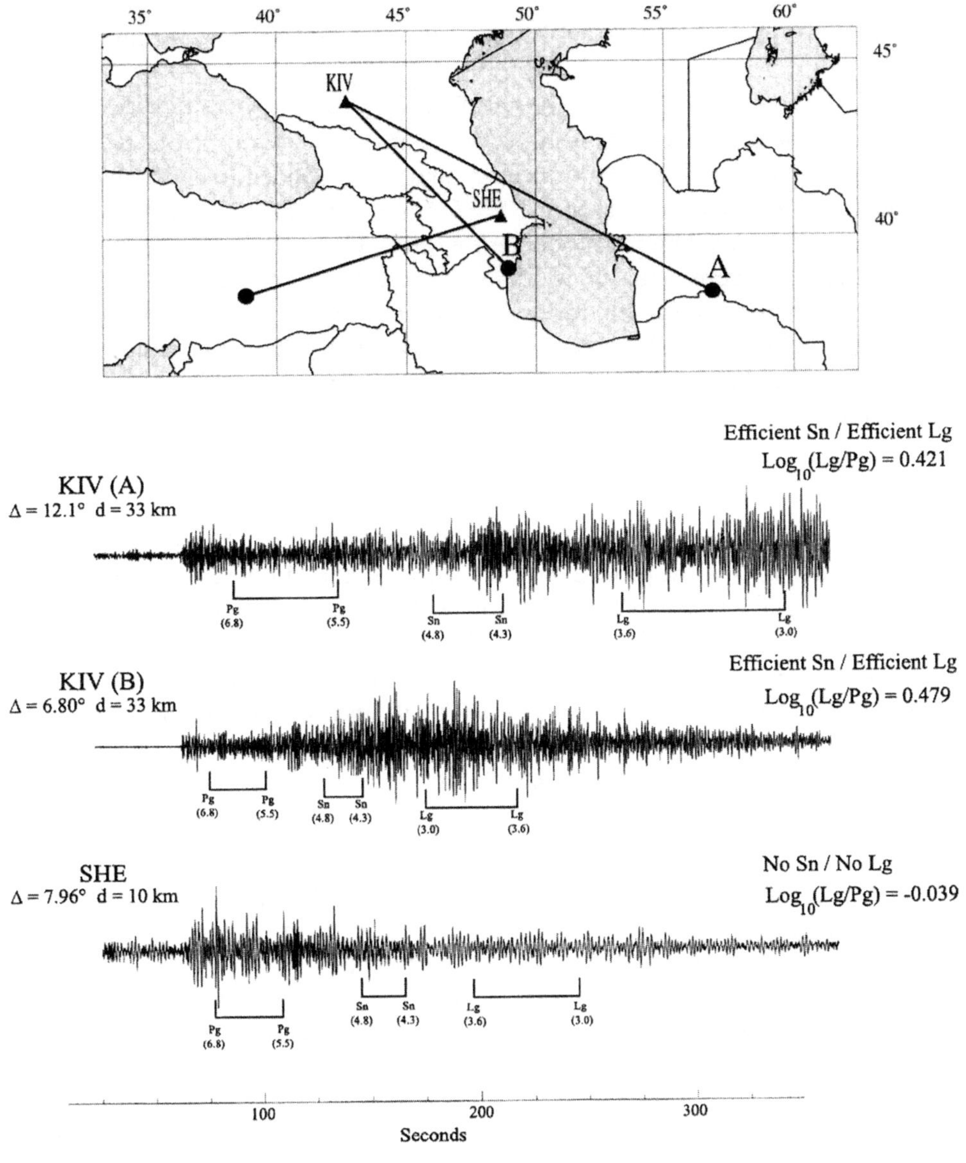

Figure 5

Examples of seismograms taken from the Caspian Sea region. Stations SHE and KIV are three component broadband stations that were a part of the temporary deployment of stations between 1993 and 1996 (MANGINO and PRIESTLEY, 1998). Station KIV is a broadband station which is part of the Global Seismographic Network (GSN). These records clearly indicate the difference in wave propagation across the north Caspian Sea and the southern Caspian Sea, as well as the *Lg* blockage in the easternmost Anatolian plateau. These seismograms also demonstrate the difference in *Sn* propagation across the Caspian Sea (station KIV) as opposed to the paths that cross the Anatolian plateau (station SHE). All seismograms are bandpassed filtered (1–4 Hz). The solid triangles are seismic stations, and the solid circles are earthquakes.

ratios to the *Lg* group velocity window and found that if we vary the group velocity windows by 0.2 km/s the calculated RMS *Lg*/*Pg* ratios did not vary by more than 10 percent.

In order to further verify that the calculated *Lg*/*Pg* ratios are representative of our seismic waveforms, we visually inspected all of the waveforms in our study. Each event-station pair was assigned an efficiency that corresponded to either efficient *Lg*, inefficient *Lg*, or absent or blocked *Lg*. We found a fairly high degree of correlation between our visual inspections and the *Lg*/*Pg* ratios. Generally the *Lg*/*Pg* ratios close to one (Fig. 6) corresponded to weak or inefficient *Lg* propagation while *Lg*/*Pg* ratios of less than one corresponded to blocked *Lg* (Fig. 7). Seismograms with very efficient *Lg* phases usually had *Lg*/*Pg* ratios larger than 5.0 although some seismograms did have anomously high *Pg* amplitudes for paths crossing large sedimentary basins, which decreased the amplitude of the *Lg*/*Pg* ratios significantly.

Method

We have applied tomographic techniques to make more objective and quantitative measurements of the propagation efficiencies in the Middle East. Also, we have a very large data set (>4400 raypaths) in which mapping by visual inspection is difficult (Fig. 8). In order to tomographically map the regional variation of our *Lg*/*Pg* ratios we derived a linear relationship between amplitude ratios and path length. Following the derivation of PHILLIPS *et al.* (1999) and the starting model of CONG *et al.* (1996), we assume that seismic phase amplitudes (a_{ijk}) can be modeled using the following relation for a given frequency band:

$$a_{ijk}(\omega) = L_{ij}^{-\beta_k} s_{jk}(\omega) c_{ik}(\omega) \exp[\alpha_k(\omega) L_{ij}] \ , \tag{2}$$

where β_k is the geometrical spreading exponent for the k-th phase, s_{jk} is the station response for the j-th station and for the k-th phase, c_{ik} is the source scaling term for the i-th seismic source and the k-th phase, L_{ij} is the total raypath length, and α_k is the spatial average attenuation coefficient and is assumed to be constant over a given frequency band. If we take the logarithm of the amplitude ratio of two phases (1 and 2) and then discretize the spatial attenuation factor:

$$\Delta A_{ij} = \Delta S_j + \Delta C_i + \Delta\beta \log_{10} L_{ij} + \log_{10} e \sum_l \Delta m_l l_{ijl} \ , \tag{3}$$

▶

Figure 6
Examples of waveforms recorded by a variety of stations in the Anatolian plateau and the northern Arabian plate. Station ANTO is a three-component broadband GSN station while BNN and DYB are short-period single component stations. Stations QASN and SLMH are also single component short period. All of these data show waveform shapes typical of regional paths crossing the Anatolian plateau (weak *Lg* and little or no *Sn*). All seismograms are bandpassed filtered (1–4 Hz). The gray lines are faults, the solid triangles are seismic stations, and the solid circles are earthquakes.

where

$$\Delta A_{ij} = \log_{10}(a_{ij1}/a_{ij2}), \quad \Delta C_i = \log_{10} c_{i1} - \log_{10} c_{i2} \ ,$$
$$\Delta S_j = \log_{10} s_{j1} - \log_{10} s_{j2}, \quad \text{and} \quad \Delta m = \alpha_2 - \alpha_1 \ .$$

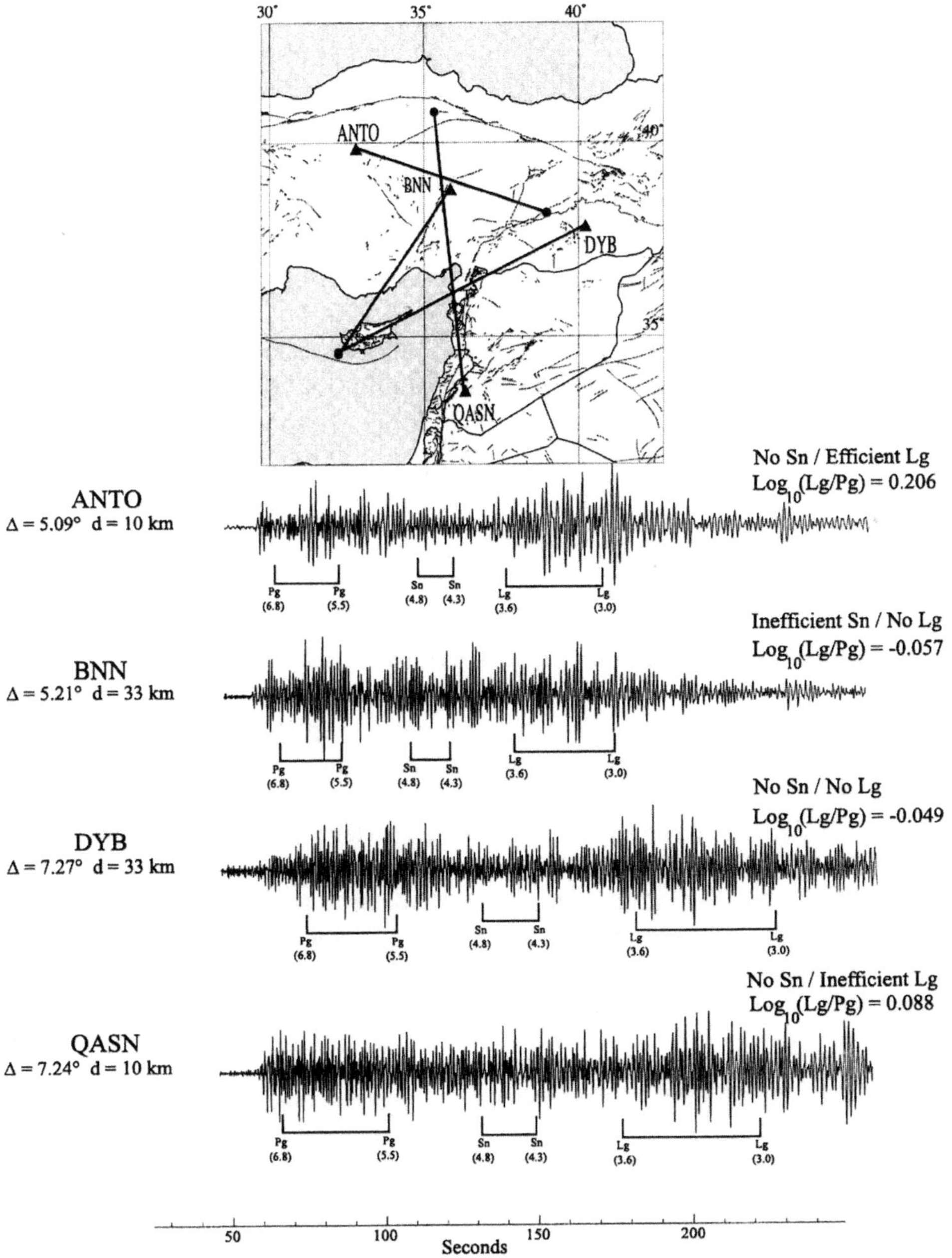

ΔA_{ij} is the logarithm of the phase amplitude ratio, ΔS_j is the logarithm ratio in instrument response for each phase, ΔC_i is the differential source scaling term, $\Delta\beta$ is the differential geometric spreading factor between each phase, Δm_{il} are the model spatial attenuation factors for the l-th cell and i-th ray, and l_{ijl} the length of the ray in the l-th cell. If we assume $s_{j1} = s_{j2}$ and a linear and constant relationship independent of frequency between *Pg* and *Lg* source scaling (i.e., $\Delta C_i = \Delta C$):

$$\Delta A_{ij} = \Delta C + \Delta\beta \log_{10} L_{ij} + \log_{10} e \sum_l \Delta m_l l_{ijl} \ . \tag{4}$$

We would expect ΔS_j to be zero as long as the seismometer response is flat for all stations over the frequency band of interest. In order to linearize this system of equations we expand equation (4) to obtain:

$$\Delta A_{ij} = \Delta C + \Delta\beta \log_{10} L_{ij} + \log_{10} e\, \Delta\mathbf{M} L_{ij} + \log_{10} e \sum_l m_l l_{ijl} \tag{5}$$

where $\Delta\mathbf{M}$ is the mean spatial attenuation parameter and Δm_l are the mean corrected spatial attenutaion parameters. We can then solve for ΔC, $\Delta\beta$, and $\Delta\mathbf{M}$ and subtract this portion of equation (5) from ΔA_{ij}:

$$\Delta \acute{A}_{ij} = \log_{10} e \sum_l \Delta m_l l_{ijl} \tag{6}$$

where

$$\Delta \acute{A}_{ij} = \Delta A_{ij} - \Delta C - \Delta\beta \log_{10} L_{ij} - \log_{10} e\, \Delta\mathbf{M} L_{ij} \ .$$

In practice we found that including $\log_{10} e\, \Delta\mathbf{M}$ made little difference for our data set. Equation (6) is a linear set of equations which relates the spatial change in the difference in *Lg* and *Pg* attenuation to our corrected *Lg*/*Pg* amplitude ratios.

In order to solve the linear system of equations we used the LSQR method of PAIGE and SAUNDERS (1982). This method has the advantage of not introducing any hidden scaling into the solution as well as providing an objective regularization scheme (NOLET, 1988). We spatially regularized our results using Laplacian damping which has been used successfully elsewhere to minimize inversion artifacts (e.g., MEYERS *et al.*, 1998). We tested a variety of Laplacian damping weights and selected one which minimizes tomographic artifacts while maintaining our spatial resolution.

▶

Figure 7
Four seismograms demonstrating the strong *Sn* attenuation occurring between eastern Jordan to northwestern Saudi Arabia. Stations (BDRN and SALA) in western Syria record a very clear *Sn* while stations in the Arabian shield (UQSK and TAIF) record no *Sn*. All seismograms are bandpassed filtered (1–4 Hz). The solid triangles are seismic stations and the solid circles are earthquakes. Note the *Sn* to *Lg* converted phases with group velocities of 3.5 to 4.0 km/s for station UQSK and TAIF.

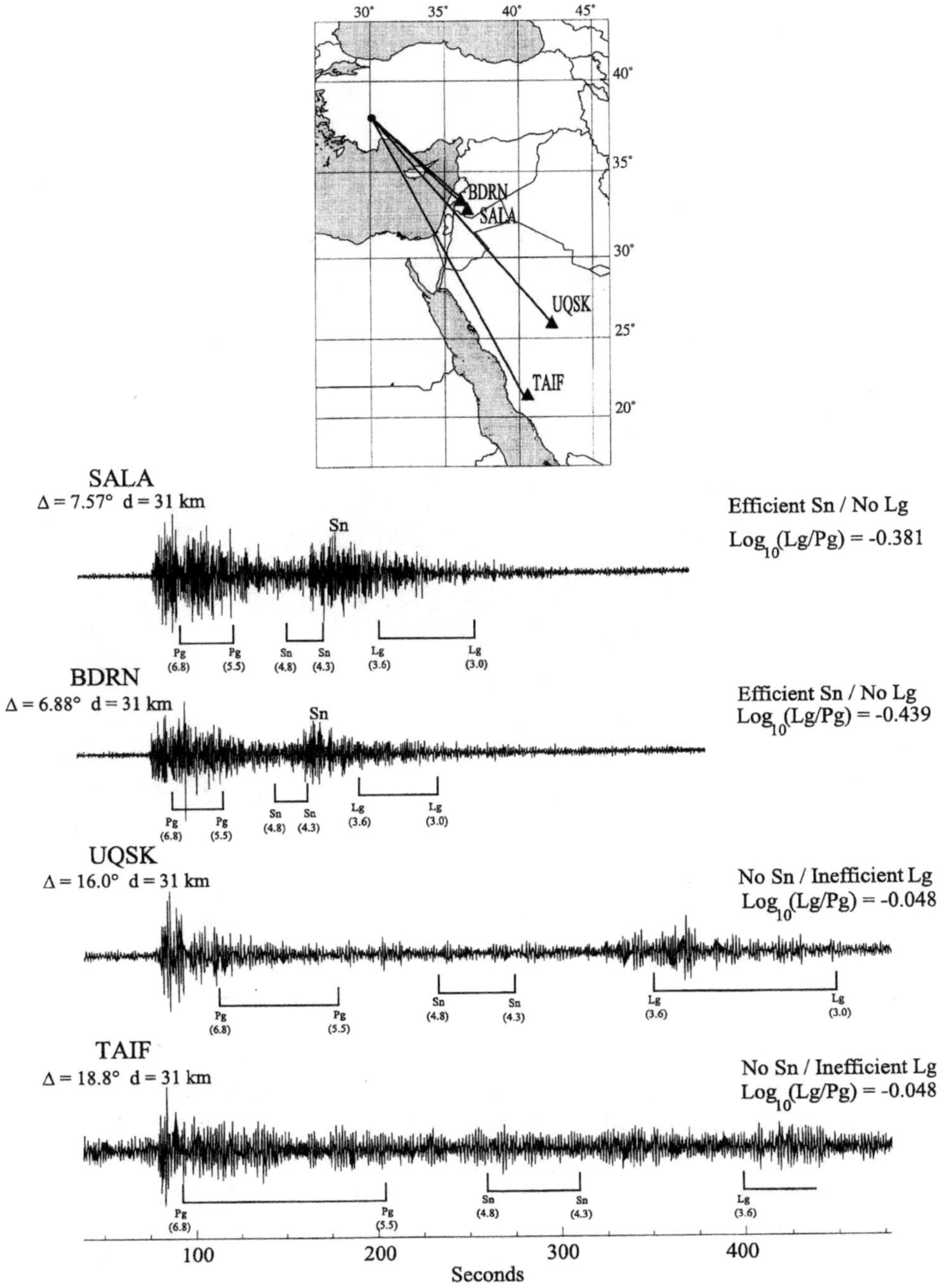

30°
35°
40°
45°
40°
35°
30°
25°
20°
BDRN
SALA
UQSK
TAIF
SALA
Δ = 7.57° d = 31 km
Efficient Sn / No Lg
Log10(Lg/Pg) = -0.381
Sn
Pg (6.8)
Pg (5.5)
Sn (4.8)
Sn (4.3)
Lg (3.6)
Lg (3.0)
BDRN
Δ = 6.88° d = 31 km
Efficient Sn / No Lg
Log10(Lg/Pg) = -0.439
Sn
Pg (6.8)
Pg (5.5)
Sn (4.8)
Sn (4.3)
Lg (3.6)
Lg (3.0)
UQSK
Δ = 16.0° d = 31 km
No Sn / Inefficient Lg
Log10(Lg/Pg) = -0.048
Pg (6.8)
Pg (5.5)
Sn (4.8)
Sn (4.3)
Lg (3.6)
Lg (3.0)
TAIF
Δ = 18.8° d = 31 km
No Sn / Inefficient Lg
Log10(Lg/Pg) = -0.048
Pg (6.8)
Pg (5.5)
Sn (4.8)
Sn (4.3)
Lg (3.6)
100
200
300
400
Seconds

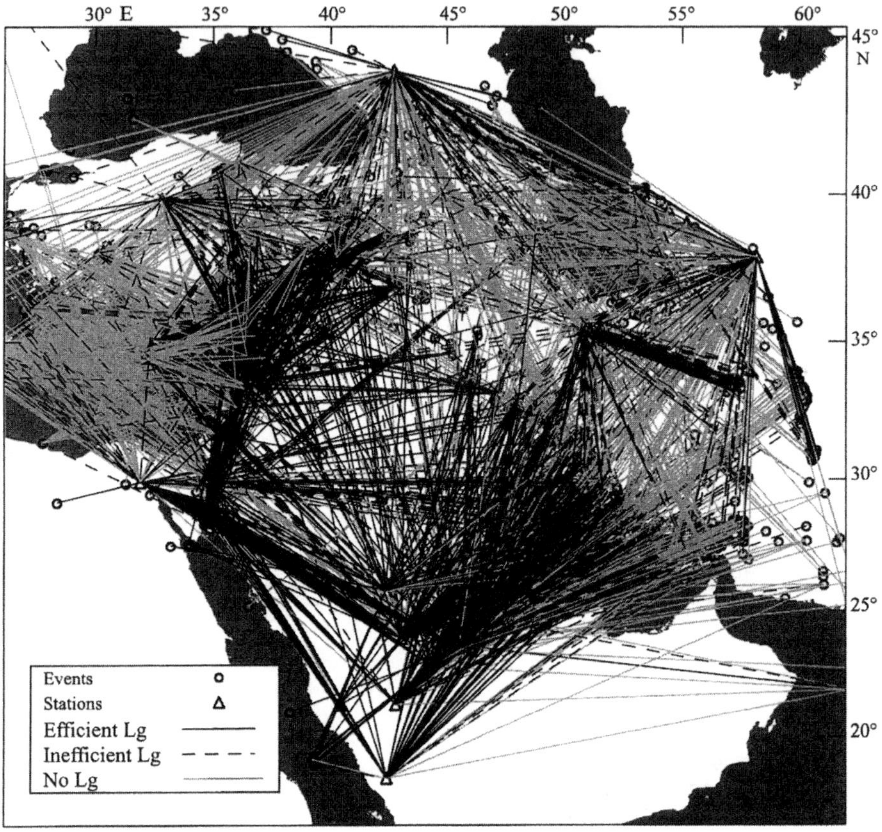

Figure 8
A map showing raypaths for all 4400 event-station pairs used in our *Lg* analysis. The efficiencies shown (efficient, inefficient, blocked) are taken from our visual inspection of each seismogram. Approximately 4200 raypaths are shown. *Lg* attenuation caused by the Iranian plateau, Mediterranean, Caspian and Black Sea can be clearly seen.

We have tested for bias or errors caused by the various instrument and station types by including the ΔS_j term in our inversion (equation (3)). We found that this term did not significantly affect our results although we did observe, for those isolated stations, some path-station corrections trade-off. We found that the attenuation structure near those stations tended to be more "damped" and of smaller amplitude than when the station corrections were not included.

Owing to the difference in geometrical spreading between *Pg* and *Lg*, many studies have shown a strong *Lg*/*Pg* ratio distance dependence (e.g., FAN and LAY, 1998). We have included this in our inversion ($\Delta\beta$) by solving the linear *Lg*/*Pg* ratio dependence on the logarithm of the path length. We found only a weak distance dependence in the Middle East, primarily because of substantial blockage and attenuation that both *Lg* and *Pg* undergo in the Middle East (Fig. 9). The resulting model of spatial attenuation is representative of our observed spatial variation in the *Lg*/*Pg* amplitude ratios in the Middle East neglecting possible source effects (Fig. 10). The amplitude of *Pg*, in particular, could potentially vary with respect to the source mechanism; however, any potential bias in our tomographic model should be lessened by redundant raypaths.

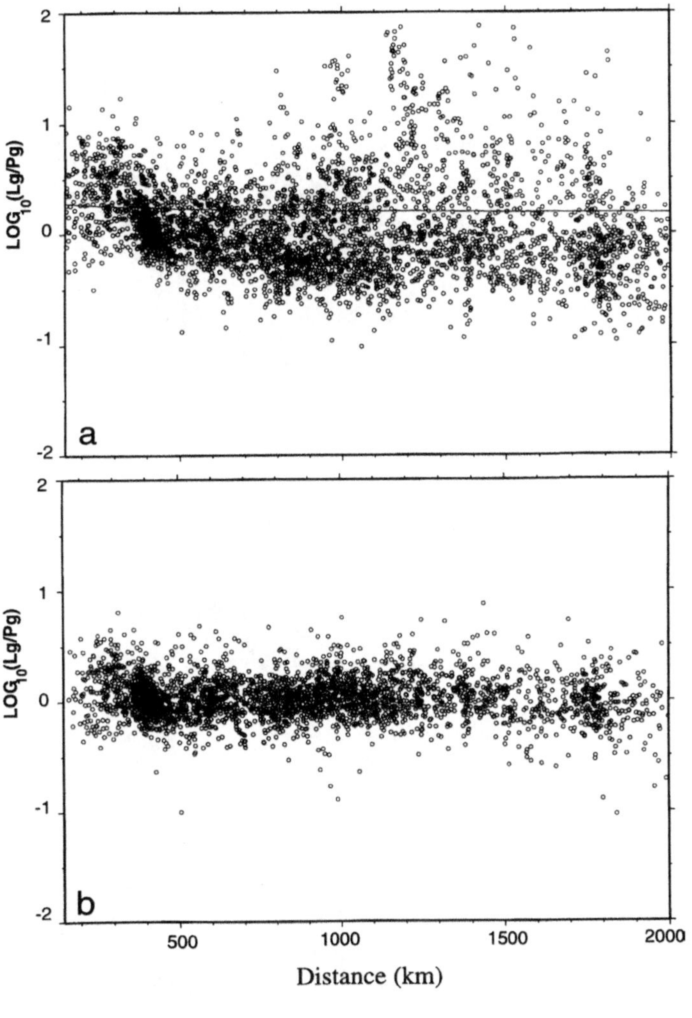

Figure 9
Plots showing all *Lg*/*Pg* amplitude ratios (versus distance) calculated for each ray-path shown in Figure 8. The line in panel (a), is the best fit for the logarithm of the distance. Panel (b) shows the residuals after 12 iterations using a Laplacian damped LSQR algorithm. There was a 50 percent reduction in the variance of the *Lg*/*Pg* ratios using our tomographic model. 95% of all *Lg*/*Pg* ratios fell within ±0.4.

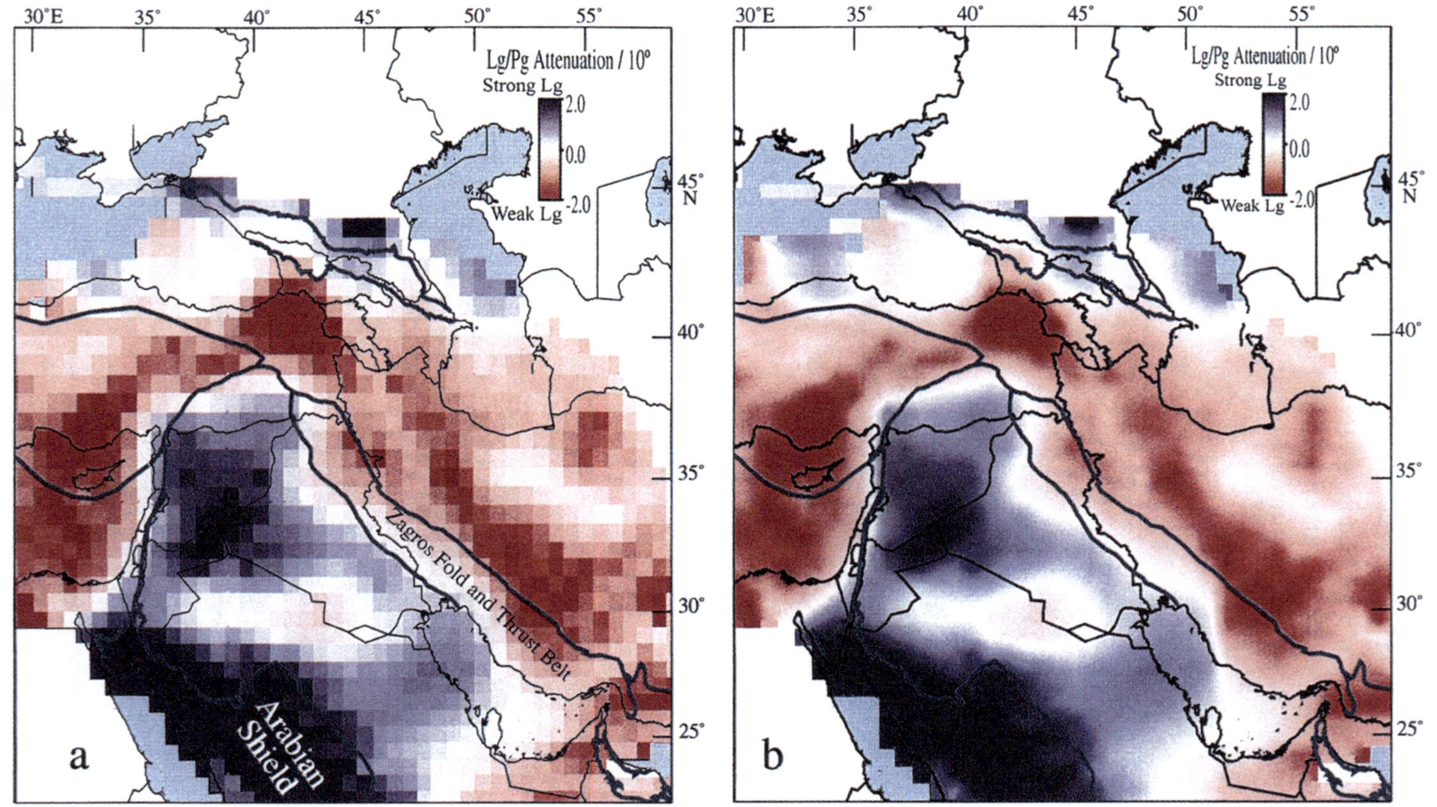

Figure 10

Maps showing Lg/Pg tomographic model for frequencies between 1 and 4 Hz. The clearest feature is the generally efficient Lg propagation within the Arabian plate and the weak or blocked Lg propagation within the Anatolian and Iranian plateaus. Panel (a) shows the raw tomographic model while panel (b) is the linearly interpolated model. In panel (a) and (b) the green lines show the major tectonic boundaries in the Middle East.

Resolution and Uncertainty

In order to test the resolution and stability of our tomographic model we performed several resolution and error tests. We would expect to obtain the highest resolution and most reliable solutions in regions of dense ray coverage. Generally we have good ray coverage, except at the model edges, throughout most of our model (Fig. 11). However, due to lack of stations and sources in Iraq and eastern Jordan we have limited resolution and a less reliable attenuation model for this subregion.

A standard test applied in most tomographic studies is to test model resolution using synthetic checkerboards or spikes to generate a synthetic data set with the same raypaths as the actual data set. We have also added Gaussian noise to the synthetic data set with amplitudes equal to $\pm 15\%$ of the maximum synthetic Lg/Pg ratio. This synthetic data set is then inverted and compared to the original synthetic input model. We employed this technique using a variety of synthetic model geometries

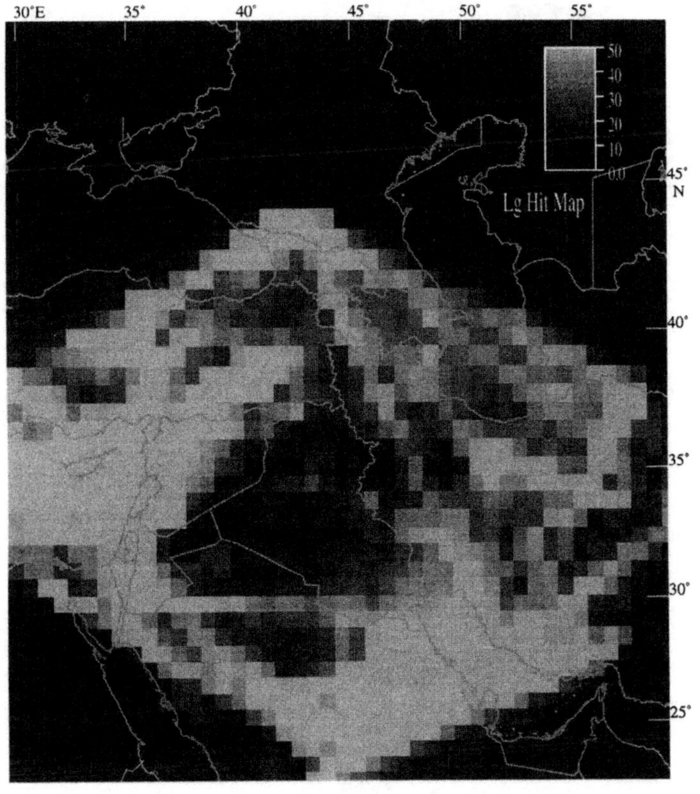

Figure 11
A map showing the hit counts for each cell in the tomographic model shown in Figure 10. Note the smaller hit counts in western and central Iraq. This is a region where the resolution is the poorest and the uncertainties are the largest.

ranging from checkerboards to vertical and horizontal stripes, as well as spike tests. The results of one spike test are shown in Figure 12. We used the same Laplacian damping weights and number of inversion iterations as were used with our original tomographic inversion. Considerable "smearing" can be seen, especially in those regions with limited ray coverage. In particular, dramatic distortion of the synthetic anomalies in Saudi Arabia is occurring due to the multitude of raypaths which travel across the Mediterranean into Saudi Arabia (Fig. 12b). Since there is very limited station coverage in this region and even fewer earthquakes, it is not surprising to see this type of artifact. The Gulf region also has relatively poor resolution. However, we have fairly good resolution in Anatolia, the eastern Mediterranean, and Syria.

We also tested the stability of our tomographic model by using a bootstrap resampling technique. This test should detect any regions of the model which are

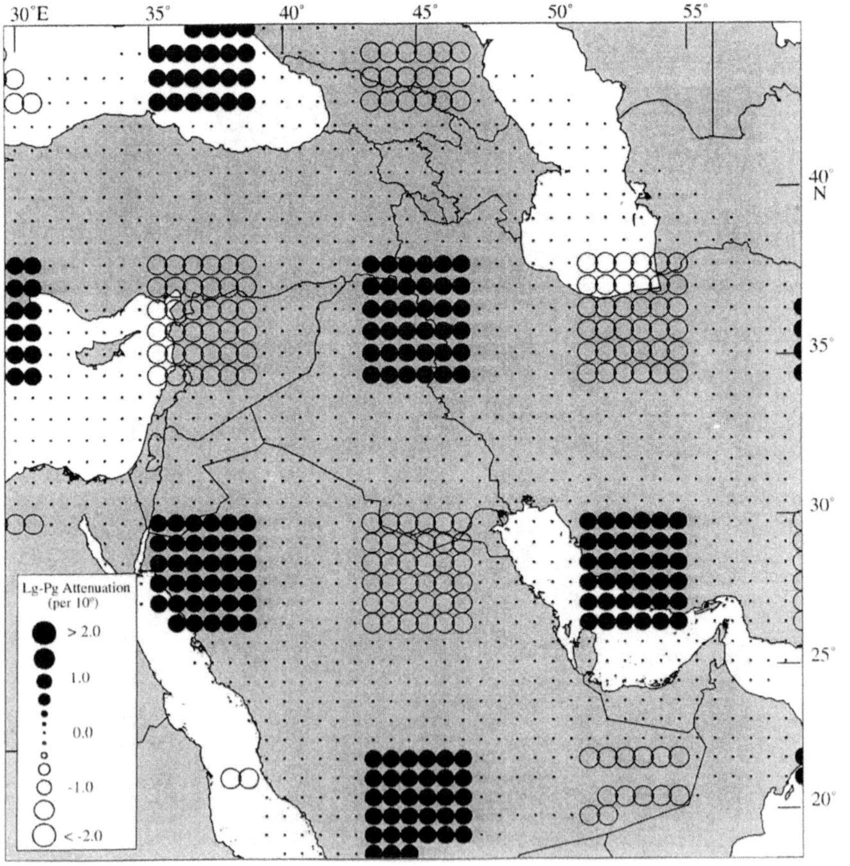

Figure 12a

A map showing one of the test models we have used for determining the resolution of our *Lg-Pg* differential attenuation model. This model contains spikes with alternating positive and negative anomalies of 2×10^{-3}.

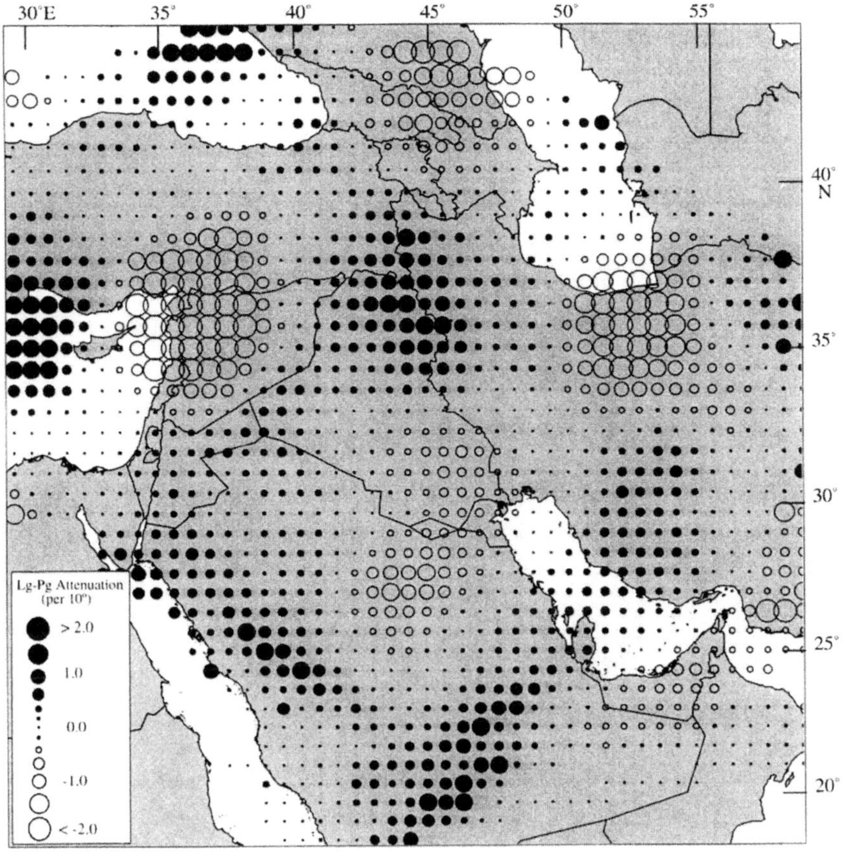

Figure 12b
A map showing the results of our resolution test for the starting model shown in Figure 12a. Note the abundant smearing that occurs in the Arabian/Persian Gulf and in northwestern Saudi Arabia. These are regions where our resolution is the poorest.

dependent on only a few inconsistent observations (HEARN and NI, 1994). We randomly resampled our raypaths until we reached the same number of observations as in our original inversion. This resampled data set is then inverted; this procedure is repeated 100 times, giving us a distribution from which a variance can be calculated. We found that the bootstrap variances converged after approximately 100 bootstrap iterations. These variances are shown in Figure 13. For our *Lg*/*Pg* model we generally have very small uncertainties; only those cells on the edge of the model (e.g., in the Red Sea and easternmost Egypt) exceed variances of more than 0.1. These results also suggest that our data set is not overly sensitive to source focal mechanisms since variations in *Lg*/*Pg* amplitude ratios due to source type should be apparent in our bootstrap uncertainty estimates as long as the distribution of source focal mechanisms is not entirely systematic.

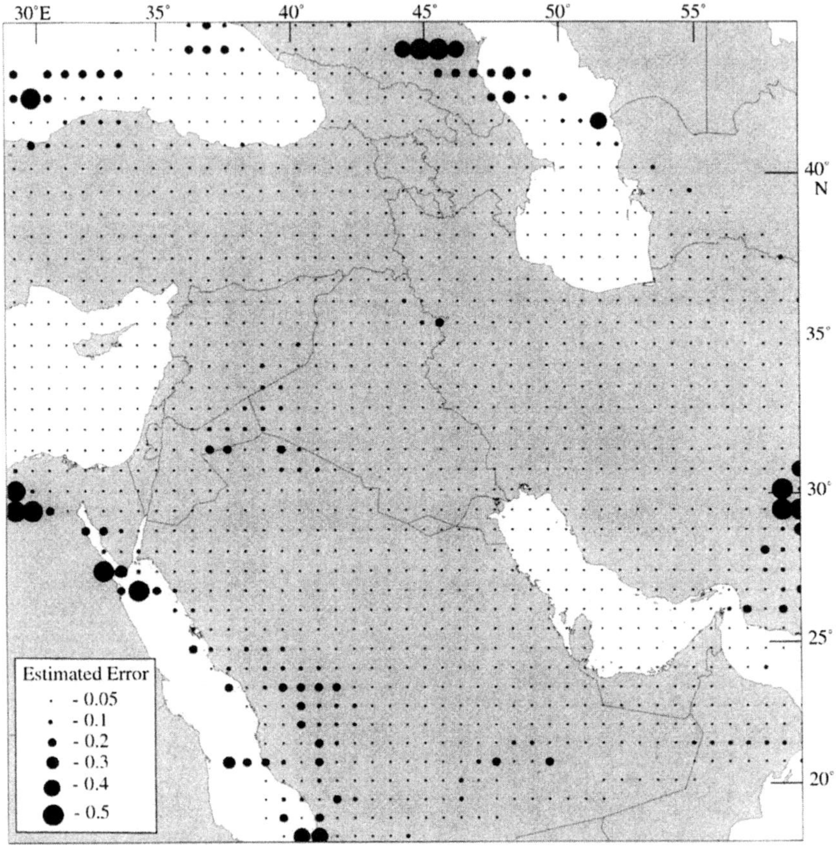

Figure 13
A map showing the bootstrap uncertainty estimates for the *Lg*/*Pg* tomographic model shown in Figure 10. These uncertainty estimates were calculated after 100 bootstrap iterations. Generally our solution was very stable. Only on the edge of our model do we find uncertainties significantly larger than 0.10.

Results

The most striking result of our *Lg*/*Pg* tomography is that regions of small *Lg*/*Pg* ratios mainly correspond with continental collision zones in the Middle East (Figs. 1 and 10). The regions near the Bitlis suture and the Zagros fold and thrust belt are both delineated very clearly in our model as zones of *Lg* blockage. Within both the Iranian and Anatolian plateaus, *Lg* propagation with respect to *Pg* is generally inefficient or weak. We do observe some inefficient *Lg* phases within the central Iranian plateau. Within the entire Anatolian plateau, however, *Lg* is inefficient. In the Lesser Caucasus and the extreme western part of the greater Caucasus there is a large region where *Lg* is either blocked or inefficient (Figs. 5 and 10). Within the south Caspian Sea basin *Lg* is also blocked; however, the northern Caspian Sea appears to have fairly efficient *Lg* propagation.

We also observe large regions of very efficient *Lg* propagation. Within the Arabian shield we observe very efficient *Lg* propagation (Figs. 4 and 10) which is consistent with the observations of MELLORS *et al.* (1999) and with the *Lg* coda *Q* models of CONG and MITCHELL (1998). Our tomographic image very clearly outlines the Arabian shield where we have very large (~2.0) *Lg*-*Pg* differential spatial attenuation while we have more modest differential attenuation (~0.5) in the Arabian platform and Mesopotamian foredeep. Another large region of very efficient *Lg* propagation is an area in western Iraq and eastern Jordan. This anomaly, however, is somewhat suspect since this is a region of relatively poor ray coverage (Fig. 13). We also found that the *Lg* to *Sn* converted phases, such as those shown in Figure 7, did not have a major impact upon our tomographic model. This is due primarily to the small amplitudes of such converted phases as well as the numerous stations and events that we have used in our inversion.

Another important issue regarding *Lg* propagation is the frequency dependence of the *Lg*/*Pg* ratios. Different crustal structure and rheologies can have different responses to different seismic frequencies. In order to investigate the frequency dependence of *Lg*/*Pg* ratios in the Middle East we applied a number of narrow bandpass filters to our seismogram database and tomographically mapped the resulting *Lg*-*Pg* differential attenuation. Figure 14a shows the same image shown in Figure 10 for reference. This image corresponds to a frequency band of 1 to 4 Hz. Figures 14b and 14c show progressively higher frequency bands. In general there is no appreciable change in the models until the frequency band of 2 to 4 Hz. Many of the larger model attenuation anomalies, seen in Figures 14a, 14b, and 14c, have been reduced considerably. There is also more *Lg* attenuation and less *Pg* attenuation in the Mesopotamian foredeep and less in the Lesser Caucasus than in the other lower frequency models.

We obtained a 50 percent variation reduction in our ratio residuals (Fig. 9) for the frequency band between 1 and 4 Hz. We obtained similar variance reductions for the other frequency bands that we analyzed, except for the 2–4 Hz band which had a variance reduction of 48 percent. This smaller variance reduction is due to the poor data quality (low signal-to-noise ratio) for this frequency band. After applying path corrections obtained from our model, the resulting 95% of our corrected residual fell within ± 0.4 of zero (Fig. 9), indicating that we have successfully removed a large portion of the path effects for *Lg*/*Pg* ratios in the Middle East.

Sn Propagation in the Middle East

In order to map *Sn* attenuation in the Middle East we reviewed and determined the *Sn* propagation efficiencies into three categories for over 4200 seismograms. The

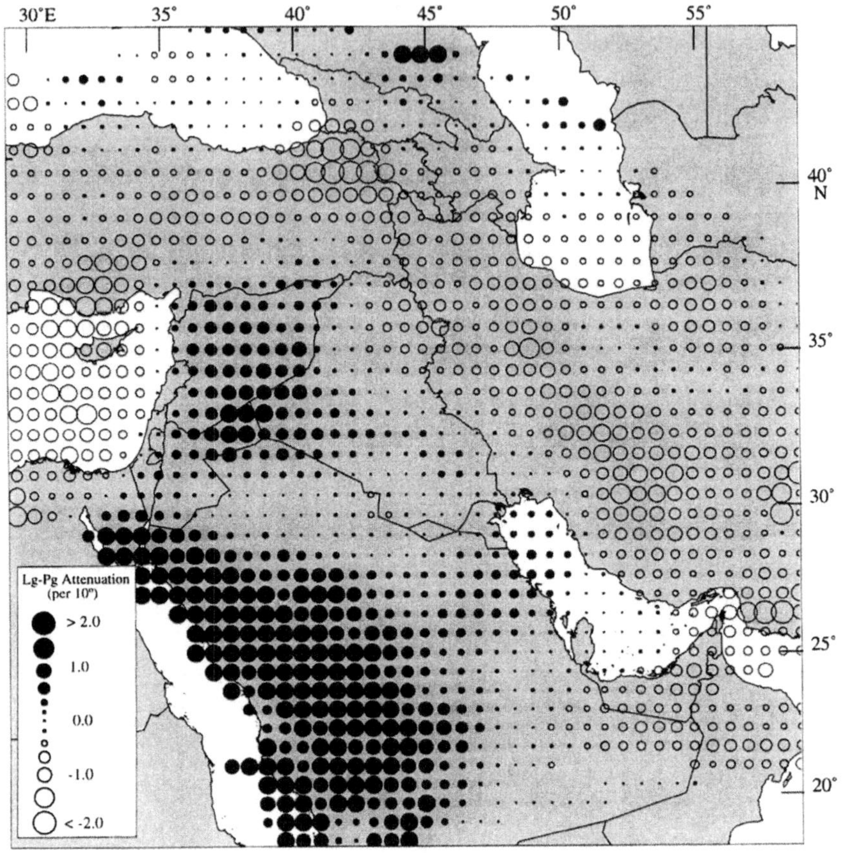

Figure 14a
A map showing a tomographic model of *Lg*/*Pg* amplitudes for the 1 to 4 Hz frequency band. This is the same model as that shown in Figure 10.

results are shown in Figure 15. These results reveal generally efficient *Sn* propagation in oceanic lithosphere and generally inefficient *Sn* propagation in the Anatolian and Iranian plateaus. However, in order to map the details of *Sn* propagation in the Middle East we have again used tomography. Unlike our study of *Lg*, because of the instability of the *Sn*/*Pn* ratios as well as the subtle nature of continental *Sn* in tectonically active regions, we chose to tomographically map qualitatively assigned *Sn* propagation efficiencies rather than amplitude ratios. This strategy allowed us to assign equal weights to the relatively weak *Sn* signals observed in the continents and the very strong *Sn* observed for oceanic paths because we are primarily interested in regions of *Sn* blockage and not the differences in *Sn* propagation in oceanic and continental lithosphere.

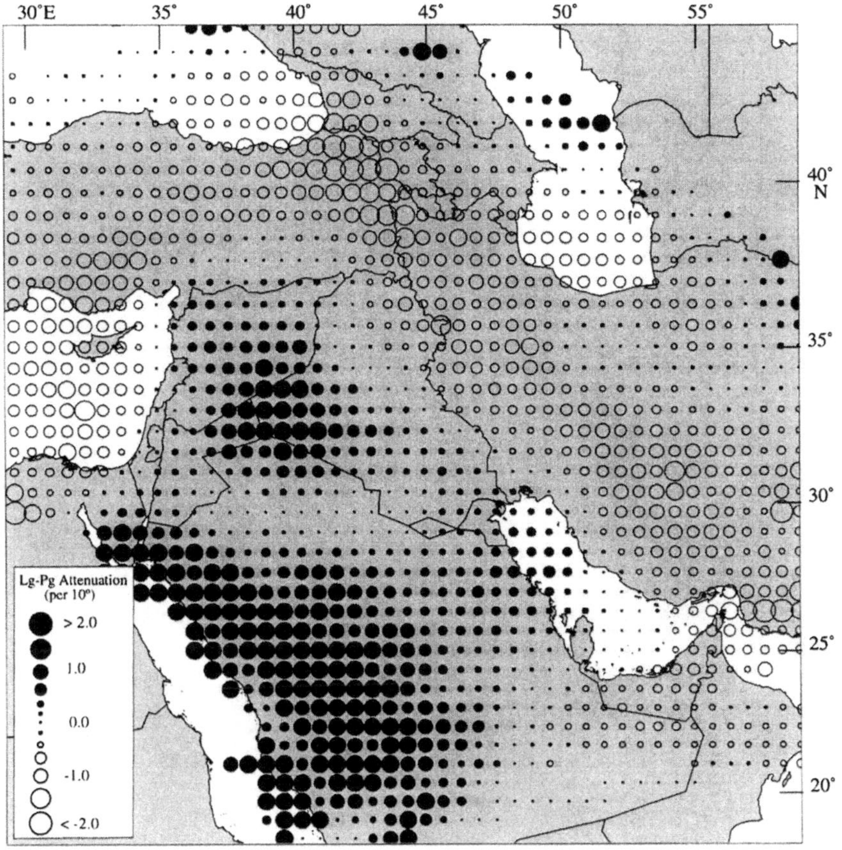

Figure 14b
A map showing a tomographic model of *Lg*/*Pg* amplitudes for the 0.5 to 2 Hz frequency band. This model is similar to the model shown in Figure 14a indicating the small amount of energy contained in the 2 to 4 Hz band. This also indicates that the short-period instrument responses have no significant effect on our *Lg*/*Pg* ratio calculations.

We assigned each seismogram to one of three categories: no *Sn*, efficient *Sn*, and weak or inefficient *Sn*. Although these categories are qualitative, we used strict criteria when evaluating each of the regional seismograms. If the seismogram showed no evidence of a discernable high-frequency *Sn* wavetrain, we categorized that path as "no *Sn* present." If some *Sn* wavetrain could be observed, regardless of its strength or amplitude, we designated it "efficient *Sn*." If there were some ambiguous signal in the seismogram that potentially could be an *Sn* signal, it was classified as "weak or inefficient *Sn*." This category was seldom used since we could usually clearly delineate, with the use of bandpass filters and spectrogram analysis, between those seismograms with *Sn* and those which did not contain an *Sn* signal. We tended to use the "inefficient *Sn*" category for those seismograms at shorter distances where regional phases cannot always be easily identified, especially for continental paths.

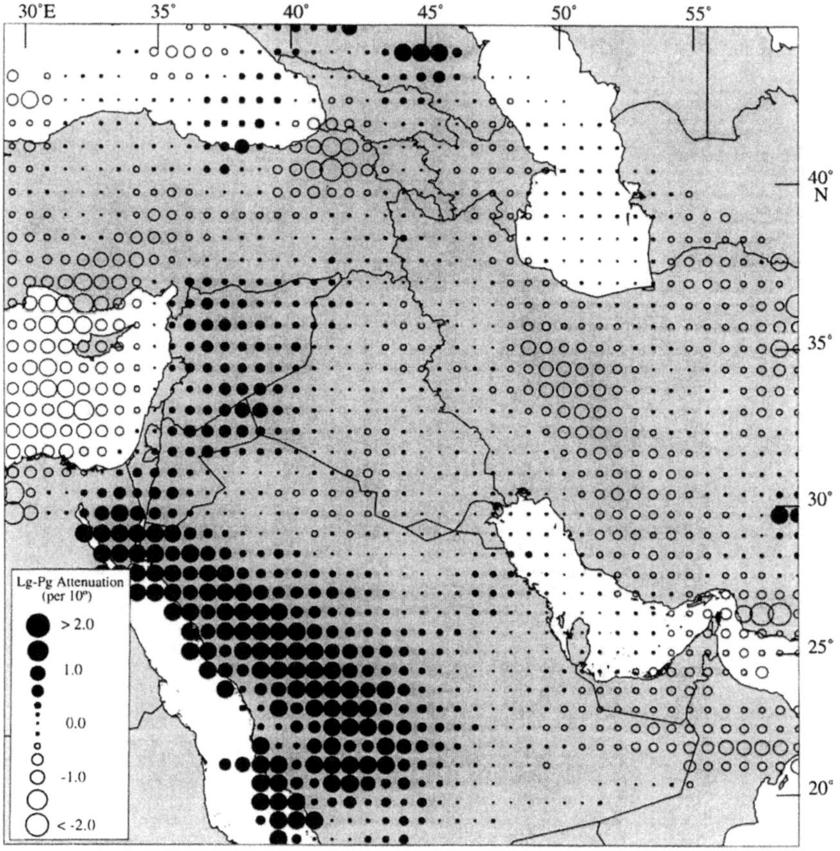

Figure 14c
A map showing a tomographic model of Lg/Pg amplitudes for the 2 to 4 Hz frequency band. This model is different than Figure 14a. We begin to see Lg attenuation in eastern Iraq and all along the deep sediments that correspond to the Mesopotamian foredeep.

Sn Efficiency Tomography Method

The method we used to tomographically image Sn propagation is similar to that used in our Lg/Pg study. However, we have chosen not to use amplitude ratios since we have found them to be primarily a function of whether the path was oceanic or continental. This is because continental Sn is generally a subtle phase which is not easily measured using amplitude ratios. Taking the logarithm of equation (2), neglecting the geometrical spreading term, discretizing the spatial attenuation factor, and substituting $\pi f m_l / V_{Sn}$ for α we obtain:

$$\log_{10}\left(\frac{a_{ij}^0}{a_{ij}}\right) = \log_{10} e \frac{\pi f}{V_{Sn}} \sum_l m_l l_{ijl} \quad (7)$$

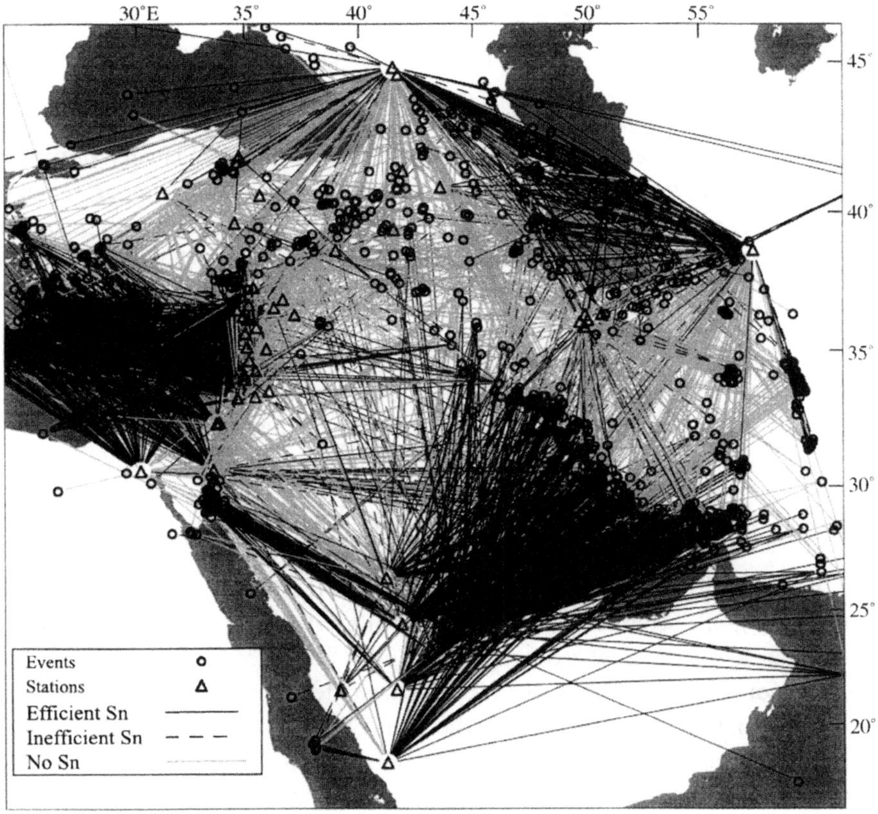

Figure 15
A map showing all 4200 *Sn* raypaths used in this study. The waveform for each raypath has been visually inspected using a variety of bandpass filters to identify if *Sn* is present or not. *Sn* attenuation for most continental paths in the northern Middle East is very clear.

where

$$a_{ij}^0 = s_j(\omega)c_i(\omega)$$

where f is the frequency of the *Sn* signal (1 Hz), V_{Sn} is the velocity of *Sn* (4.6 km/s) and all variables and indices are the same as was used in equations (2) through (5). If we assume, to first order, that the effect of geometrical spreading will not play a major role in *Sn* blockage, we can discretize the data vector and use discrete wave propagation efficiencies rather than amplitudes or amplitude ratios.

$$\hat{A}_{ij}^{dsc} = \log_{10} e \frac{\pi f}{V_{Sn}} \sum_l m_l l_{jjl} \tag{8}$$

where

$$\hat{A}_{ij}^{dsc} = 1, 0.5, \quad \text{or} \quad 0.0 \ .$$

We assigned the propagation efficiencies as follows: 1 for no or blocked *Sn*, 0.5 for inefficient or weak *Sn*, and 0.0 for an efficient *Sn*. Equation (8) is very similar to the method of SAUNDERS *et al.* (1988), who used local analog seismograms to map regions of high attenuation. By choosing these three numbers for our propagation efficiencies we have defined both our model parameter as the reciprocal of the extinction path length and the necessary, albeit idealized, attenuation that defines blocked, inefficient and efficient *Sn*.

For

$$\hat{A}_{ij}^{dsc} = 1\colon \ a_{ij} = 0.1 \ a_{ij}^{0} \ (90\% \text{ attenuation})$$
$$\hat{A}_{ij}^{dsc} = 0.5\colon \ a_{ij} = 0.3 \ a_{ij}^{0} \ (70\% \text{ attenuation})$$
$$\hat{A}_{ij}^{dsc} = 0.0\colon \ a_{ij} = a_{ij}^{0} \ (0\% \text{ attenuation}) \ .$$

Using these definitions of efficient, inefficient and blocked *Sn*, our model parameters in equation (8), m_l, becomes the reciprocal of the extinction path length. Because we have created a set of idealized criteria defining our propagation efficiencies, it is important not to overinterpret our model parameters. In practice, because we have chosen to define efficient *Sn* as an indication of no attenuation at all, our results represent in effect the maximum extinction path length for *Sn*. We have allowed our model parameters to fall below zero if the data require it. We have interpreted the negative extinction path lengths as an indication that our assumption of no attenuation or very small attenuation for efficient *Sn* paths is incorrect for those regions. For the most part, only a few cells decreased significantly beneath zero.

Solving this linear set of equations (equation (8)) allows us to objectively and quantitatively map our qualitative *Sn* efficiencies (Fig. 16). We chose our Laplacian damping parameter using the same technique we used for our *Lg*/*Pg* tomography. In order to minimize the inversion artifacts for our *Sn* tomography we required larger weighting for the Laplacian damping. This is in part due to the discrete nature of our database as well as the longer path lengths that were required for our *Sn* observations although generally we had adequate ray coverage throughout most of our model (Fig. 17).

▶

Figure 16

Maps showing the *Sn* efficiency tomographic model for all the rays shown in Figure 15. Blue indicates *Sn* spatial attenuation of 5×10^{-4} km^{-1} or lower (efficient *Sn*) while red indicates a spatial attenuation of 1.5×10^{-3} km^{-1} or greater (no *Sn*). White color corresponds to regions of inefficient *Sn* propagation or *Sn* attenuation that only occurs for path lengths of approximately 1000 km or greater. The thick green lines show the major tectonic boundaries in the Middle East.

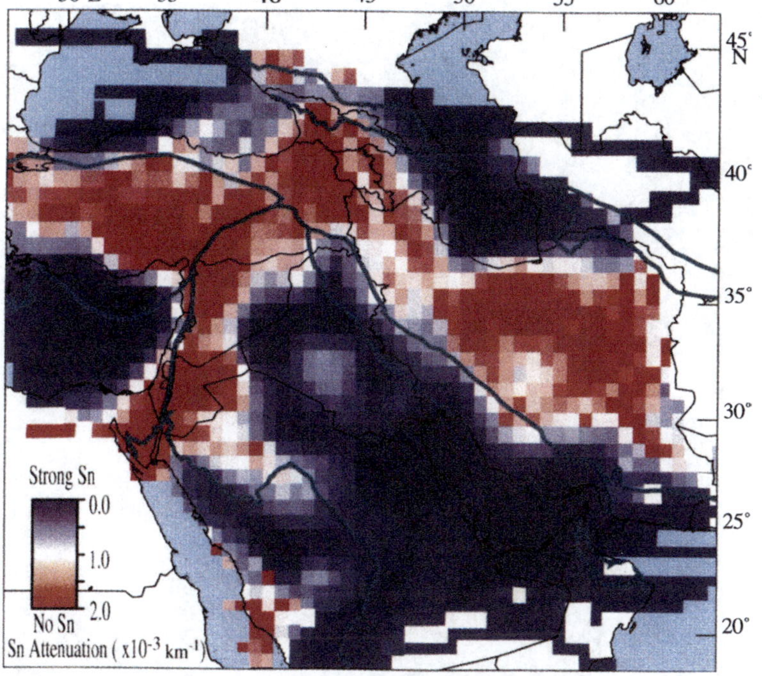
30°E
35°
40°
45°
50°
55°
60°
45°
N
40°
35°
30°
25°
20°
Strong Sn
0.0
1.0
2.0
No Sn
Sn Attenuation (x10-3 km-1)

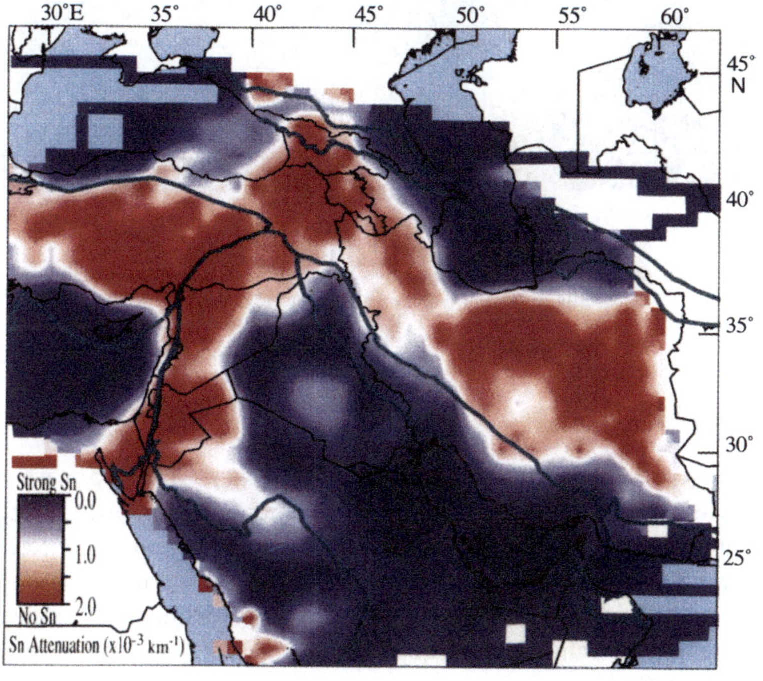
30°E
35°
40°
45°
50°
55°
60°
45°
N
40°
35°
30°
25°
Strong Sn
0.0
1.0
2.0
No Sn
Sn Attenuation (x10-3 km-1)

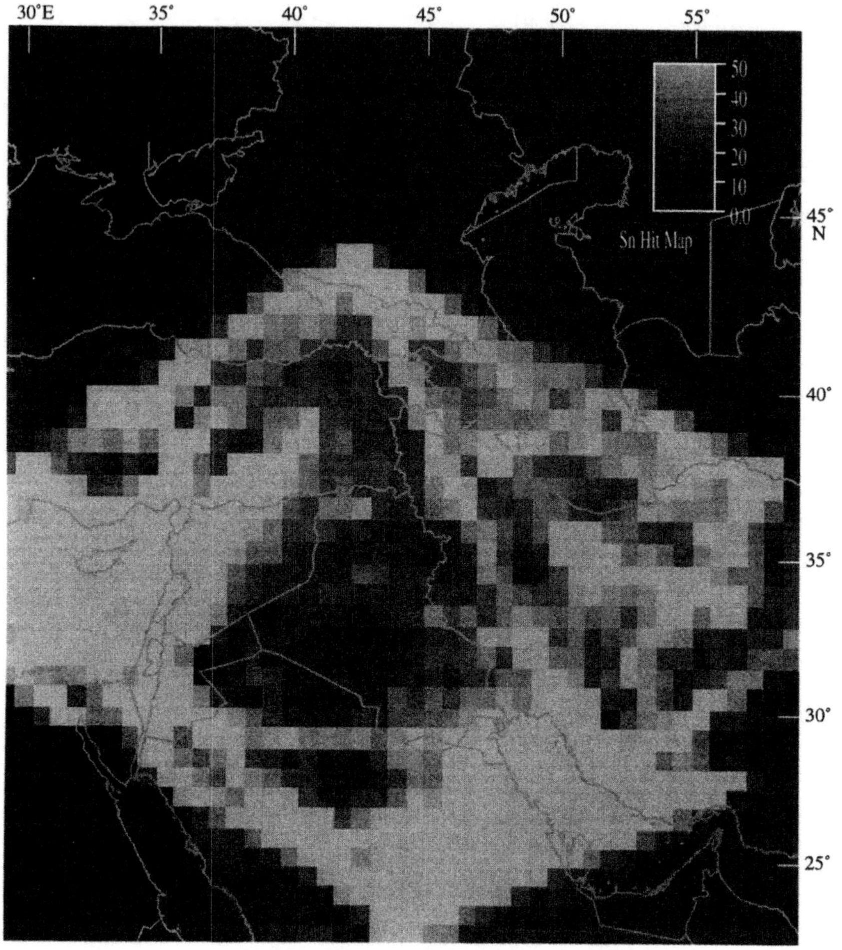

Figure 17
A map showing the hit count map for the tomographic model shown in Figure 16. Note the low number of hit counts in Iraq and some paths of northeastern Turkey.

Resolution and Uncertainties

In order to estimate the resolution of our *Sn* efficiency tomography we performed resolution tests as in our analysis of *Lg*/*Pg* ratio tomography. In order to reproduce a discrete synthetic data set we used a criterion to convert the non-discrete synthetic amplitude which is produced by propagating rays through a synthetic test model. We created a synthetic set of propagation efficiencies by defining an extinction path length. Assuming a minimum frequency for *Sn*, the extinction path length corresponds to a Q value:

$$Q = \frac{-\pi L^{\mathrm{ext}} f}{V_{Sn} \ln(0.10)} \tag{10}$$

for example $V_{Sn} = 4.6$ km/s, $f = 1$ Hz, $L^{\mathrm{ext}} = 250$ km, $Q = 75$ (Fig. 18). For the synthetic tests we did not create any inefficient *Sn* synthetic data, only efficient and blocked *Sn*, in order to make conservative estimates of our resolution. We also added ten percent noise by randomly switching 10% of our synthetic efficiencies from 1.0 to 0.0 or from 0.0 to 1.0. After creating a discrete synthetic set of *Sn* propagation efficiencies analogous to our actual data set, we used the Laplacian damped LSQR algorithm to solve for the input test model for a variety of extinction path lengths (Fig. 18). We used a variety of synthetic models to test our *Sn* model resolution including checkerboards, vertical strips, horizontal strips, and spike tests. Generally we observed the same resolution problems with all of the test models.

The apparent resolution is not as good as that of the *Lg*/*Pg* ratio tomography and depends upon our choice of the extinction path length. Our resolution tests

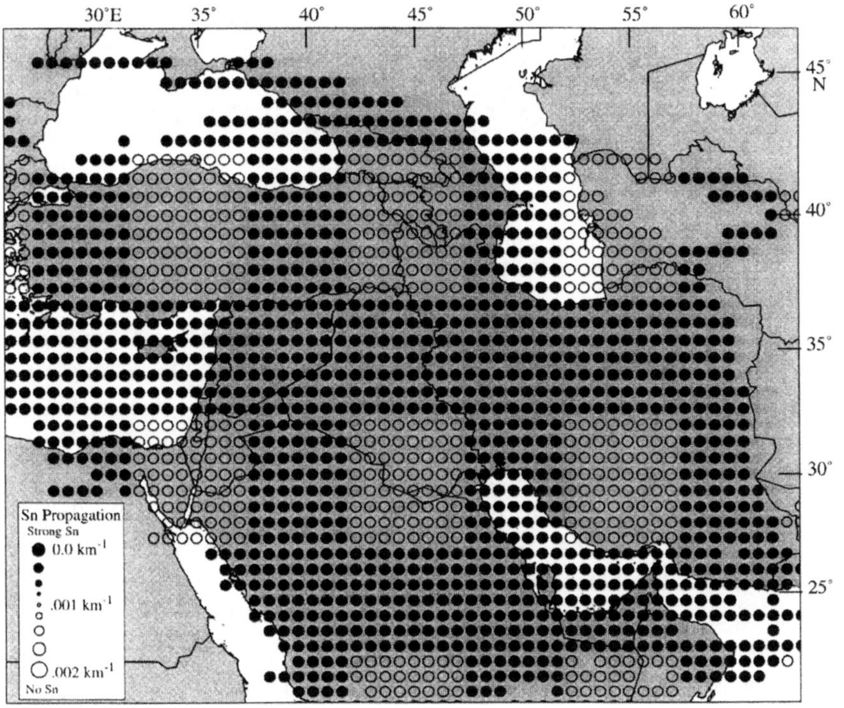

Figure 18a
A map showing the test model used to determine the resolution of our *Sn* efficiency tomography. This model contain "spikes" of regions with *Sn* extinction path lengths of 500 km surrounded by regions with no *Sn* attenuation (infinite extinction path length). Besides this test model we used a variety of other model geometries to test our model resolution.

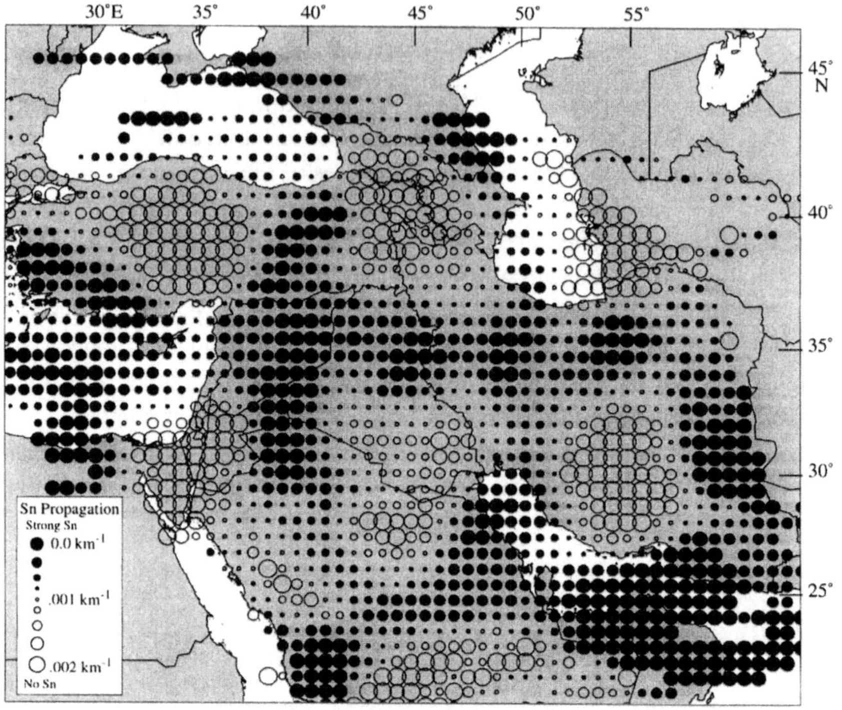

Figure 18b
A map showing the results of our resolution spike test using an extinction path length of 250 km in this test model. We observe east to west smearing of anomalies in northwestern Saudi Arabia as well as north-south smearing along the Red Sea and the Gulf of Aqaba.

demonstrate that conservatively only 200 to 300 km sized features, depending upon the degree of attenuation (i.e., extinction path length), can be resolved with our technique. This is not surprising given the large path length (greater than 300 km) that is required to observe *Sn*. We are able to image the original input model reasonably well in the eastern Mediterranean and Anatolian plateaus. However, in southern Arabia and Iran there is "smearing" in the east-west direction. This is mainly caused by the lack of stations and sources in central and southern Arabia. We also see north-south distortion of the test anomalies in the southern Zagros mountains. Examining Figure 18d we see that anomalies with extinction paths exceeding 500 km are not as well imaged as those anomalies with extinction paths of 250 km. In general, we cannot resolve attenuation anomalies where *Sn* is eliminated in 700 km or greater.

In order to estimate the uncertainties we employed the bootstrap resampling technique that we used for *Lg*/*Pg* tomography. By randomly resampling our original data set and reinverting for our model *Sn* propagation efficiencies 100 times, we were able to estimate the variance of each model parameter (Fig. 19).

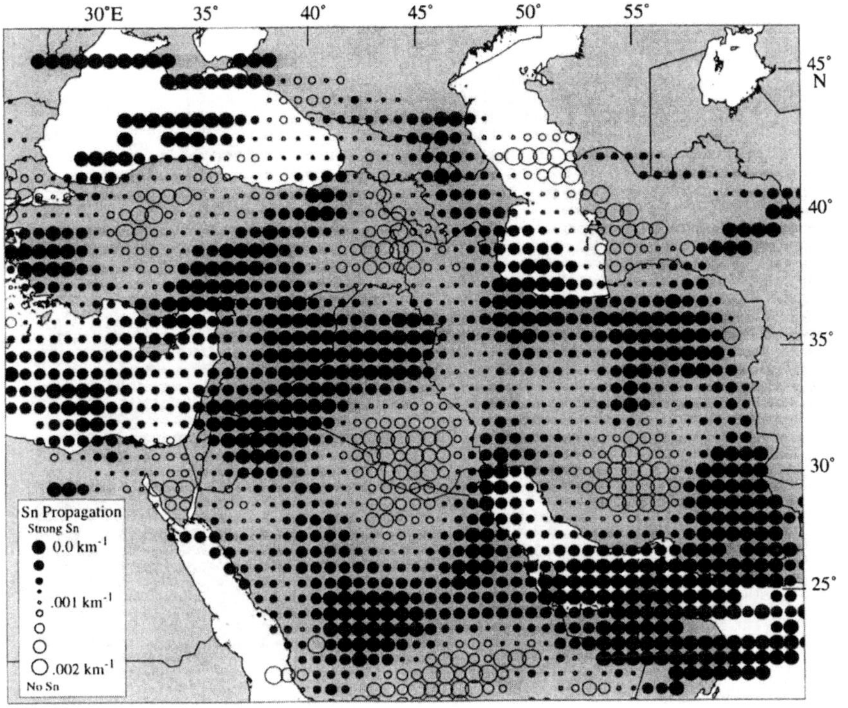

Figure 18c
A map showing the results of our resolution spike test using an extinction path length of 500 km. We still observe considerable amounts of smearing even with the less-attenuating test model.

We found that in general our estimated uncertainties were fairly minor. The exception is in Iraq and northern Saudi Arabia where we found substantial uncertainties in the model *Sn* efficiencies, indicating that these values are not very reliable. This is due, for the most part, to the lack of ray coverage in this part of our model (Fig. 17).

Results

The most obvious feature in our model is weak or no *Sn* propagation in Turkey and most of Iran, and efficient *Sn* propagation in the oceanic lithosphere as well as in most of the Arabian plate (Fig. 16). In both the Anatolian and the Iranian plateaus, *Sn* is heavily attenuated. The inefficient *Sn* zone in the Iranian plateau is bounded on the west by the Zagros fold and thrust belt and extends all the way to the East Iranian ranges (~60°E longitude) and as far south as the Makran in southern Iran. Throughout most of the Anatolian plateau and into the Anatolian block *Sn* propagation is inefficient. Similarly, in the Lesser Caucasus *Sn* is heavily attenuated; however, *Sn* propagation is efficient in the eastern half of the Greater Caucasus.

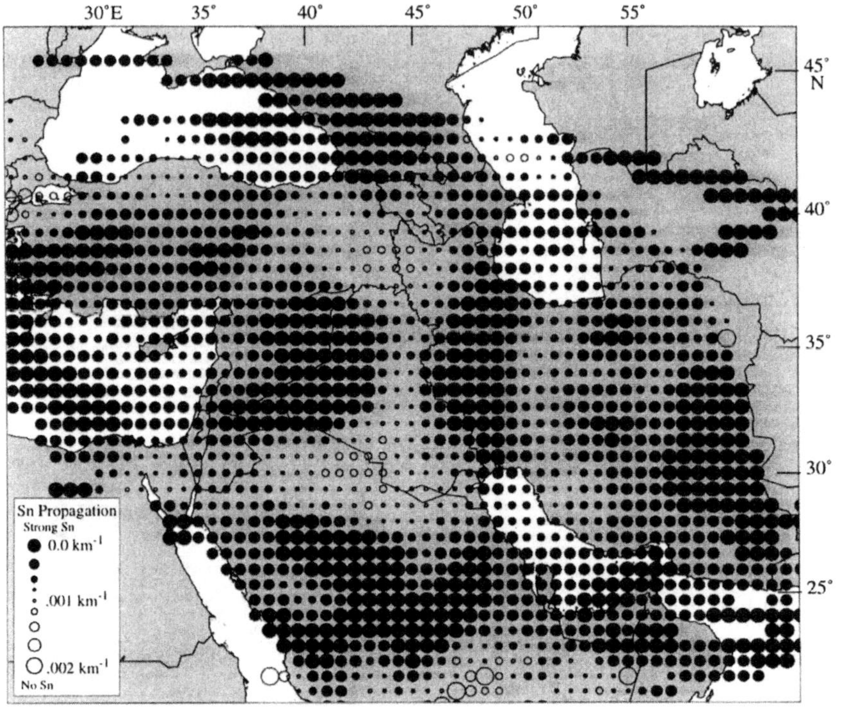

Figure 18d
A map showing the results of our resolution spike test using an extinction path length of 750 km. Note that we no longer can resolve the anomalies in central Turkey since the majority of our rays are not sufficiently long to be attenuated.

Our tomographic models indicate that the zone of inefficient *Sn* is not confined to the Anatolian plateau, but extends to the south across the Bitlis suture and into western Syria. Our resolution tests and uncertainty estimates indicate that this is neither an artifact nor a product of "smearing" of the inefficient *Sn* region within the Anatolian plateau. We also observe a large region of inefficient *Sn* that includes the southern Dead Sea fault in Jordan, Palestine, and Israel, as well as the Gulf of Aqaba. This region of *Sn* attenuation also extends approximately 500 km to the east, well into the northwestern segment of the Arabian plate. Some of this could be an artifact, as was demonstrated by our resolution tests in Figure 18. This feature could be due to smearing of the anomaly along the Dead Sea fault or perhaps the blurring together of two separate anomalies in the region; one in Jordan, Palestine, and Israel and the other in northern Saudi Arabia. Another region with similar resolution problems is the eastern Black Sea where we only have two stations in the Greater Caucasus recording a large number of events in the Anatolian plateau. This has caused the inefficient *Sn* zone in the Anatolian

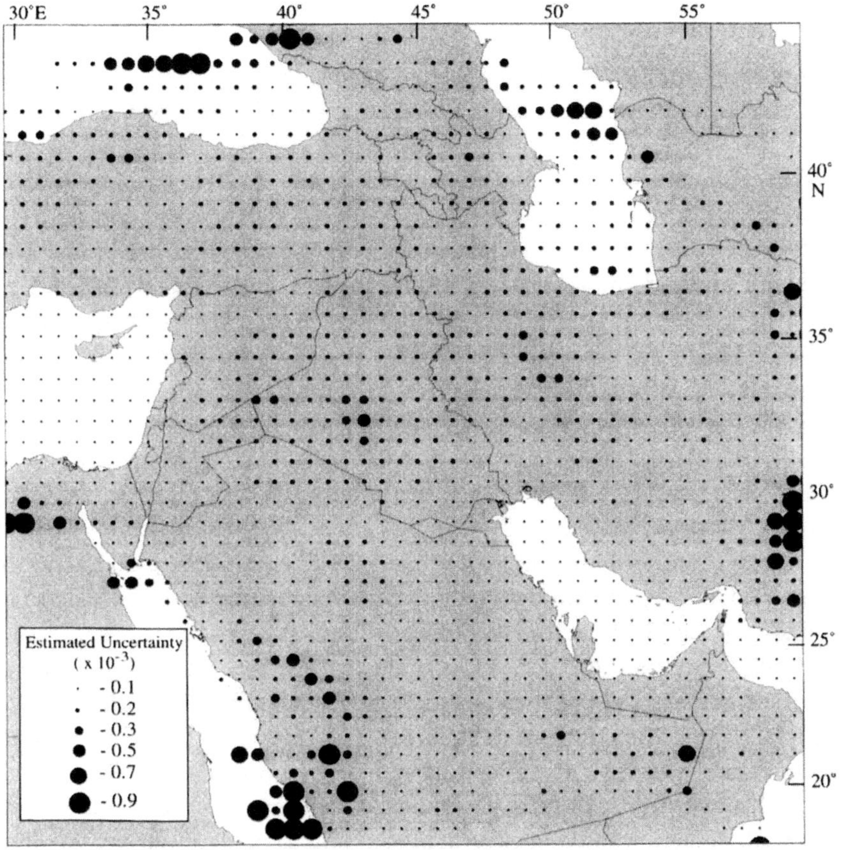

Figure 19
A map showing bootstrap uncertainty estimates for our *Sn* efficiency tomography. A bootstrap uncertainty of 0.5 corresponds to approximately 50% error for either most efficient *Sn* or blocked *Sn* anomalies shown in Figure 16. In general our tomographic model is fairly stable; however, in central Iraq and northern Saudi Arabia we have fairly large uncertainties approaching 50% of the anomalies shown in Figure 16.

plateau to "bleed" into the eastern Black Sea where we have scant model resolution.

Along the Red Sea coast in Saudi Arabia we observe another region of *Sn* attenuation which is based mostly on observations at two stations along the Red Sea coast (Figs. 4 and 16). To the northeast of this anomaly we found a region of efficient *Sn* propagation in the northern Arabian shield between the region of no *Sn* propagation along the Red Sea coast and inefficient *Sn* propagation in northern Saudi Arabia. Throughout most of southern and central Arabia, including Iraq and the Persian Gulf, *Sn* was found to be efficient. We have also found that *Sn* is efficient in the Koppeh Dagh region and farther to the west in the eastern Caucasus and in the Caspian Sea.

Discussion and Conclusions

The results of our *Lg*/*Pg* tomography are fairly consistent with prior studies of *Lg* in the Middle East. However, because of the considerably more waveform data used in this study we were able to accurately map the boundaries of *Lg* and *Sn* attenuation. The blockage of *Lg* phases across the Zagros mountains has been demonstrated by several different studies including KADINSKY-CADE *et al.* (1982), RODGERS *et al.* (1997), CONG and MITCHELL (1997), and MELLORS *et al.* (1999). In the Greater Caucasus, however, our results suggest a somewhat different picture from that suggested by KADINSKY-CADE *et al.* (1982) and RODGERS *et al.* (1997). We found that the region of complete *Lg* blockage resides in the easternmost edge of the Anatolian plateau in northeastern Turkey rather than in the Greater Caucasus, where for the most part we observe relatively efficient *Lg* propagation. Furthermore, we have found a very good correlation with inefficient *Lg* propagation and the Arabian plate boundary and the major tectonic provinces (Figs. 20 and 10). Prior studies (e.g., MITCHELL *et al.*, 1997; RODGERS *et al.*, 1997) found less of a correlation with *Lg* attenuation, probably because of the lack of resolution. Local network data from Syria and Turkey used in our study significantly improved our ability to resolve the geometry of *Lg* blockage and attenuation.

The boundaries marking regions of high *Lg* attenuation, relative to *Pg*, and regions of efficient *Lg* propagation correspond remarkably well with the boundary between the Eurasian and Arabian plates. This correlation is probably due to changes in crustal and sediment thickness across these boundaries. There is evidence for a crustal root beneath the Zagros mountain belt (e.g., SNYDER and BARAZANGI, 1986) as well as significantly thick sediments. This type of rapid crustal thickness variation has been demonstrated to block *Lg* (KENNETT, 1986). However, there is an approximately 100 to 200 km offset between the location of the crustal root and the location of minimum *Lg*-*Pg* differential attenuation. This might be due to the lack of seismic stations in the Iranian plateau which has led to most raypaths traveling from east to west in this region. This bias in our data set might cause smearing of the region of *Lg* blockage eastwards.

We observe similar types of *Lg* blockage in easternmost Anatolia. This may be a result of a sudden increased crustal thickness in eastern Turkey or perhaps to intrinsic attenuation in the crust caused by partial melt. We do not observe any *Lg* blockage for paths crossing the Dead Sea fault system. Any potential Moho topography and structural complexity associated with the Dead Sea fault are

▶

Figure 20

Maps showing several independent data sets which may be correlated with both our *Lg*/*Pg* and *Sn* efficiency tomography results. Panel (a) shows both the topography of the Middle East as well as Holocene volcanoes and Neogene volcanic fields. Panel (b) shows the *Pn* velocity map of HEARN and NI (1994). Regions of high topography generally have high *Lg* attenuation relative to *Pg*. We also see a fairly good correlation between low *Pn* velocities shown in panel (b) and high *Sn* attenuation.

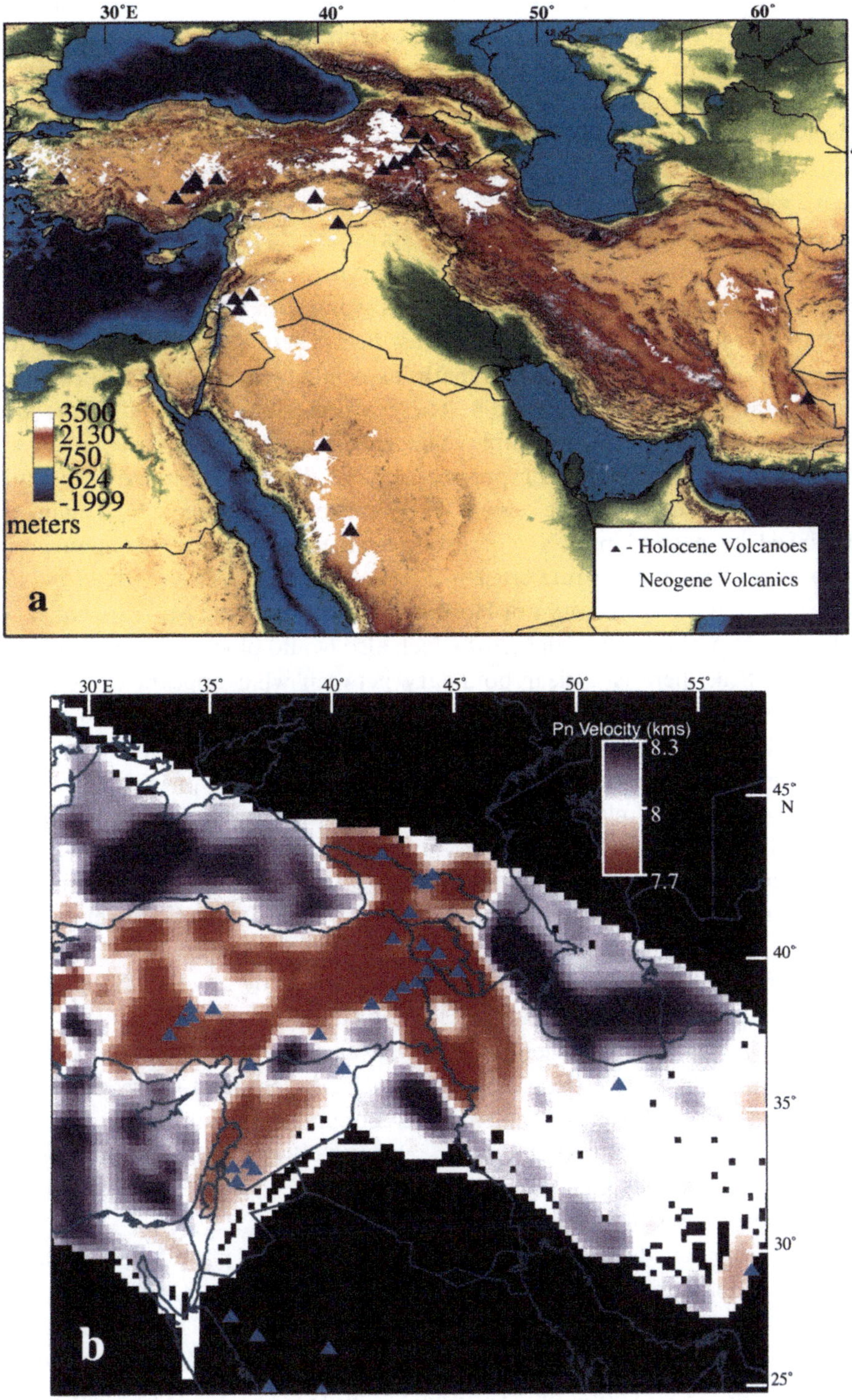
30°E
40°
50°
60°
40° N
30°
20°
3500
2130
750
-624
-1999
meters
a
▲ - Holocene Volcanoes
Neogene Volcanics
30°E
35°
40°
45°
50°
55°
Pn Velocity (kms)
8.3
8
7.7
45° N
40°
35°
30°
25°
b

apparently insufficient to disrupt *Lg* propagation, unlike the other major plate boundaries in the Middle East. The importance of sediment thickness on *Lg* propagation can be seen from the patterns of observed *Lg*/*Pg* amplitude ratio variation in Saudi Arabia. In the west, where no sedimentary rocks are present, *Lg* clearly propagates more efficiently than in the Arabian platform where sediment thickness ranges between 5 and 15 km (HUSSEINI, 1988). There is a region of similarly efficient *Lg* propagation in central and western Iraq which correlates with the Rutbah uplift as mapped by LOVELOCK (1984) and BREW *et al.* (1997).

We observe a small region in central Iran where *Lg* propagation is somewhat efficient, especially for lower frequencies, contrasting with the complete *Lg* blockage we observe across the Zagros fold and thrust belt. There is very little known about the crustal structure of this region. Raypaths associated with efficient *Lg* propagation in central Iran are shown in Figure 8. These observations suggest that the crustal structure in central Iran is less variable than other parts of the Iranian plateau where we do not observe as efficient *Lg* propagation. In the Makran region of southern Iran, however, we see very large regions of complete *Lg* blockage which again could be produced by a combination of crustal thickness variations in Oman and the Makran along with the oceanic crust in the Gulf of Oman (RAVAUT, *et al.* 1997). However, there is a large amount of Holocene volcanism in the Makran which could be indicative of partial melt in the crust which also would block *Lg*. North of Iran, in the Caspian Sea, there is a clear boundary between what appears to be remnant oceanic crust in the south Caspian (ZONENSHAIN and LEPICHON, 1986) and what appears to be thicker crust in the north Caspian Sea.

When interpreting uppermost mantle structure using wave propagation, efficient *Sn* typically implies a relatively cold/thick mantle lithosphere, such as most old oceanic lithosphere, where uppermost mantle attenuation is small. Poor *Sn* propagation, which is indicative of high uppermost mantle attenuation, implies hotter uppermost mantle and perhaps the presence of partial melt. However, it is not necessary to have partial melt in the uppermost mantle to substantially (5 to 10 times) lower *Q* (KAMPFMANN and BERCKHEMER, 1985; SATO *et al.*, 1989); therefore inefficient *Sn* propagation could also be an indication of simply hot lithosphere as opposed to a region of asthenospheric upwelling. Attempts at modeling *Sn* wavetrains have indicated that a negative velocity gradient is required in the uppermost mantle (lithosphere-asthenosphere boundary) in order to produce an effective waveguide for *Sn* (STEPHENS and ISACKS, 1977). If this velocity gradient were reduced or altered, then *Sn* propagation could conceivably be attenuated or perhaps in some cases blocked.

To first order, our tomographically mapped regions of poor *Sn* propagation in the Middle East agree with the prior work of KADINSKY-CADE *et al.* (1982), RODGERS *et al.* (1997), and MELLORS *et al.* (1999). However, there are important differences to note. In general, we map larger regions of poor *Sn* propagation than KADINSKY-CADE *et al.* (1982) and RODGERS *et al.* (1997). This is especially evident when examining the region of inefficient *Sn* in Iran, where we found that *Sn* does not propagate across the entire Iranian plateau. Similarly *Sn* does not propagate within the Anatolian plateau. Furthermore, we found a very large region of poor *Sn* propagation along the southern Dead Sea region that extends into eastern Jordan. A portion of this inefficient *Sn* zone could be due to resolution problems in our tomographic model. However, given the extensive *Sn* attenuation that we observe at stations in Saudi Arabia, Egypt, and southern Israel, this zone of uppermost mantle attenuation must be fairly substantial. We would suggest that most of western Jordan and western Syria is underlain by this zone of uppermost mantle attenuation. The region of *Sn* attenuation along the Red Sea coast agrees well with the study of MELLORS *et al.* (1999), with the exception that along the northeastern Red Sea coast we have observed efficient *Sn* propagation that extends in the Arabian shield for approximately 500 km.

Comparison of our *Sn* attenuation with the *Pn* velocity tomography of HEARN and NI (1994) can eliminate some of the ambiguity in interpreting individual observations. Generally there is a fairly good correlation between lower *Pn* velocities (7.6–7.8 km/s) and our regions of *Sn* attenuation, with a few notable exceptions (Fig. 20). For the most part central and eastern Turkey as well as the Lesser Caucasus have generally low *Pn* velocity (~7.75 km/s) and inefficient *Sn* propagation. These observations are an indication that there is partial melt in the uppermost mantle beneath the Anatolian plateau and the northwestern corner of the Iranian plateau. This observation is confirmed, to some degree, by the large amount of Quaternary and active volcanism in the Plateau (Fig. 20). However, we also find very important differences between the two tomographic models. The most striking difference is the region of high *Pn* velocity (8.2 km/s) south of the Bitlis suture in southeastern Turkey where we also see inefficient *Sn* propagation. Although this could be due to a lack of resolution by both tomographic models, this is the region with the best ray coverage for our *Sn* attenuation tomography. This observation is also interesting since there is evidence of Quaternary volcanism south of the Bitlis suture. Perhaps it is an indication of strong lateral heterogeneity or azimuthal seismic anisotropy affecting the *Pn* velocity measurements.

Another important result of our study is the large area of *Sn* attenuation in the Iranian plateau, especially since the study of HEARN and NI (1994) did not find particularly low *Pn* velocities (7.9–8.1 km/s). Again, this might be the bias towards *Sn* blockage in the Iranian plateau caused by the predominantly long raypaths in this region. Similarly this observation might indicate a lack of partial melt in the lithospheric mantle although perhaps the lithosphere is still hot and not fully recovered from the substantial volcanism in the late Cenozoic. This hypothesis is also consistent with the lack of extensive Holocene volcanism within the Iranian plateau (Fig. 20).

One of the most interesting regions of *Sn* attenuation in the Middle East occurs in Israel, Palestine, and most of Jordan. It is possible that this anomaly is artificially enlarged due to the lack of stations and sources in this region. However, complete elimination of *Sn* waves traveling from the Eastern Mediterranean Sea into Saudi Arabia (see Figs. 4 and 15) suggests that this zone is substantial (i.e., 200–400 km). There is little evidence for lower uppermost mantle velocities in the Gulf of Aqaba (HOFSTETTER *et al.*, 1991) and the Dead Sea fault system (GINZBURG and MAKRIS, 1979). However, few measurements of uppermost mantle velocity have been made farther east in Jordan and Saudi Arabia. Based on the Quaternary volcanism in this region (Fig. 20), CAMP and ROOBOL (1992) have proposed a broad zone of upwelling asthenosphere in western Arabia. Our observations suggest that the lithosphere is not hot or thinned across the entire Arabian shield.

Results from regional wave propagation tomography offer new and important insight into the attenuation structure of the crust and uppermost mantle. However, further resolution and reliable criteria must be developed to eliminate some of the ambiguity in the interpretation of the behavior of regional waveforms. However, in order to unambiguously determine the path effects on *Lg* propagation we need to better account for source contributions to *Lg* excitation. Even though we have not fully accounted for the source effects, given our data redundancy our *Lg*-*Pg* differential attenuation model can be used to correct *Lg*/*Pg* ratios for the path effects, which is critical to reliably discriminating earthquakes from explosions in the Middle East, given the dramatic variations in *Lg* and *Pg* propagation.

Acknowledgments

We thank the Syrian National Seismological Center of the General Establishment of Geology and Mineral Resources (Damascus, Syria) and the Kandilli Observatory and Earthquake Research Institute of Bogaziçi University (Istanbul, Turkey) for providing us data for this study and for the maintenance of their networks. We would like to thank Scott Phillips for providing special insight into our *Lg*-*Pg* tomographic method. We also are grateful to Christine Orgren and Francisco Gomez for their useful comments and discussions. We would especially like to thank Brian Mitchell for his reivew of our manuscript. The Cornell researchers are supported by a grant from the Defense Threat Reduction Agency of the Department of Defense No. DSWA01-97-1-0006. INSTOC contribution No. 257.

REFERENCES

BARKA, A., and REILINGER, R. (1997), *Active Tectonics of the Eastern Mediterreanean Region: Deduced from GPS, Neotectonic and Seismicity Data*, Annali Di Geofisica *40*, 587–610.

BAUMGARDT, D. R. (1996), *Investigations of Lg Blockage and Transportability of Regional Discriminants in the Middle East*, Sci. Report 1, PL-TR-96-2294, ENSCO Inc., Springfield, Virginia.

BAUMGARDT, D. R., and DER, Z. (1997), *Investigation of the Transportability of the P/S Ratio Discriminat to Different Tectonic Regions*, Sci. Report 1, PLT-TR-94-229, ENSCO, Inc., Springfield, Virginia.

BERBERIAN, M., and KING, G. (1981), *Towards a Paleogeography and Tectonic Evolution of Iran*, Can. J. Earth Sci. *18*, 33–69.

BOSTOCK, M., and KENNETT, B. L. N. (1990), *The Effect of 3-D structure on Lg Propagation Patterns*, Geophys. J. Int. *101*, 355–365.

BOUCHON, M. (1982), *Complete Synthesis of Seismic Crustal Phases at Regional Distances*, J. Geophys. Res. *87*, 1,735–1,741.

BREW, G., LITAK, R., SEBER, D., BARAZANGI, M., AL-IMAM, A., and SAWAF, T. (1997), *Basement Depth and Sedimentary Velocity Structure in the Northern Arabian Platform, Eastern Syria*, Geophys. J. Int. *128*, 617–631.

CAMP, V. C., HOOPER, P. R., ROOBOL, M. J., and WHITE, D. L. (1987), *The Madinah Eruption, Saudi Arabia: Magma Mixing and Simultaneous Extrusion of Three Basaltic Chemical Types*, Bull. Volcanol. *49*, 489–508.

CAMP, V. E., and ROOBOL, M. J. (1992), *Upwelling Asthenosphere Beneath Western Arabia and its Regional Implications*, J. Geophys. Res. *97*, 15,255–15,271.

CONG, L. L., XIE, J. K., and MITCHELL, B. (1996), *Excitation and Propagation of Lg from Earthquakes in Central Asia with Implications for Explosion/Earthquake Discrimination*, J. Geophys. Res. *101*, 27,779–27,789.

CONG, L., and MITCHELL, B. (1998), *Lg Coda Q and its Relation to the Geology and Tectonics of the Middle East*, Pure appl. geophys. *153*, 563–585.

DERCOURT, J., ZONENSHAIN, L., RICOU, L., KAZMIN, V., LE PICHON, X., KNIPPER, A., GRANDJACQUET, C., SBORTSHIKOV, I., GEYSSANT, J., LEPVRIER, C., PECHERSKY, D., BOULIN, J., SIBUET, J., SAVOSTIN, L., SOROKHTIN, O., WESTPHAL, M., BAZHENOV, M., LAUER, J., and BIJU-DUVAL, B. (1986), *Geological Evolution of the Tethys Belt from the Atlantic to the Pamirs since the Liasic*, Tectonophysics *123*, 241–315.

DEWEY, J. F., HEMPTON, M. R., KIDD, W. S. F., and SENGOR, A.M.C., *Shortening of Continental Lithosphere: The Neotectonics of Eastern Anatolia, a Young Collision Zone* (eds. Coward, M. P. *et al.*, 1986) Geological Society Special Publications *19*, 3–36.

DOTDUYEV, S., I. (1986), *Nappe Structure of the Greater Caucasus Range*, Geotectonics *20*, 420–430.

FAN, G., and LAY, T. (1998), *Statistical Analysis of Irregular Wave Guide Influences on Regional Seismic Discriminants in China*, Bull. Seismol. Soc. Am. *88*, 74–88.

FREUND, R. (1970), *The Geometry of Faulting in Galilee*, Isr. J. Earth Sci. *19*, 200–227.

GARFUNKEL, Z. (1981), *Internal Structure of the Dead Sea Leaky Transform (Rift) in Relation to Plate Kinematics*, Tectonophysics *80*, 81–108.

GINZBURG, A., and MAKRIS, J. (1979), *Gravity and Density Distribution in the Dead Sea Rift and Adjoining Areas*, Tectonophysics *54*, 17–25.

GHALIB, H. A. A. (1992), *Seismic Velocity Structure and Attenuation of the Arabian Plate*, Ph.D. Dissertation, Saint Louis University, pp. 260–288.

GREENWOOD, W. R., ANDERSON, R. E., FLECK, R. J., and ROBERTS, R. J., (1980), *Precambrian Geologic History and Plate Tectonic Evolution of the Arabian shield.* Saudi Arabian Directorate for Miner. Resources Bull., pp. 24–35.

HARTSE, H., FLORES, R., and JOHNSON, P. (1998), *Correcting Regional Seismic Discriminants for Path Effects in Western China*, Bull. Seismol. Soc. Am. *88*, 596–608.

HEARN, T., and NI, J. (1994), *Pn Velocities Beneath Continental Collision Zones: The Turkish-Iranian Plateau*, Geophys. Int. *117*, 273–283.

HOFSTETTER, A., FELDMAN, L., and ROTSTEIN, Y. (1991), *Crustal Structure of Israel: Contraints from Teleseismic and Gravity Data*, Geophys. J. Int. *104*, 371–379.

HUESTIS, S., MOLNAR, P., and OLIVER, J. (1973), *Regional Sn Velocites and Shear Velocity in the Uppermost Mantle*, Bull. Seismol. Soc. Am. *63*, 469.

HUSSEINI, M. (1988), *The Arabian Infracambrian Extensional System*, Tectonophysics *148*, 93–103.

JACKSON, J. A., and MCKENZIE, D. (1988), *The Relationship Between Plate Motions and Seismic Moment Tensors, and the Rates of Active Deformation in the Mediterranean and Middle East*, Geophys. J. *93*, 45–73.

KADINSKY-CADE, K., BARAZANGI, M., OLIVER, J., and ISACKS, B. (1981), *Lateral Variations in High Frequency Seismic Wave Propagation at Regional Distances across the Turkish and Iranian Plateaus*, J. Geophys. Res. *86*, 9,377–9,396.

KADINSKY-CADE, K., and BARAZANGI, M. (1982), *Seismotectonics of Southern Iran: The Oman Line*, Tectonics *1*, 389–412.

KAMPFMANN, W., and BERCKHEMER, H. (1985), *High Temperature Experiments on the Elastic and Anelastic Behavior of Magmatic Rocks*, Phys. Earth Planet. Int. *40*, 223–247.

KENNETT, B. L. N. (1986), *Lg Waves and Structural Boundaries*, Bull. Seismol. Soc. Am. *76*, 1,133–1,141.

LOVELOCK, P. E. R. (1984), *A Review of the Tectonics of the Northern Middle East Region*, Geol. Mag. *121*, 577–587.

INNOCENTI, F., MAZZUOLI, R., PASQUARE, G., RADICATI DI BROZOLO, F., and VILLARI, L. (1982), *Tertiary and Quaternary Volcanism of the Erzurum-Kars Area (Eastern Turkey): Geochronological Data and Geodynamic Evolution*, J. Volcanol. Geotherm. Res. *13*, 223–240.

MANGINO, S., and PRIESTLEY, K. (1998), *Crustal Structure in the Caspian Sea Region*, Geophys. J. Int. *133*, 630–648.

MCNAMARA, D., Owens, T., and WALTER, W. (1995), *Observations of Regional Phase Propagation Across the Tibetan Plateau*, J. Geophys. Res. *100*, 22,215–22,229.

MCKENZIE, D. (1972), *Active Tectonics of the Meditereanean*, Geophys. J. R. Astr. Soc. *30* (2), 109–185.

MCKENZIE, D. (1976), *The East Anatolian Fault: A Major Structural in Eastern Turkey*, Earth Planet. Sci. Lett. *29*, 189–193.

MELLORS, R., VERNON, F., CAMP, V., AL-AMRI, A., GHALIB, A., and AL-DAIL, M. (1999), *Regional Waveform Propagation in the Saudi Arabian Peninusla*, Geophys. Res. Lett., in press.

MEYERS, S., BECK, S., ZANDT, G., and WALLACE, T. (1998), *Lithospheric-scale Structure across the Bolivian Andes from Tomographic Images of Velocity and Attenuation for P and S Waves*, J. Geophys. Res. *103*, 21,233–21,252.

MITCHELL, B. J., PAN, Y., XIE, J., and CONG, L. (1997), *Lg coda Q Variation across Eurasia and its Relation to Crustal Evolution*, J. Geophys. Res. *102*, 22,767–22,779.

MOLNAR, P., and OLIVER, J. (1969), *Lateral Variations of Attenuation in the Uppermost Mantle and Discontinuities in the Lithosphere*, J. Geophys. Res. *74*, 2,648–2,683.

NI, J., and BARAZANGI, M. (1983), *High Frequency Seismic Wave Propagation Beneath the Indian Shield, Himalayan Arc, Tibetan Plateau and the Surrounding Regions: High Uppermost Mantle Velocities and Efficient Sn Propagation Beneath Tibet*, Geophys. J. R. Astr. Soc. *72*, 665–689.

NOLET, G., *Seismic wave propagation and seismic tomography*. In *Seismic Tomography* (ed. NOLET, G., 1988), pp. 1–23.

NUTTLI, O. (1980), *The Excitation and Attenuation of Seismic Crustal Phases in Iran*, Bull. Seismol. Soc. Am. *70*, 469–480.

PAIGE, C. C., and SAUNDERS, M. A. (1982), *LSQR: An Algorithm for Sparse Linear Equations and Sparse Least Squares*, ACM Trans. Math. Software *8*, 43–71.

PEARCE, J. A., BENDER, J. F., DE LONG, S. E., KIDD, W. S. F., LOW, P. J., GUNER, Y., SAROGLU, F., YILMAZ, Y., MOORBATH, S., and MITCHELL, J. G. (1990), *Genesis of Collision Volcanism in Eastern Anatolia, Turkey*, J. of Volc. and Geotherm. Res. *44*, 189–229.

PHILLIP, H., CISTERNAS, A., GVISHIANI, A., and GORSHKOV, A. (1989), *The Caucnasus: An Actual Example of the Initial Stages of Continental Collision*, Tectonophysics *161*, 393–398.

PHILLIPS, W. S., HARTSE, H. E., TAYLOR, S., VELASCO, A. A., and RANDALL, G. E. (1999), *Regional Phase Amplitude Ratio Tomography for Seismic Verification*, submitted to Pure appl. geophys.

POWERS, R. W., RAMIREZ, L. F., REDMOND, C. P., and ELBERG, E. L. (1966), *Geology of the Arabian Peninsula: Sedimentary Geology of Saudi Arabia*, U.S. Geol. Surv. Prof. Pap., 560–D, 91–96.

PRESS, F., and EWING, M. (1952), *Two Slow Surface Waves across North America*, Bull. Seismol. Soc. Am. *42*, 219–228.

QUENNELL, A. M. (1959), *Tectonics of the Dead Sea Rift*, Int. Geol. Congr., 20th, Mexico, 1956 –Assoc. Sewrv. Geol Afr., 385–405.

RAVAUT, P., BAYER, R., HASSANI, R., ROUSSET, D., and AL YAHYA'EY, A. (1997), *Strcuture and Evolution of the Northern Oman Margin: Gravity and Seismic Constraints over the Zagros-Makran-Oman Collision Zone*, Tectonophysics *279*, 253–280.

REILINGER, R., MCCLUSKY, S., ORAL., B., KING, R., TOKSÖZ, N., BARKA, A., KINIK, I., LENK, O., and SANLI, I. (1997), *Global Positioning System Measurements of Present-day Crustal Movements in the Arabia-Africa-Eurasia Plate Collision Zone*, J. Geophys. Res. *102*, 9,983–9,999.

RODGERS, A., NI, J., and HEARN, T. (1997), *Propagation Characteristics of Short-period Sn and Lg in the Middle East*, Bull. Seismol. Soc. Am. *87*, 396–413.

SATO, H., SACKS, S., MURASE, T., MUNCIL, G., and FUKUYAMA, H. (1989), *Qp Melting Temperature Relation in Peridotite at High Pressure and Temperature: Attenuation Mechanism and Implications for the Mechanical Properties of the Uppermost Mantle*, J. Geophys. Res. *94*, 10,647–10,661.

SEARLE, R. C. (1975), *The Dispersion of Surface Waves across the Afar*. In *Afar Depression of Ethiopa* (eds. A. Pilgerand and A. Roesler) (Schweizerbart, Stuttgart, Germany) pp. 113–120.

SEBER, D., and MITCHELL, B. (1992), *Attenuation of Surface Waves across the Arabian Peninsula*, Tectonophysics *204*, 137–150.

SEBER, D., VALLVE, M., SANDVOL, E., STEER, D., and BARAZANGI, M. (1997), *Middle East Tectonics: Applications of Geographic Information Systems (GIS)*, GSA Today *7*, 1–6.

SENGOR, A. M. C., and KIDD, W. S. F. (1979), *Post-collisional Tectonics of the Turkish-Iranian Plateau and a Comparison with Tibet*, Tectonophysics *55*, 361–376.

SNYDER, D., and BARAZANGI, M. (1986), *Deep Crustal Structure and Flexure of the Arabian Plate Beneath the Zagros Collisional Mountain Belt as Inferred from Gravity Observations*, Tectonics *5*, 361–373.

STEPHENS, C., and ISACKS, B. (1977), *Toward an Understanding of Sn: Normal Modes Love Waves in an Oceanic Structure*, Bull. Seismol. Soc. Am. *67*, 69–78.

WIER, S. (1982), *Inefficient Propagation of Sn Under Southeastern China*, Geophys. J. R. Astr. Soc. *71*, 151–158.

ZHANG, T. R., and LAY, T. (1995), *Why the Lg Phase Does not Traverse Oceanic Crust*, Bull. Seismol. Soc. Am. *85*, 1665–1678.

ZONENSHAIN, L., and LEPICHON, X. (1986), *Deep Basins of the Black Sea and Caspian Sea as Remanants of Mesozoic Back-Arc Basins*, Tectonophysics, *123*, 181–211.

(Received July 14, 1999, revised January 19, 2000, accepted February 1, 2000)

Pure appl. geophys. 158 (2001) 1165–1188
0033–4553/01/071165–24 $ 1.50 + 0.20/0

Pure and Applied Geophysics

Mapping Crustal Heterogeneity Using *Lg* Propagation Efficiency Throughout the Middle East, Mediterranean, Southern Europe and Northern Africa

D. E. McNamara[1] and W. R. Walter[2]

Abstract—In this paper we describe a technique for mapping the lateral variation of *Lg* characteristics such as *Lg* blockage, efficient *Lg* propagation, and regions of very high attenuation in the Middle East, North Africa, Europe and the Mediterranean regions. *Lg* is used in a variety of seismological applications from magnitude estimation to identification of nuclear explosions for monitoring compliance with the Comprehensive Nuclear-Test-Ban Treaty (CTBT). These applications can give significantly biased results if the *Lg* phase is reduced or blocked by discontinuous structure or thin crust. Mapping these structures using quantitative techniques for determining *Lg* amplitude attenuation can break down when the phase is below background noise. In such cases *Lg* blockage and inefficient propagation zones are often mapped out by hand. With our approach, we attempt to visually simplify this information by imaging crustal structure anomalies that significantly diminish the amplitude of *Lg*. The visualization of such anomalies is achieved by defining a grid of cells that covers the entire region of interest. We trace *Lg* rays for each event/station pair, which is simply the great circle path, and attribute to each cell a value equal to the maximum value of the *Lg*/*P-coda* amplitude ratio for all paths traversing that particular cell. The resulting map, from this empirical approach, is easily interpreted in terms of crustal structure and can successfully image small blockage features often missed by analysis of raypaths alone. This map can then be used to screen out events with blocked *Lg* prior to performing *Q* tomography, and to avoid using *Lg*-based methods of event identification for the CTBT in regions where they cannot work.

For this study we applied our technique to one of the most tectonically complex regions on the earth. Nearly 9000 earthquake/station raypaths, traversing the vast region comprised of the Middle East, Mediterranean, Southern Europe and Northern Africa, have been analyzed. We measured the amplitude of *Lg* relative to the *P-coda* and mapped the lateral variation of *Lg* propagation efficiency. With the relatively dense coverage provided by the numerous crossing paths we are able to map out the pattern of crustal heterogeneity that gives rise to the observed character of *Lg* propagation. We observe that the propagation characteristics of *Lg* within the region of interest are very complicated but are readily correlated with the different tectonic environments within the region. For example, clear strong *Lg* arrivals are observed for paths crossing the stable continental interiors of Northern Africa and the Arabian Shield. In contrast, weakened to absent *Lg* is observed for paths crossing much of the Middle East, and *Lg* is absent for paths traversing the Mediterranean. Regions that block *Lg* transmission within the Middle East are very localized and include the Caspian Sea, the Iranian Plateau and the Red Sea. Resolution is variable throughout the region and strongly depends on the distribution of seismicity and recording stations. *Lg*

[1] USGS, Golden, CO, USA.
[2] Lawrence Livermore National Laboratory, Livermore, CA, USA.

Correspondence to: Daniel E. McNamara, USGS, Box 25046, MS 966, Denver, CO 80225, E-mail: mcnamara@ghtmail.cr.usgs.gov propagation is best resolved within the Middle East where regions of crustal heterogeneity on the order of 100 km are imaged (e.g., South Caspian Sea and Red Sea). Crustal heterogeneity is resolvable but is poorest in seismically quiescent Northern Africa.

Key words: *Lg* regional phase seismology, attenuation, CTBT, nuclear explosion.

Introduction

The seismic phase *Lg* often contains the largest amplitudes in regional continental seismograms and is thus used for many purposes in seismological investigations including, magnitude estimation, source spectra and moment determination, and for discriminating explosions from earthquakes for monitoring the Comprehensive Nuclear-Test-Ban Treaty (CTBT). However, because this phase is strongly influenced by the crustal waveguide it also exhibits extreme variations in its amplitude relative to other seismic phases. Indeed it is well known that the phase can be completely absent from regional seismograms that traverse regions of thin crust (e.g., ZHANG and LAY, 1995) or very strong attenuation. Traditional quantitative amplitude measurement techniques such as attenuation Q estimation or attenuation tomography will fail in cases where *Lg* is completely absent due to the impossibility of measuring the amplitude of a phase that is below the background noise. Such regions, that never show efficient *Lg* propagation, are often mapped out by hand, based on examining all available regional seismograms (e.g., RODGERS *et al.*, 1997). In this paper we apply a more quantitative approach to such mapping by plotting the maximum *Lg* amplitude normalized by the *P-coda* in cells across a region. Our main goal in this paper is to map *Lg* characteristics such as *Lg* blockage, efficient *Lg* propagation, and regions of very high attenuation in a vast area that includes the Middle East, North Africa, Europe and the Mediterranean.

The regional mapping aspect of our work expands upon the previous work of numerous scientists (e.g., KADINSKY-CADE *et al.*, 1981, RODGERS *et al.*, 1997) by presenting a method of graphically representing *Lg* propagation characteristics. Our approach maps empirical observations of *Lg*/*P-coda* amplitude ratios and successfully images alternating regions of efficient, inefficient and absent *Lg* transmission. This is achieved by tracing *Lg* rays through a grid of cells that covers the region of interest. We then attribute to each cell the maximum value of the *Lg*/*P-coda* amplitude ratio for all paths traversing that particular cell. The resulting map successfully images *Lg* propagation characteristics often missed by simple analysis of individual raypaths alone. Our mapping technique is also an expansion of two methods previously presented by CAMPILLO *et al.* (1993) and KENNETT *et al.* (1985). We will briefly describe these previous methods in a later section and demonstrate how our empirical amplitude ratio approach contributes to a better assessment of *Lg* propagation characteristics.

This study has two main goals: To map out crustal heterogeneity as illuminated by *Lg* propagation efficiency and to determine regions of blockage or severe attenuation for CTBT monitoring and as a pre-screen for *Q* tomography.

Crustal structure. We intend to understand how the various geologic crustal structures of this tectonically complex region affect the amplitude of high-frequency (0.5–5 Hz) *Lg* arrivals. *Lg* is commonly thought to be generated by a superposition of higher-mode surface waves (OLIVER and EWING, 1957; EWING *et al.*, 1957; KNOPOFF *et al.*, 1973; MITCHELL, 1995) or, as its group velocity implies, *Lg* propagates as multiply reflected shear waves trapped within the crust (GUTENBERG, 1955; PRESS and EWING, 1952; HERRIN and RICHMOND, 1960; BOUCHON, 1982). *Lg* is of tectonic and geologic interest because it is commonly observed that lateral crustal heterogeneity plays a significant role in shaping the characteristics of the *Lg* signal (RUZAIKIN *et al.*, 1977; KENNETT *et al.*, 1985; KENNETT, 1986; MCNAMARA *et al.*, 1996; RODGERS *et al.*, 1997). Consequently, *Lg* carries information about the average crustal shear-wave velocity and apparent attenuation along its path and is sensitive to varying tectonic environments.

CTBT. A second motivation for this study is that *Lg* plays an important role in event discrimination efforts for the CTBT. It is commonly observed that explosions are less efficient at producing *Lg* than earthquakes and *P*/*Lg* amplitude ratios are often used to discriminate the two event types (e.g., WALTER *et al.*, 1995; HARTSE *et al.*, 1997). An accurate understanding of the factors that affect the variation of *Lg* amplitude and propagation characteristics is important to enable effective discrimination between naturally occurring earthquakes and nuclear explosions.

Factors Affecting Lg Propagation

For naturally occurring earthquakes, *Lg* is a seismic wavetrain that travels with a group velocity of about 3.5 km/s and is generally the largest amplitude arrival on all three components of motion on regional short-period seismograms. The absence of *Lg* on the regional seismogram can be caused by several factors. First, for small magnitude events, *Lg* amplitude can be well beneath background noise levels. Second, high attenuation (low *Q*) can reduce *Lg* amplitude to below noise levels. It is generally observed that *Lg* attenuation is higher for regions with active tectonism than for stable continental interiors (AKI, 1980a, b; MCNAMARA *et al.*, 1996). Third, crustal structure plays an important role in determining the amplitude and propagation of *Lg*. The presence of *Lg* is often used to infer the existence of continental crust because *Lg* quickly loses energy in the thin waveguide provided by oceanic crust (KENNETT, 1986; ZHANG and LAY, 1995). Also, *Lg* propagation is affected by variations of the crustal waveguide along its path due to scattering along faults and fractures in tectonically active regions and/or when the waveguide is not continuous, as in the case of abrupt crustal thickness variations (AKI, 1980a;

McNamara *et al.*, 1996). Finally, source mechanism can determine the amplitude of *Lg*. Pure explosion sources theoretically do not generate shear energy and would not generate significant *Lg* amplitude arrivals on regional seismograms.

Motivation for our study can be observed in Figure 1. The top panel shows three separate regional seismograms, recorded at the short-period center element of the Sonseca array in central Spain (station ESDC), from three separate sources. The bottom panel shows the region of interest in our study with continental areas of Africa, Europe and the Middle East and the deep ocean basins (> 1 km) of the Atlantic and Indian Oceans, and the Mediterranean, Black and Red Seas. From

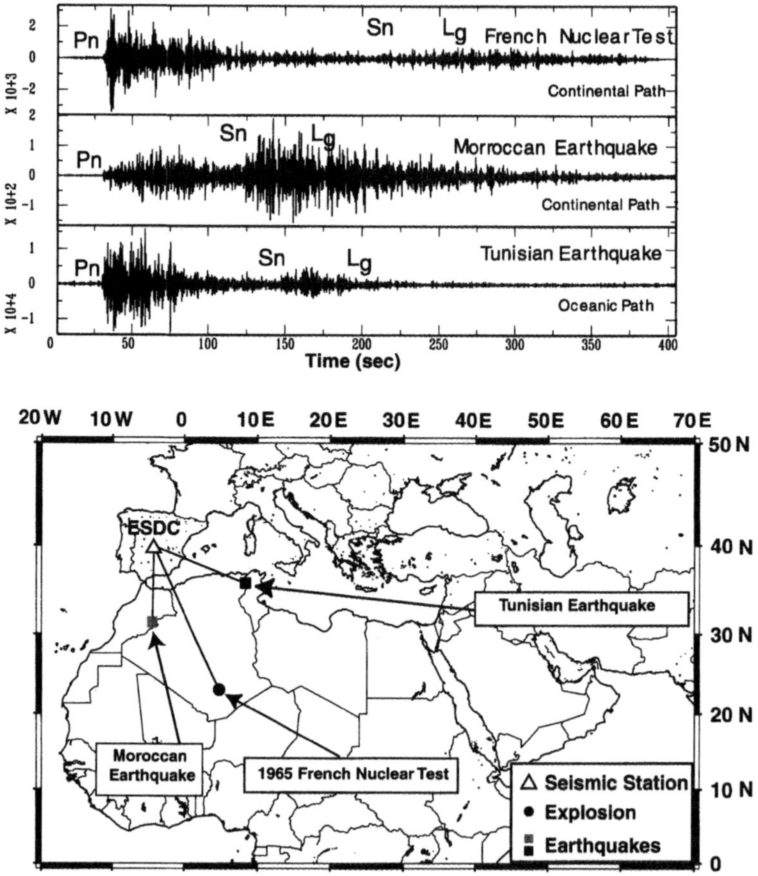

Figure 1
The top panel shows three separate regional seismograms recorded in Spain at station ESDC, from three separate sources bandpass filtered from 1–5 Hz. The sources are a 1965 French nuclear test, an earthquake from Morocco with a dominantly continental propagation path and an earthquake from Tunisia with a dominantly oceanic propagation path. The bottom panel illustrates the region of interest in our study with continental areas of Africa, Europe and the Middle East and the deep ocean basins (> 1 km) of the Atlantic and Indian Oceans, and the Mediterranean, Black and Red Seas. Note the relative amplitude variations between *Pn* and *Lg* within the three seismograms.

previous, qualitative studies (KADINSKY-CADE *et al.*, 1981; RODGERS *et al.*, 1997) we anticipate significant lateral variation of *Lg* propagation throughout the region. For example, it is well established that the stable continental interior of Northern Africa has extremely high values of *Lg* coda *Q* (XIE and MITCHELL, 1990; MITCHELL, 1995). For this reason we expect the continental interiors within the region, such as Northern Africa and the Arabian Peninsula, to be imaged as having very efficient *Lg* propagation. Conversely, it is also well established that *Lg* does not propagate across deep ocean basins (ZHANG and LAY, 1995). For this reason we expect that the deep ocean basins of the Pacific and Indian Oceans will be imaged as having inefficient to absent *Lg* propagation. In addition there is considerable previous work demonstrating that the oceanic crust underlying the deep seas within the region, block *Lg* propagation. For example, SHAPIRO *et al.* (1996) have shown that the deep portions of the Mediterranean Sea do not efficiently propagate *Lg* and the Black and Caspian Seas have been shown to block *Lg* transmission (KADINSKY-CADE *et al.*, 1981; RODGERS *et al.*, 1997; MANGINO and PRIESTLEY, 1998). We have the opportunity to compare the numerous previous results to our maps.

The top panel of Figure 1 clearly demonstrates that significant variation of *Lg* amplitudes are observed within the region of interest. Note the relative amplitude variations between *Pn* and *Lg* within the three seismograms. As expected the 1965 French nuclear test, with a dominantly continental path, shows very weak to absent *Lg* energy. Such weak *Lg* amplitudes relative to *Pn* are a source characteristic of explosions and are often used to discriminate them from earthquakes (e.g., WALTER *et al.* 1995; HARTSE *et al.*, 1997). In contrast, an earthquake from eastern Morocco, with a similar path traversing the continental crust beneath the Alboran Sea, shows relatively large *Lg* amplitude relative to *Pn*. Finally, the seismogram from an earthquake in northern Tunisia manifests weak to absent *Lg* energy and closely resembles the seismogram from the French nuclear test. This is expected for earthquakes crossing structure that blocks or severely attenuates *Lg*, such as the oceanic path in this case. The Tunisian earthquake clearly demonstrates the difficulties encountered in discriminating between naturally occurring earthquakes and nuclear explosions for the CTBT, if regions that block or severely attenuate *Lg* are not well established and accounted for. With this in mind we map the lateral propagation characteristics of *Lg* in the vast region of the Tethys collision zone. This is a complex tectonic region resulting from the convergence of the African and Arabian plates with southern Eurasia. The region includes the Middle East, the Mediterranean Sea, Southern Europe and Northern Africa.

Waveform Data

We collected data from nearly 4000 moderate to large earthquakes ($m_b > 4.0$) recorded at over 100 seismic stations, sparsely distributed throughout the entire region. The distribution of earthquakes and recording stations is shown in Figures

2a, b and c. We used only crustal events with reported depths less than 40 km that occurred between 1990–1997. Event locations, origin times and magnitudes were obtained from the United States Geological Survey's (USGS) preliminary determination of epicenters (PDE) catalogs. As shown in Figure 2, seismicity across the region is not uniform. Earthquakes are concentrated in specific tectonic regions such as the subduction zones of the Mediterranean Sea and Southern Europe, the Caucasus Mountains, the Zagros thrust of western Iran, the Turkish-Iranian Plateau, to the east in the Hindu Kush and Pamirs continental collision zones, surrounding the Caspian Sea and also the active spreading centers of the Red Sea and East African Rift. Large stable continental interiors such as Northern Africa and the Arabian Peninsula have little seismicity and consequently few recording stations.

Data sources include 3-component broadband stations operated within permanent networks (WWSSN, MEDNET, GEOSCOPE, GSN), and broadband, 3-component stations from various portable deployments (Saudi Arabian PASSCAL array (VERNON *et al.*, 1996), Caspian Sea (MANGINO and PRIESTLEY, 1998)). Generally, these stations operated using Streckheisen STS2 or Guralp CMG3 and/or 40T sensors recorded at either 20 or 40 samples/sec. Such instruments are broadband and have a flat velocity response ranging from about 0.02–0.01 Hz to 10–30 Hz. Additional waveform data were obtained from vertical-component short-period stations operated in permanent and temporary arrays (Sonseca, Bilbasi). These instruments generally are operated as an array of borehole sensors and have a peak response of 1 Hz.

A large majority of waveform data was obtained through the IRIS data management center and is currently archived at Lawrence Livermore National Laboratory. The remainder of the data used in this study was obtained from individual network operators. Many stations were included that operated at different times, also the region of interest is very large. Consequently all stations do not record events shown in Figure 2.

Amplitude Measurement and Data Selection Criteria

To obtain amplitude measurements we first integrated the vertical component, raw velocity seismogram to displacement. Selection of the vertical component is based on two reasons. First, we generally observe that *Lg* energy is distributed nearly evenly across all three components of motion (MCNAMARA *et al.*, 1996). Second, all single component array stations, included in this study, recorded only vertical motion. The broadband stations operate using three-components. However, for consistency between stations, all amplitude measurements were made on the vertical component of motion. Since regional arrivals, such as *Lg*, are generally best observed at shorter periods, the broadband displacement seismograms were then filtered with a two-pole, two-pass, Butterworth filter using a passband from 0.5 to 5.0 Hz. It was not necessary to filter the 1 Hz short-period seismograms. Each filtered and short-period seismogram was then smoothed about the mean of a 10-sample moving window. Next, we determined the seismogram envelope $E(t)$ using:

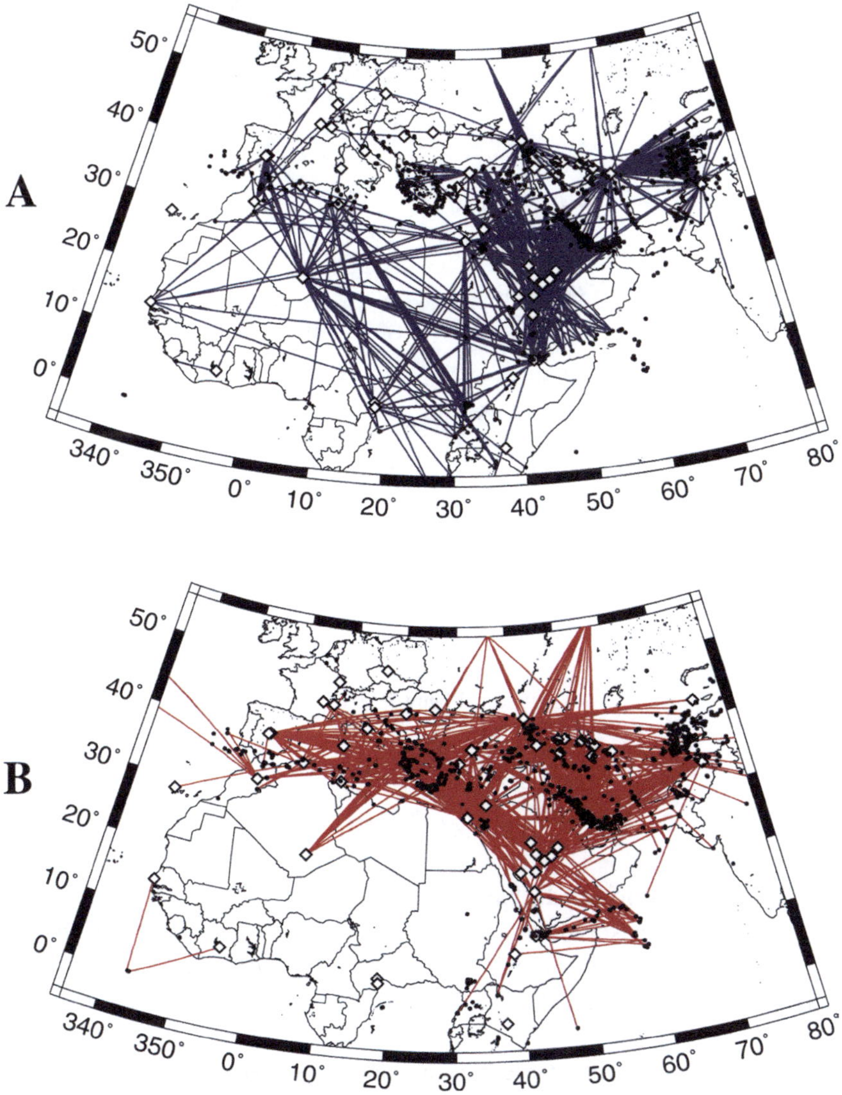

Figure 2
Maps showing the distribution of 3868 earthquakes (black circles), 108 stations (white diamonds) used in this study. Also shown are 3547 color coded raypaths reduced from 8457 after the data selection criteria were applied. (a) Raypaths with efficient *Lg* propagation. (b) Raypaths with absent *Lg* propagation. (c) Raypaths with inefficient to weak *Lg* propagation.

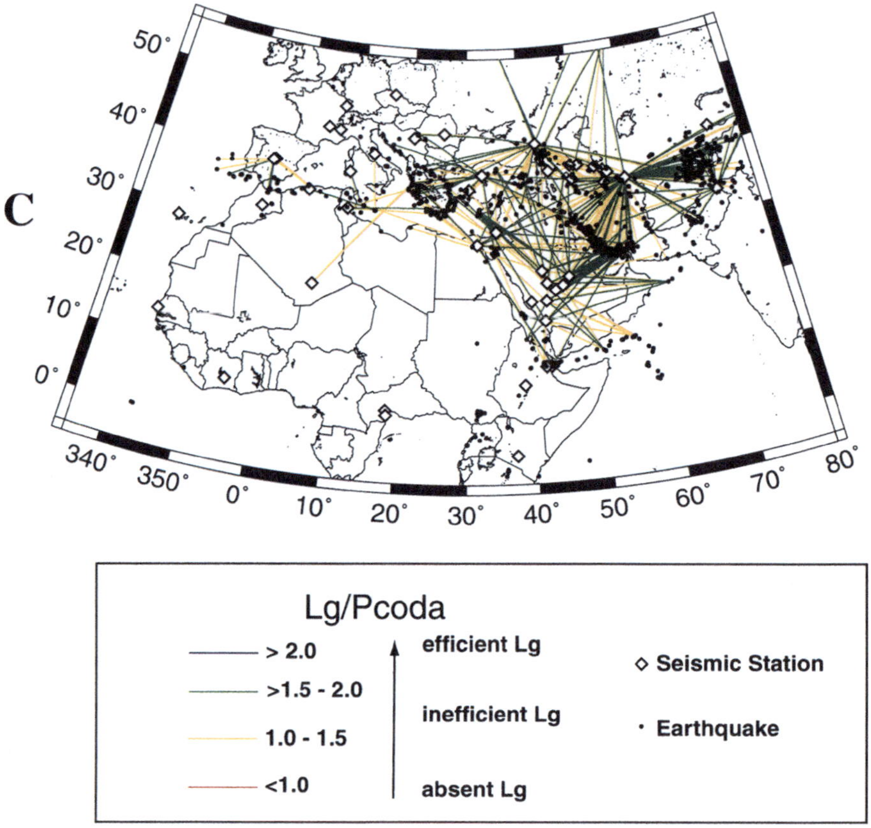

Figure 2C

$$E(t) = [A(t)^2 + H(t)^2]^{0.5} \tag{1}$$

where $A(t)$ is the smoothed, bandpassed time series and $H(t)$ is its Hilbert transform. We then measured root-mean-squares (RMS) amplitudes of the envelope function within the following group velocity windows: pre-event noise (origin time-10 km/s), *P-coda* (5.8–4.8 km/s) and *Lg* (3.6–3.0 km/s) (Fig. 3). In some cases, due to possible errors such as event location and/or depth, these windows were inappropriate and the data were not included. We also encountered numerous examples of *Sn* to *Lg* conversion for several raypaths traversing the Mediterranean Sea (SHAPIRO *et al.*, 1996). In these cases the *Lg* phase was always significantly faster than 3.6 km/s, rendering our velocity window inappropriate. Amplitude measurements, using these data, were also not included in further analysis of *Lg* propagation efficiency.

After visually inspecting all seismograms and obtaining amplitude measurements in the three-group velocity windows, amplitude ratios were computed. The *P-coda/pre-event noise* mean amplitude ratio was used to assess the overall signal strength of the seismogram. For inclusion in the propagation efficiency analysis we required a *P-coda/noise* ratio of 2 or greater. The *Lg/P-coda* ratio proved to be most useful for quantitatively assessing the propagation of *Lg*. We selected the *P-coda* window in order to isolate scattered energy arriving after the main regional *Pn* and *Pg* arrivals and before *Sn*. We found that normalizing *Lg* amplitude by the *P-coda* rather than *Pn* and/or *Pg* was more stable, since it is not as sensitive to source depth, focal mechanism radiation pattern and upper mantle attenuation. Using the *P-coda* amplitude, rather than the individual regional phase amplitudes (*Pn*, *Pg*) to form the ratio, allows for isolation and more accurate measure of amplitude variation of the *Lg* phase. Finally, normalizing with amplitude ratios is also required because of the inclusion of numerous instrument types with uncertain response information and to remove source-size effects.

Mapping the Lateral Variation of Lg Propagation Efficiency

Upon initial inspection of the regional seismograms, it becomes readily apparent that *Lg* amplitude can vary dramatically throughout the region of interest. For example, Figure 3 demonstrates the wide range of *Lg/P-coda* amplitude ratios that can be expected. For an earthquake in Morocco, recorded in southern Algeria, (92.297.09.11.09 at TAM) the *Lg* arrival is over 17 times larger than the energy contained within the *P-coda* (Fig. 3). This is likely due to the high Q homogeneous nature of the stable interior continental crust. In comparison, for an event of similar magnitude with a raypath of similar length, in the tectonically complex region of the Caucasus Mountains between the Caspian and Black Seas (90.215.11.57.13 at KIV), the average amplitude of the *P-coda* is three times larger than *Lg* (Fig. 3). This strongly suggests that the crustal structures associated with the Caucasus Mountains severely inhibit *Lg* transmission.

Raypath maps. As an initial step in understanding the spatial distribution of *Lg* propagation we map event/station raypaths, color coded, according to the measured *Lg/P-coda* amplitude ratio (Fig. 2). We are able to observe *Lg* within a wide range of path lengths, from hundreds to thousands of kilometers. We eliminated paths with epicentral distances less than 150 km because it was difficult to get adequate separation of *Lg* from other regional *S* phases. The upper path length bound was more difficult to determine because significant *Lg* energy was observed for several paths as long as 6000 km traversing Northern Africa. We defined four categories of *Lg* propagation efficiency based on the *Lg/P-coda* amplitude ratio. *Lg* is considered efficient when the *Lg/P-coda* ratio is greater than 2.0 and is mapped with blue lines in Figure 2a. *Lg* is considered absent when the *Lg/P-coda* ratio is less than 1.0 and

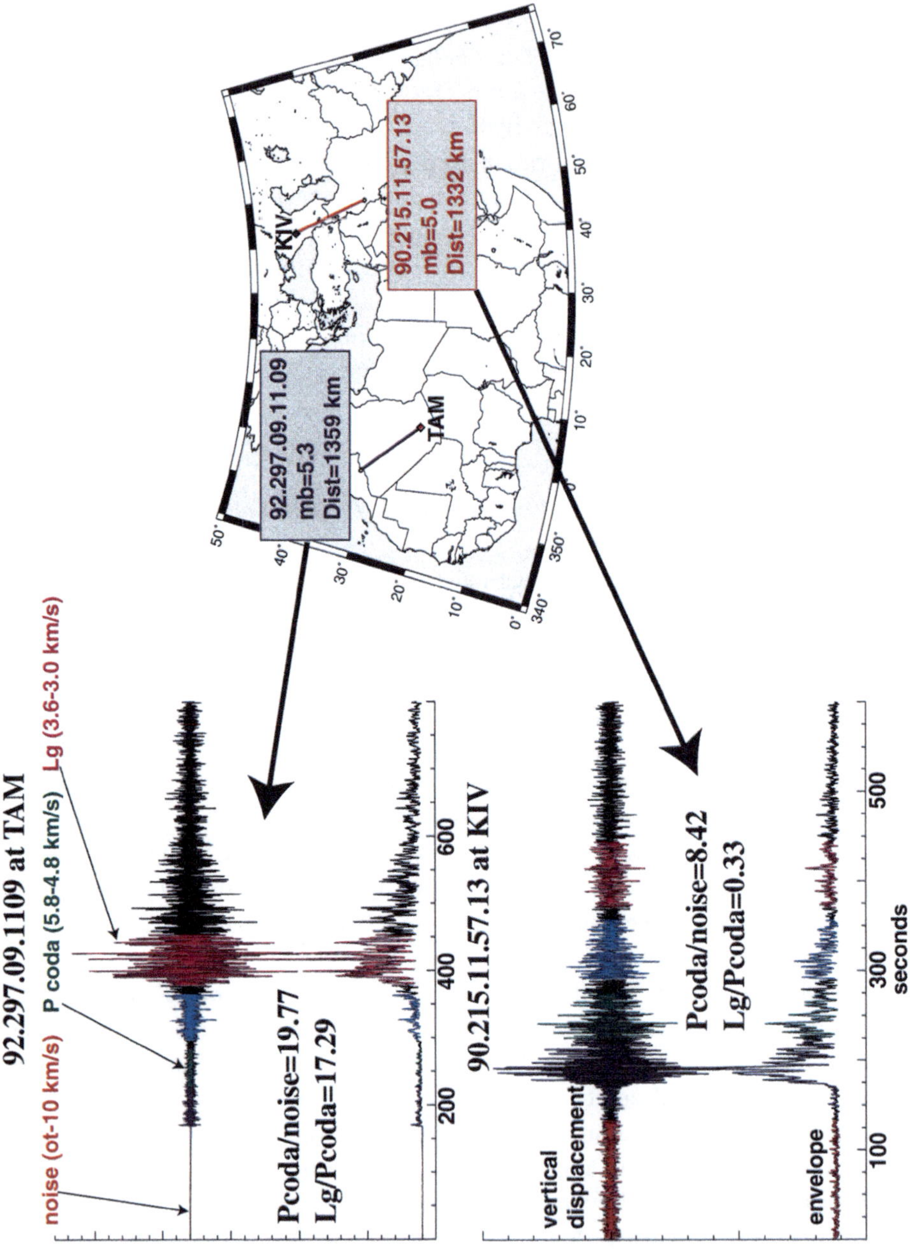
92.297.09.11.09
mb=5.3
Dist=1359 km
90.215.11.57.13
mb=5.0
Dist=1332 km
KIV
TAM
92.297.09.1109 at TAM
noise (ot-10 km/s)
P coda (5.8-4.8 km/s)
Lg (3.6-3.0 km/s)
Pcoda/noise=19.77
Lg/Pcoda=17.29
90.215.11.57.13 at KIV
Pcoda/noise=8.42
Lg/Pcoda=0.33
vertical displacement
envelope
seconds

is mapped with red lines in Figure 2b. Weak and inefficient *Lg* propagation (*Lg*/*P-coda* = 1.0 < 2.0) is mapped in Figure 2c.

The raypath maps show significant variation of *Lg* propagation efficiency throughout the region. We observe strong *Lg* in the stable continental interiors of Northern Africa and the Arabian Peninsula, and clearly absent *Lg* for most paths traversing the Mediterranean Sea. By visual inspection of the raypath maps it is difficult however, to completely identify the regions where *Lg* is absent and even more difficult to interpret the regions of weak and inefficient propagation. For this reason we now present our method to more clearly image the lateral variation of *Lg* propagation efficiency.

Maximum Lg/P-coda cell maps. To better constrain the spatial distribution of *Lg* propagation characteristics we attempt to visually simplify this information by imaging crustal structures that efficiently propagates *Lg* and significantly diminishes the amplitude of *Lg*. The imaging of such amplitude anomalies is achieved by defining a grid of cells that covers the entire region of interest. We trace *Lg* rays for each event/station pair, which is simply the great circle path, and attribute to each cell a value equal to the maximum value of the *Lg*/*P-coda* amplitude ratio for all paths traversing that particular cell. If all rays traversing a particular cell do not contain significant *Lg* energy (*Lg*/*P-coda* < 1.0), the cell is colored red (Fig. 4). Cells where only absent or inefficient *Lg* paths are present are colored from orange to yellow depending on the maximum value (Fig. 4). Cells that allow efficient to strong *Lg* propagation are colored from green to dark blue (Fig. 4). By mapping the maximum *Lg*/*P-coda* ratio, traversing the cell, we are able to negate the bias from inefficient or absent *Lg* that later propagate into regions of efficient *Lg* propagation. This method allows us to more accurately resolve the boundaries of regions of different propagation characteristics by reducing smearing, and reduces inefficient *Lg* propagation artifacts.

Figure 4 shows a map using a regularly defined grid, with 111.14 km cells, covering the entire region of interest. Raypath lengths range from 150–6000 km, unless the *Lg*/*P-coda* ratio is less than 1.0. In this case, the maximum allowable raypath length was reduced to 2000 km. The elimination of long paths with absent *Lg* was done to minimize normal attenuation effects and the effects of off-azimuth

◄

Figure 3

We show the bandpassed filtered (0.5–5.0 Hz), high signal-to-noise seismograms and envelope functions for two earthquakes with dramatically different *Lg* propagation characteristics. We also present examples of the velocity windows used for amplitude measurement: pre-event noise (0–10 km/s), *P-coda* (5.8–4.8) km/s and *Lg* (3.6–3.8 km/s). Event 92.297.09.11.09 (*mb* = 5.3) occurred in Morocco and was recorded in Southern Algeria (GEOSCOPE station TAM) at a distance of 1359 km. The seismogram and envelope functions show the characteristic efficient *Lg* propagation that occurs in the continental crust of Northern Africa (*Lg*/*P-coda* = 17.29). Event 90.215.11.57.13 (*mb* = 5.0) occurred in western Iran and was recorded at GSN station KIV at a distance of 1332 km. The seismogram and envelope functions indicate that along this path through the Caucasus Mountains *Lg* propagation is absent (*Lg*/*P-coda* = 0.33).

propagation. It also reduced inefficient *Lg* artifacts in regions of efficient propagation. After applying the data selection criteria discussed in the previous sections, 8257 paths were reduced to 3547 high-quality *Lg*/*P-coda* amplitude measurements. These high-quality measurements were then used to construct the map shown in Figure 4.

Mapping event/station paths, color coded to represent the *Lg* amplitude observations, is the typical technique used to display this type of information. However, in our case, with several thousand observations, we find mapping the maximum *Lg*/*P-coda* by cell provides a clearer and more easily interpretable view of the regional variation of *Lg* propagation. In effect, we are mapping the large-scale, first-order *Lg* blockage and propagation efficiency features. The most significant observation from our data set is that *Lg* propagation efficiency varies considerably and regions that do not effectively transmit *Lg* energy can be roughly correlated to tectonic structures. Features shown in Figure 4 are consistent with previous studies and clearly identify regions of inefficient to absent *Lg*, such as the Mediterranean Sea as well as regions of efficient to strong *Lg* propagation, such as Northern Africa and the Arabian Peninsula (ROGERS *et al.*, 1997).

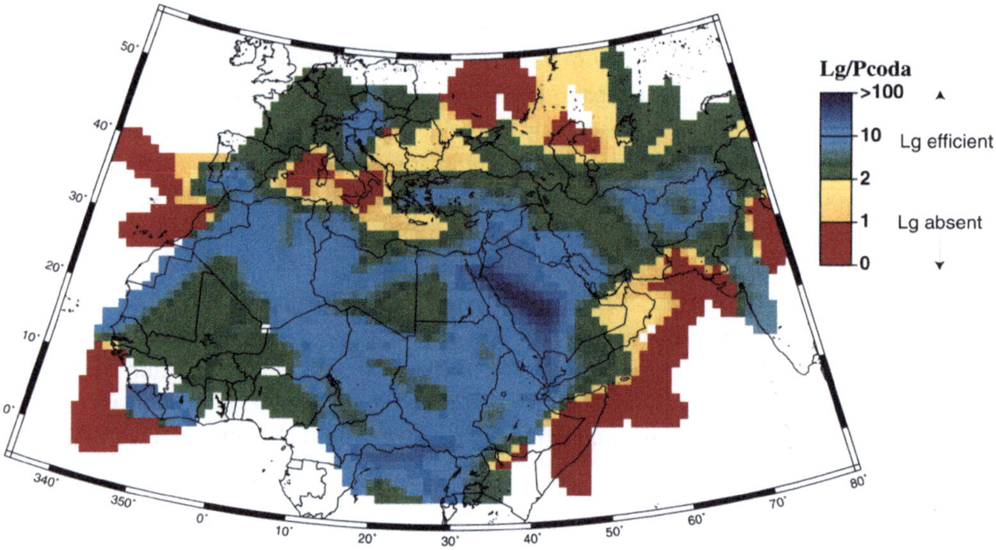

Figure 4

A map showing the results of our empirical *Lg* propagation efficiency mapping technique. Cells are color coded by the maximum *Lg*/*P-coda* value that traverses a particular cell. In this example the entire region is divided into 111.14 km cells and 3547 raypaths are used to construct the map. The maximum path length was 6000 km unless *Lg* was absent from the trace (Lg/P-coda < 1.0). In this case the maximum allowable path length was reduced to 2000 km. Only large-scale features are imaged in this map. For example, the Mediterranean Sea propagates *Lg* inefficiently while continental interiors such as Northern Africa and the Arabian Peninsula show very efficient *Lg* propagation.

Resolution. A first-order measure of data set resolution can be obtained by inspecting the raypath distribution within the region (Figs. 2a, b, c). Regions with both high raypath density and good azimuthal distribution will have the best resolution. Figure 5 is a map showing the number of raypaths traversing each cell. Raypath density is greatest in the Middle East, including Iran, Iraq, the Caucasus and southern Caspian region. The Arabian Peninsula has relatively good raypath density while Northern Africa, the Mediterranean and Europe are the poorest resolved regions. In these areas, where raypath density and azimuthal distribution is limited, as in Northern Africa, streaking occurs and the colored cells will resemble the event/station ray path map (Fig. 4). Only larger scale features, on the order of hundreds of kilometers (i.e., Mediterranean Sea), are resolvable. We will show, in later sections, that by applying further data selection criteria, to reduce multipathing, we can significantly increase the resolution of our data set. In areas of good raypath coverage we are able to image *Lg* propagation features on the order of 100 km and less (i.e., Caspian Sea and Red Sea).

Previous Lg propagation efficiency mapping methods. Numerous researchers have attempted to use the propagation characteristics of *Lg* to map crustal heterogeneity in a variety of regions throughout the world. For many years now, these studies have relied on qualitatively accessing the amplitude of *Lg* and then analyzing individual propagation paths in search of blockage features. For example RUZAIKIN *et al.* (1977) observed that the Tibetan Plateau does not effectively propagate *Lg* energy by

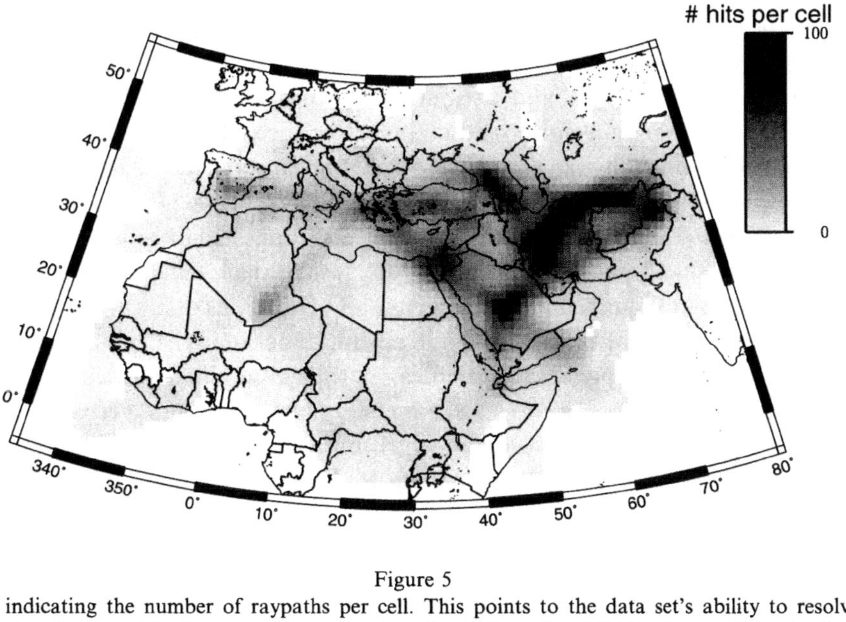

Figure 5
A map indicating the number of raypaths per cell. This points to the data set's ability to resolve *Lg* propagation features. The best coverage is in the seismically active Middle East and in the Arabian Peninsula. All shaded cells have at least 3 raypath hits.

simple visual inspection of *Lg* amplitudes. Later studies added more data (NI and BARAZANGI, 1983) and a slightly more quantitative approach, by measuring amplitude ratios (MCNAMARA *et al.*, 1996), to further refine the *Lg* propagation features in the Tibetan Plateau. A similar history exists in the Middle East. Early raypath mapping results of KADINSKY-CADE *et al.* (1981) were updated and improved by the more quantitative work of RODGERS *et al.* (1997). Qualitative raypath mapping has been an effective method for understanding first-order *Lg* propagation features.

Attempts to quantitatively assess the propagation quality of *Lg*, and then map it in a detailed grid, are relatively uncommon. One early example is KENNETT *et al.* (1985) who mapped crustal heterogeneity in the North Sea basin with a back-projection inversion scheme. *Lg* amplitude was assessed and assigned a code from 1–5 based on its amplitude relative to *Sn*. Their scheme included intermediate *Lg* amplitudes, as was an extension of the classification system defined earlier by GREGERSEN (1984). Heterogeneity values were then assigned to cells within the gridded region, based on the *Lg* classification codes of all paths traversing each cell. While KENNETT *et al.* (1985) represents an improvement over qualitative approaches, we found it limited in two ways. First, the *Lg* amplitude classification system is significantly less quantitative than a simple measure of normalized amplitude. Second, *Sn* amplitude is sensitive to source depth, radiation pattern and upper mantle *Qs*. Consequently, an assessment of crustal heterogeneity is significantly biased by the inclusion of *Sn* since the variations of *Lg* amplitude and crustal *Q* are not easily isolated.

Using a method more similar to ours, CAMPILLO *et al.* (1993) mapped the variation of *Lg* amplitudes in the southwestern region of the Alpine Range. They divided the region into a grid and attributed to each cell the mean value of the *Lg*/*Pn* amplitude ratios computed for all paths traversing the cell, weighted by their length in the particular cell. This approach represents an improvement, but is limited for two reasons. First, unlike the scattered energy of the *P-coda*, *Pn* amplitude is very sensitive to source depth, radiation pattern and upper mantle *Qp*, and velocity gradient. Consequently, the inclusion of *Pn* rather than the P-coda significantly biases an assessment of crustal heterogeneity, and variations of *Lg* amplitude are not easily isolated. Second, the mean amplitude ratio, rather than the maximum, will bias amplitude values in regions of efficient *Lg* propagation unrealistically low. This occurs when a path with weak to absent *Lg* energy traverses into a region of high *Q* or efficient *Lg* propagation. By using the maximum amplitude ratio we are essentially removing the effect an absent *Lg* (small *Lg*/*P-coda*) propagation path in regions of strong *Lg* (large *Lg*/*P-coda*) propagation energy. This is not the case when these weak or absent *Lg* amplitudes are included to compute the mean amplitude ratio assigned to a particular cell. Therefore, mapping the maximum *Lg*/*P-coda* value traversing a given cell is a quantitative and more accurate representation of the crustal heterogeneity responsible for the observed *Lg* amplitudes.

Technique limitations. It is important to note that our technique has several limitations. For example, we assume that any attenuating or blocking structure is isotropic. Specifically, the azimuth of propagation does not effect Lg propagation efficiency. While this is a simple assumption pertaining to the structure of the crust, it appears to be adequate enough to determine the first-order propagation characteristics of the region. Extracting more subtle features such as anisotropic behavior simply requires analysis of individual raypaths, in addition to the maximum *Lg*/*P-coda* cell maps.

Second, the selection of our *P-coda* velocity window (5.8–4.8 km/s) may allow some contamination from slowly propagating direct *Pg* arrivals. Our goal was to normalize *Lg* amplitudes with the signal closest to the phase (*Lg*) using a narrow window uncontaminated by other direct phases (*Pn*, *Pg*, *Sn*) yet long enough to provide a good RMS amplitude. While the window bound of 5.8 km/s may allow some leakage of *Pg* energy, the bulk of the window contains scattered *P-coda*.

Our approach also does not account for the effects of ray-bending and multipathing. Because the raypath of a regional phase, such as *Lg*, will bend around low velocity structures, the lateral extent of regions that attenuate *Lg* transmission may be underestimated. The longer the raypath the more serious the effect of ray-bending. This effect can be observed in both the Red Sea and Caspian Sea. Both are known to severely attenuate the transmission of *Lg*. However they do not appear as blocking regions when long raypaths are included in the low-resolution cell map (Fig. 4). To eliminate the effects of ray-bending and the consequent reduction of anomaly area in the next section, we will show, individual sub-regions with a finer grid and shorter, regional, raypaths.

Finally, our approach requires very high quality data and is heavily dependent on accurate amplitude ratio measurements. Since we use only the single maximum amplitude ratio to define the value of a particular cell, rather than an average of all paths traversing the cell, a single poor-quality amplitude measurement can seriously degrade the reliability of the image within that region. By applying our strict data selection criteria we have removed a considerable amount of erroneous paths, resulting in a data set of high quality *Lg*/*P-coda* amplitude ratios.

Regional Observations Lg Propagation Efficiency

We will now discuss the efficiency of *Lg* propagation within four sub-regions, the Western Mediterranean and Alboran Seas (Fig. 6), the Red Sea and Arabian Peninsula (Fig. 7), the Caspian Sea (Fig. 8) and the Eastern Mediterranean and Black Seas (Fig. 9). In this section we will compare our results with those of previous qualitative studies and demonstrate that our technique successfully images heterogeneous crustal structure responsible for the observed *Lg* amplitude variations. We will also demonstrate how we can enhance image features and improve resolution by

A

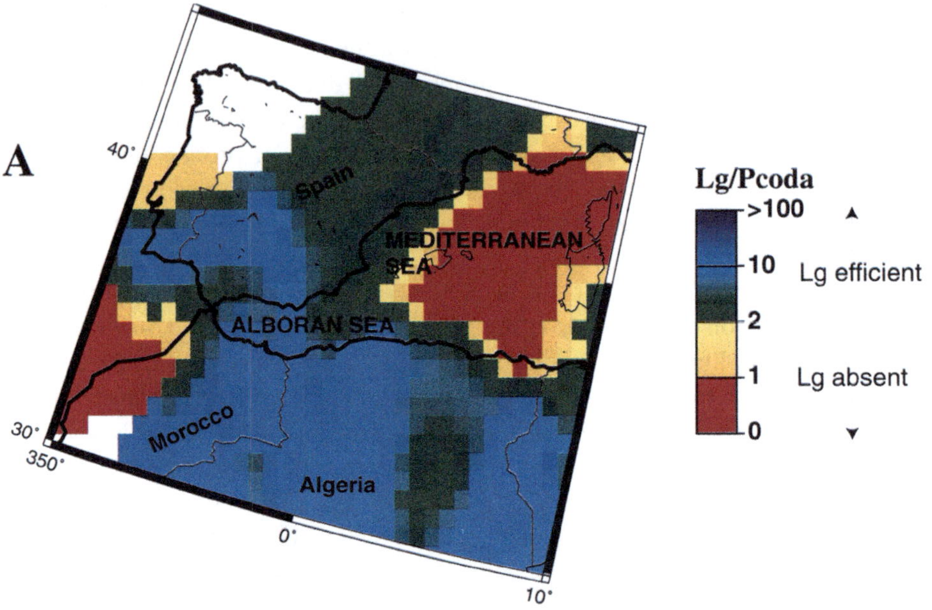

B

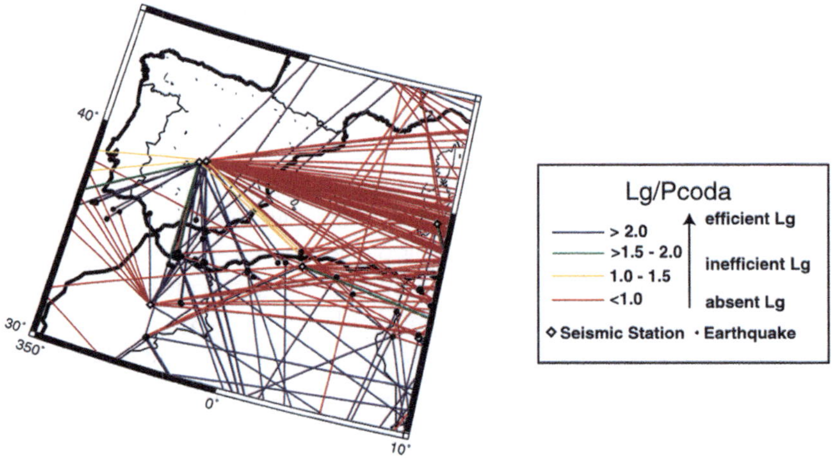

Figure 6

(a) Regional *Lg* propagation map showing the region surrounding the western Mediterranean and Alboran Seas. A finer grid (55.7 km cells) and shorter, regional raypaths lengths are used to improve resolution and allow the imaging of small-scale *Lg* propagation features. Small-scale features, such as efficient *Lg* propagation across the Alboran Sea and the sharp boundary between the Mediterranean Sea and continental Africa and Europe, are well imaged. (b) 135 regional raypaths used to construct the map in Figure 6a (maximum path length = 1500 km, maximum with absent *Lg* = 1000 km. The area containing the raypaths ranges in latitude from −20.0° to 20.0° and longitude from −20.0° to 20.0° and is slightly larger than the mapped area.

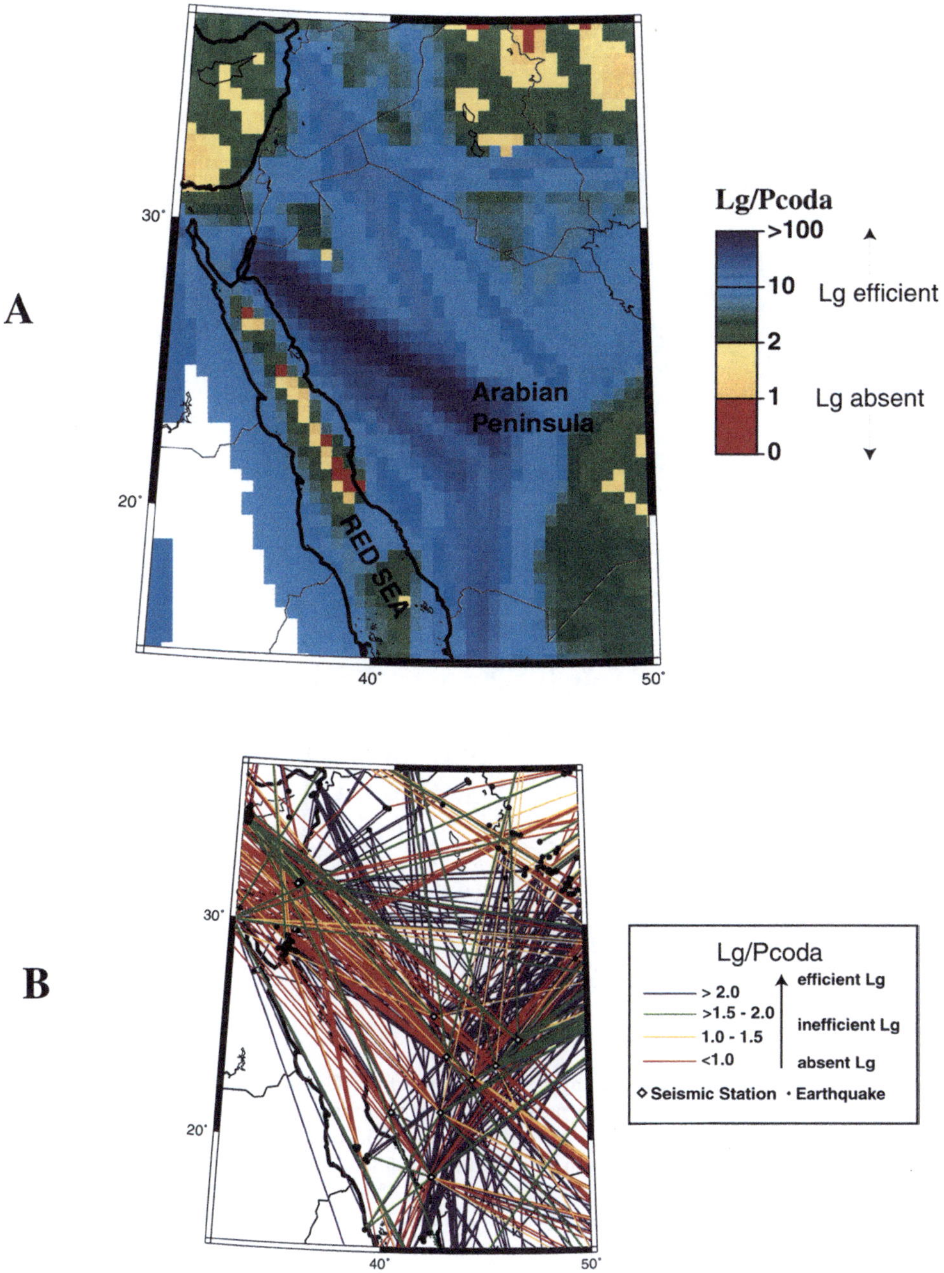

Figure 7

(a) Regional *Lg* propagation map delineating a narrow zone of absent to inefficient *Lg* propagation down the center of the Red Sea and the very efficient propagation region of the Arabian Peninsula. A finer grid (55.7 km cells) and shorter, regional raypaths lengths are used to improve the imaging of small-scale *Lg* propagation features. (b) 1247 regional raypaths used to construct the map in figure 6a (maximum path length = 2000 km, maximum with absent $Lg = 1500$ km). The area containing the raypaths ranges in latitude from 0.0° to 40.0° and longitude from 20.0° to 60.0° and is slightly larger than the mapped area.

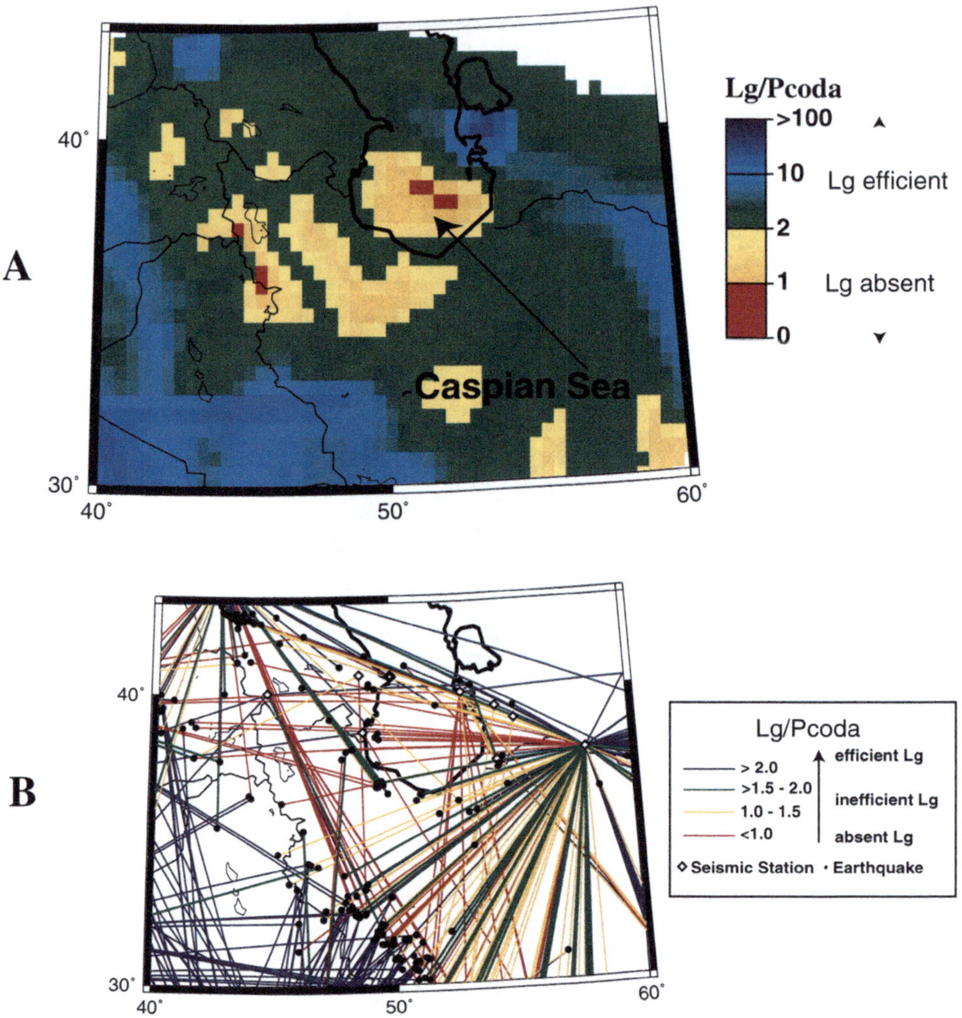

Figure 8
Regional *Lg* propagation map designating a zone of absent to inefficient *Lg* propagation in the southern Caspian Sea and Caucasus Mountains. A finer grid (55.7 km cells) and shorter, regional raypaths are used to improve the imaging of small-scale *Lg* propagation features. (b) 1009 regional raypaths used to construct the map in Figure 6a (maximum path length = 1500 km, maximum with absent *Lg* = 1000 km). The area containing the raypaths ranges in latitude from 20.0° to 50.0° and longitude from 30.0° to 70.0° and is slightly larger than the mapped area.

limiting *Lg* raypath lengths to regional distances and using a finer grid (55.7 km cells). Raypath lengths, with efficient to strong *Lg*, are limited to 1500–2000 km, depending on coverage within the particular region. Raypaths with inefficient to absent *Lg* are limited to 1000–1500 km.

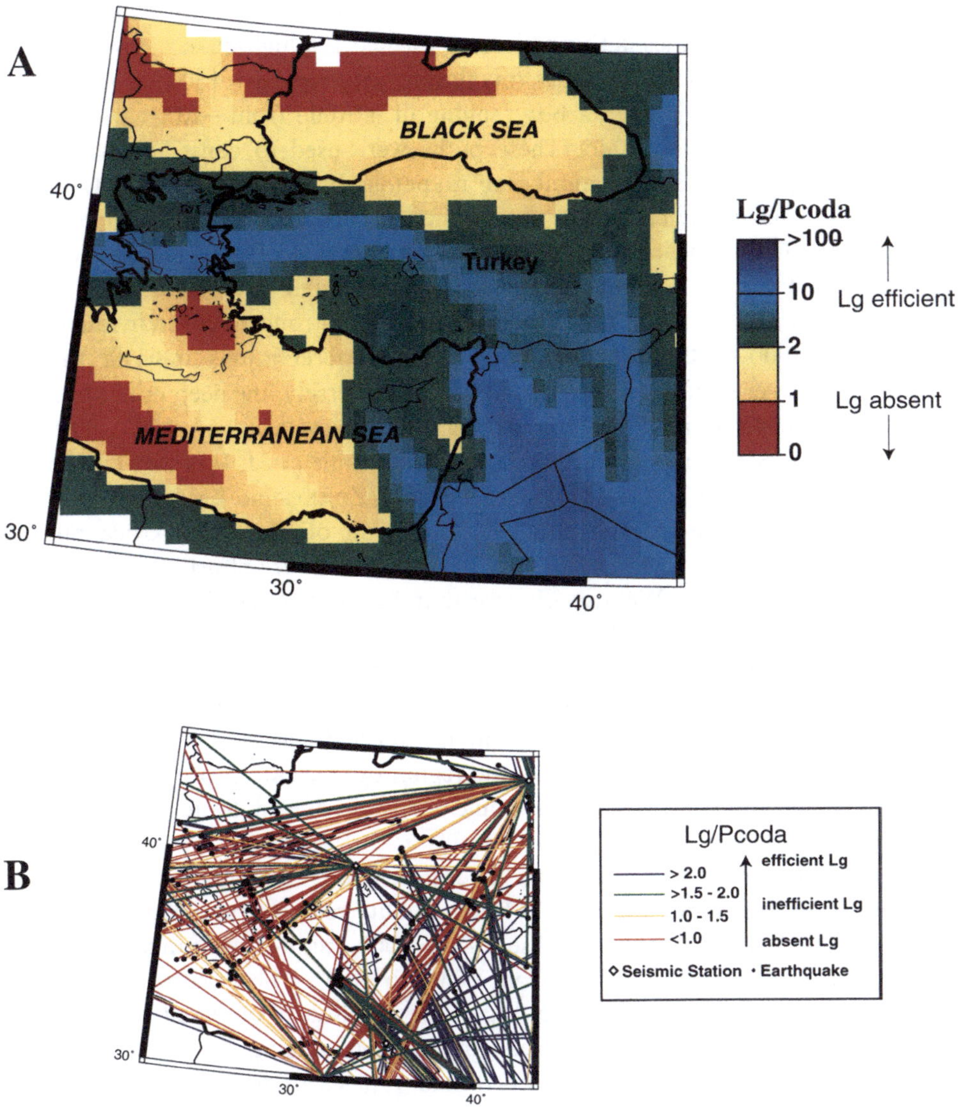

Figure 9

Regional *Lg* propagation map showing regions of absent to inefficient *Lg* propagation in the deep basins of the Mediterranean and Black Seas. *Lg* propagation is relatively efficient in Turkey, Syria, Israel, and the northern Arabian Peninsula. Also *Lg* propagation is relatively efficient in the shallow basins of the far eastern Mediterranean Sea, implying that they are underlain by continental crust. A finer grid (55.7 km cells) and shorter, regional raypaths are used to improve the imaging of small-scale *Lg* propagation features. (b) 664 regional raypaths used to construct the map in Figure 6a (maximum path length = 1500 km, maximum with absent $Lg = 1000$ km). The area containing the raypaths ranges in latitude from 20.0° to 55.0° and longitude from 15.0° to 50.0° and is slightly larger than the mapped area.

Western Mediterranean and Alboran Sea. To improve the resolution of *Lg* propagation characteristics in this region we selected 135 high-quality measurements with regional propagation distances (maximum path length = 1500 km, maximum with absent *Lg* = 1000 km) in a box ranging in latitude from −20.0° to 20.0° and longitude from −20.0° to 20.0°. These paths were used to construct the map in Figure 6. Despite the very low number of raypaths, this region clearly demonstrates how well our technique can define the boundaries between regions of different *Lg* propagation characteristics (Fig. 6). For example, it is well known that the stable continental interior of Northern Africa has extremely high values of *Lg*-coda *Q* (XIE and MITCHELL, 1990). To illustrate, we have observed strong *Lg* for individual paths as long as 5000–6000 km in this region. Such observations indicate that *Lg* propagation is very strong in Northern Africa. Conversely, the deep ocean basins of the Mediterranean do not efficiently propagate *Lg* (SHAPIRO *et al.*, 1996). The deep portions of the Mediterranean are underlain by oceanic crust. It is well known that *Lg* does not propagate across deep oceanic basins (ZHANG and LAY, 1995). Figure 6 demonstrates that the efficient propagation region of Northern Africa and the inefficient propagation region of the Mediterranean Sea are clearly delineated with our mapping technique. We also observe that *Lg* propagates efficiently across the shallow Alboran Sea. This is likely due to the fact that the Alboran Sea is underlain by continental rather than oceanic crust (SEBER *et al.*, 1996).

The pattern of crustal heterogeneity is less easily interpreted off the Atlantic Coast of Morocco. The blocking region is expected due to the transition from oceanic to continental crust, however, the region of efficient *Lg* propagation off the coast of Spain is not expected. This could likely be due to *Sn* to *Lg* conversions (SEBER *et al.*, 1994).

The Red Sea and Arabian Peninsula. To improve the resolution of *Lg* propagation features in this region we selected 1247 high-quality regional paths (maximum path length = 2000 km, maximum with absent *Lg* = 1500 km) in a box ranging in latitude from 0.0° to 40.0° and longitude from 20.0° to 60.0°. These paths were used to construct the map in Figure 7. The largest *Lg* amplitudes observed in our crustal heterogeneity mapping were observed for paths traversing the western Arabian Peninsula. As shown in Figure 7, *Lg/P-coda* ratios can be a high as 100. This observation is consistent with several previous qualitative studies of *Lg* amplitude variation (KADINSKY-CADE *et al.*, 1981; RODGERS *et al.*, 1997) and surface wave attenuation (SEBER and MITCHELL, 1992) in the Arabian Peninsula. Using a more quantitative approach, MELLORS *et al.* (1999) analyzed regional seismogram stacks for the amplitude variation of *Lg* as a function of source location and backazimuth using data recorded at portable broadband seismic stations deployed within the Arabian Peninsula (Vernon *et al.*, 1996). They noted that the largest *Lg* amplitudes were observed from events to the north in the Gulf of Aquaba and that seismograms from events located in Africa showed weak to absent *Lg* energy. This is consistent with the results of our study. The most striking feature in this map is a narrow

50–100-km wide band of inefficient *Lg* propagation. This band corresponds to the deepest portions of the northern Red Sea and indicates that raypaths traversing the Red Sea should show inefficient *Lg* propagation. We suspect that the narrow band of inefficient *Lg* propagation extends the length of the Red Sea, however, due to the limited number of earthquakes in the region and limited raypath coverage, we are unable to completely image this feature.

Caspian Sea. Previous studies that qualitatively assess *Lg* amplitude clearly show that paths traversing the southern Caspian Sea do not have significant *Lg* energy (MANGINO and PRIESTLEY, 1998; KADINSKY-CADE *et al.*, 1981; RODGERS *et al.*, 1997). When we limit raypaths lengths to include only regional distances (1009 paths, maximum path length = 1500 km, maximum with absent *Lg* = 1000 km) in a box ranging in latitude from 20.0° to 50.0° and longitude from 30.0° to 70.0°, the southern Caspian is clearly imaged as a region of weak to absent *Lg* propagation (Fig. 8). We also observe that the lesser Caucasus, adjacent to the Caspian Sea, are also imaged as a region of inefficient *Lg* propagation. This observation is evident in the seismogram in Figure 3 and is consistent with the previous qualitative studies discussed above. The region of *Lg* blockage in the Southern Caspian is likely larger than imaged due to *Lg* paths bending around regions of low velocity. We suspect that ray-bending will reduce the areal extent of the imaged blocking region. Due to limited shorter raypath coverage, we are not able to further improve the image resolution in this region without additional data. However, with our current data set and mapping method we are able to resolve *Lg* blocking structures in this region on the order of several hundred kilometers (Fig. 8).

Eastern Mediterranean and Black Seas. Finally, to improve the resolution of *Lg* propagation features in this region, we selected 664 high quality regional paths (maximum path length = 1500 km, maximum with absent *Lg* = 1000 km) in a box ranging in latitude from 20.0° to 55.0° and longitude from 15.0° to 50.0°. These paths were used to construct the map in Figure 9. Figure 9 clearly shows regions of absent to inefficient *Lg* propagation in the deep basins of the Mediterranean and Black Seas. In contrast, *Lg* propagation is relatively efficient in Turkey, Syria, Israel, and the northern Arabian Peninsula. These observations are expected and consistent with previous qualitative *Lg* amplitude studies (KADINSKY-CADE *et al.*, 1981; RODGERS *et al.*, 1997). In addition, *Lg* propagation is relatively efficient in the shallow basins of the eastern Mediterranean Sea, implying that they are underlain by continental crust.

Conclusions

The patterns of crustal heterogeneity that we have imaged, using *Lg* amplitude variations, correlate well with the major tectonic features of the region and are consistent with several previous investigations. From this we conclude that our empirical and qualitative method of mapping maximum *Lg*/*P-coda* values can successfully image relatively small-scale, detailed crustal structure. We have shown in our regional analysis that our approach can resolve *Lg* blocking features on the order of 50–100 km in regions of good raypath distribution and coverage (e.g., South Caspian Sea and Red Sea).

The inefficient propagation and blockage of *Lg* from earthquake sources, by complex tectonic features in the Middle East and Northern Africa, pose serious challenges for the utility of the *Lg* phase in CTBT monitoring. An accurate understanding of the factors that affect the variation of *Lg* amplitude and propagation characteristics is important to enable effective discrimination between earthquakes and nuclear explosions. With our mapping method, accurate determination of the expected *Lg* amplitude variations, due to crustal structure within a particular region, can be quickly obtained and applied to discrimination efforts. Also, results from our imaging can be used to screen out regional paths that traverse *Lg* blocking regions. This may be required to accurately obtain crustal Q parameters using the *Lg* phase. For example, the inclusion of paths where *Lg* is absent from the seismogram would significantly bias tomographic Q models.

The resultant set of *Lg* propagation maps from our empirical approach has been demonstrated to be easily interpreted in terms of crustal structure, and can successfully image detailed blockage features often missed by analysis of raypaths alone. Furthermore it is extremely useful for monitoring the Comprehensive Test-Ban Treaty.

Acknowledgments

The authors would like to thank J. Lahr, P. Earle and 2 anonymous reviewers for comments that improved our manuscript.

Thanks to Terri Hauk and Stan Ruppert for data acquisition and archiving. Work performed under the auspices of the U.S. Department of Energy by the Lawrence Livermore National Laboratory under contract number W-7405-ENG-48. This work was supported by DOE LLNL grant #GI 99-82 to the University of Alaska. Maps were created using GMT (Wessel and Smith, 1990).

References

Aki, K. (1980a), *Scattering and Attenuation of Shear Waves in the Lithosphere*, J. Geophys. Res. *85*, 6496–6504.

Aki, K. (1980b), *Scattering and Attenuation of Shear Waves in the Lithosphere from 0.05 to 25 Hz*, Phys. Earth Planet. Interiors *21*, 50–60.

Bouchon, M. (1982), *The Complete Synthesis of Seismic Crustal Phases at Regional Distances*, J. Geophys. Res. *87*, 1735–1741.

EWING, W., JARDETZKY, W., and PRESS, F. (1957), *Elastic Waves in Layered Media* (McGraw-Hill, New York, 380 pp).

CAMPILLO, M., FEIGNIER, B., BOUCHON, M., and BETHOUX, N. (1993), *Attenuation of Crustal Waves Across the Alpine Range*, J. Geophys. Res. *98*, 1987–1996.

GUTENBERG, B. (1955), *Channel Waves in the Earth's Crust*, Geophysics *20*, 283–294.

GREGERSEN, S. (1984), *Lg Wave Propagation and Crustal Structure Differences near Denmark and the North Sea*, Geophys. J. R. Astron. Soc. *79*, 217–234.

HARTSE, H., TAYLOR, S. R., PHILLIPS, W. S., and RANDALL, G. E. (1997), *A Preliminary Study of Regional Seismic Discrimination in Central Asia with an Emphasis on Western China*, Bull. Seism. Soc. Am. *87*, 551–568.

HERRIN, E., and RICHMOND, J. (1960), *On the Propagation of the Lg Phase*, Bull. Seism. Soc. Am. *50*, 197–210.

KADINSKY-CADE, K., BARAZANGI, M., OLIVER, J., and ISAAKS, B. (1981), *Lateral Variations of High-frequency Seismic Wave Propagation at Regional Distances across the Turkish and Iranian Plateaus*, J. Geophys. Res. *86*, 9377–9396.

KENNETT, B. (1986), *Lg Waves and Structural Boundaries*, Bull. Seismol. Soc. Am. *76*, 1133–1141.

KENNETT, B., GREGERSEN, S., MYKKELTVEIT, S., and NEWMARK, R. (1985), *Mapping of Crustal Heterogeneity in the North Sea Basin via the Propagation of Lg-Waves*, Geophys. J. R. Astr. Soc. *83*, 299–306.

KNOPOFF, L., SCHWAB, F., and KAUSSEL, E. (1973), *Interpretation of Lg*, Geophys. J. R. Astron. Soc. *33*, 389–404.

MANGINO, S., and PRIESTLEY, K. (1998), *The Crustal Structure of the Southern Caspian Region*, Geophys. J. Int. *133*, 630–648.

MCNAMARA, D., OWENS, T., and WALTER, W. (1996), *Propagation Characteristics of Lg across the Tibetan Plateau*, Bull. Seismol. Soc. Am. *86*, 457–469.

MELLORS, R., (1997), *Preliminary Noise Survey and Data Report of Saudi Arabian Data, UCRL-ID-128949*, 15pp., Lawrence Livermore National Laboratory, Livermore CA.

MELLORS, R., VERNON, F., CAMP, V., AL-AMRI, A., and GHALIB, A. (1999), *Regional Waveform Propagation in the Saudi Arabian Peninsula* (in press).

MITCHELL, B. J. (1995), *Anelastic Structure and Evolution of the Continental Crust and Upper Mantle from Seismic Surface Wave Attenuation*, Rev. Geophys. *33*, 441–462.

NI, J., and BARAZANGI, M. (1983), *High frequency Seismic Wave Propagation Beneath the Indian Shield, Himalayan Arc, Tibetan Plateau and Surrounding Regions: High Uppermost Mantle Velocities and Efficient Propagation Beneath Tibet*, Geophys. J. R. Astr. Soc. *72*, 665–689.

OLIVER, J., and EWING, M. (1957), *Higher Mode Surface Waves and their Bearing on the Earth's Mantle, Bull. Seismol. Soc. Am. 47*, 187–204.

PRESS, F., and EWING, M. (1952), *Two Slow Surface Waves across North America*, Bull. Seism. Soc. Am. *42*, 219–228.

RODGERS, A., NI, J., and HEARN, T. (1997), *Propagation Characteristics of Short-period Sn and Lg in the Middle East*, *87*, 396–413.

RUZAIKIN, A., NERSESOV, I., KHALTURIN, V., and MOLNAR, P. (1977), *Propagation of Lg and Lateral Variations in Crustal Structure in Asia*, J. Geophys. Res. *82*, 307–316.

SEBER, D., BARAZANGI, M., TADILI, B., RAMDANI, M., IBENBRAHIM, A., BEN SARI, D., and OTMAN EL ALAMI, S. (1993), *Sn to Sg Conversion and Focusing along the Atlantic Margin, Morocco: Implications for Earthquake Hazard Evaluation*, Geophys. Res. Lett. *20*, 1503–1506.

SEBER, D., BARAZANGI, M., TADILI, B., RAMDANI, M., IBENBRAHIM, A., and BEN SARI, D. (1996) *Three-dimensional Upper Mantle Structure Beneath the Intraplate Atlas and Interplate Rif Mountains of Morocco*, J. Geophys. Res. *101*, 3125–3138.

SEBER, D., and MITCHELL, B. (1992), *Attenuation of Surface Waves across the Arabian Peninsula*, Tectonics *204*, 137–150.

SHAPIRO, N., BETHOUX, N., CAMPILLO, M., PAUL, A. (1996) *Regional Seismic Phases across the Ligrian Sea: Lg Blockage and Oceanic Propagation*, Phys. Earth Planet. Interiors *93*, 257–268.

VERNON, F., MELLORS, R., BERGER, J., AL-AMRI, A., and ZOLLWEG, J. (1996), Initial results from the deployment of broadband seismometers in the Saudi Arabian shield. In *Proc. 18th Annual Seism. Res.*

Symp. on Monitoring a Comprehensive Test Ban Treaty 4–6 September, 1996, (J. F. Lewkowicz, J. M. MacPhetres and D. T. Reiter, eds.), pp. 108–117.
WALTER, W. R., MAYEDA, K. M., and PATTON, H. (1995), *Phase and Spectral Ratio Discrimination between NTS Earthquakes and Explosions. Part I: Empirical Observations*, Bull. Seismol. Soc. Am. *85*, 1050–1067.
WESSEL, P., and SMITH, W. (1991), *Free Software Helps Display Data*, EOS *72*, 445–446.
XIE, J., and MITCHELL, B. (1990), *A Back-projection Method for Imaging Large-scale Lateral Variations of Lg Coda Q with Application to Continental Africa*, Geophys. J. Int. *100*, 161–11.
ZHANG, T., and LAY, T. (1995), *Why the Lg Phase does not Traverse Oceanic Crust*, Bull. Seismol. Soc. Am. *85*, 1665–1678.

(Received July 14, 1999, revised March 7, 2000, accepted March 10, 2000)

Pure appl. geophys. 158 (2001) 1189–1206
0033–4553/01/071189–18 $ 1.50 + 0.20/0

Pure and Applied Geophysics

Application of Regional Phase Amplitude Tomography to Seismic Verification

W. S. Phillips,[1] H. E. Hartse,[1] S. R. Taylor,[1] A. A. Velasco[1] and G. E. Randall[1]

Abstract—We have applied tomographic techniques to amplitude data to quantify regional phase path effects for use in source discrimination studies. Tomography complements interpolation (kriging) methods by extending our ability to apply path corrections into regions devoid of calibration events, as well as raising levels of confidence in the corrections because of their more physical basis. Our tomography technique solves for resolvable combinations of attenuation, source-generation, site and spreading terms. First difference regularization is used to remove singularities and reduce noise effects.

In initial tests the technique was applied to a data set of 1488, 1.0 Hz, P_g/L_g amplitude ratios from 13 stations for paths inside a 30° by 40° box covering western China and surrounding regions. Tomography reduced variance 60%, relative to the power-law distance correction traditionally applied to amplitude ratios. Relative P_g/L_g attenuation varied with geologic region, with low values in Tibet, intermediate values in basins and high values for platforms and older crust. Spatial patterns were consistent with previous path effect studies in Asia, especially local earthquake coda-Q. Relative spreading was consistent with expected values for P_g and L_g. Relative site terms were similar to one another, yet some tradeoff with attenuation was evident.

Tomography residuals followed systematic trends with distance, which may result from the evolution from direct to coda phases, focusing, model tradeoff or data windowing effects. Examination of the residuals using a kriging interpolator showed coherent geographical variations, indicating unmodeled path effects. The residual patterns often follow geological boundaries, which could result from attenuating zones or minor blockages that are too thin to be resolved, or that have anisotropic effect on regional phases. These results will guide efforts to reparameterize tomography models to more effectively represent regional wave attenuation and blockage. The interpolated residuals also can be combined with predictions of the tomographic model to account for path effects in discrimination studies on a station by station basis.

Key words: Discrimination, tomography, path correction, L_g blockage.

Introduction

Removing the effects of propagation path on seismic amplitudes allows us to isolate source signatures, which should lead to more effective discrimination between natural and man-made events. The Comprehensive Nuclear-Test-Ban Treaty (CTBT) requires us to monitor small events, which increases the importance of

[1] Seismic Research Center, Los Alamos National Laboratory, Los Alamos, New Mexico 87545, USA. E-mail: wsp@lanl.gov

recordings collected at regional distances. Path variations are strong at regional distances because ray paths sample the heterogeneous crust and upper mantle. Thus quantification of regional path effects is a crucial effort for extending Test Ban monitoring techniques to small events.

Path effects include intrinsic attenuation of different crust and upper mantle media, scattering or blockage by structural discontinuities, lateral refraction and multipathing. Seismograms from Tibet and from the Baikal rift, recorded at Lanzhou (LZH) in central China demonstrate the effects different paths can produce on phase amplitudes in a tectonically complex region (Fig. 1). Note the large and impulsive L_g phase from the Baikal event, compared to the lack of L_g from the Tibet event, relative to the P phases. Such behavior has been well documented in China and surrounding regions (e.g., RUZAIKIN *et al.*, 1977; McNAMARA *et al.*, 1996; RAPINE *et al.*, 1997).

Our strategy is to predict path effects on earthquake data by whatever means, and use this information to improve event identification and magnitude estimates on a station by station basis. To facilitate the event identification, corrections are expected to reduce the scatter in the earthquake population of discriminant ratios such as P_g/L_g (e.g., TAYLOR *et al.*, 1989). In addition, the corrections eliminate the effect of path on any separation between earthquake and explosion populations, and may act to increase or decrease that separation. Decreasing the separation is not necessarily detrimental; otherwise, we employ a practice of identifying events based on their path effects, which in many cases, is just a complicated way to discriminate based on event

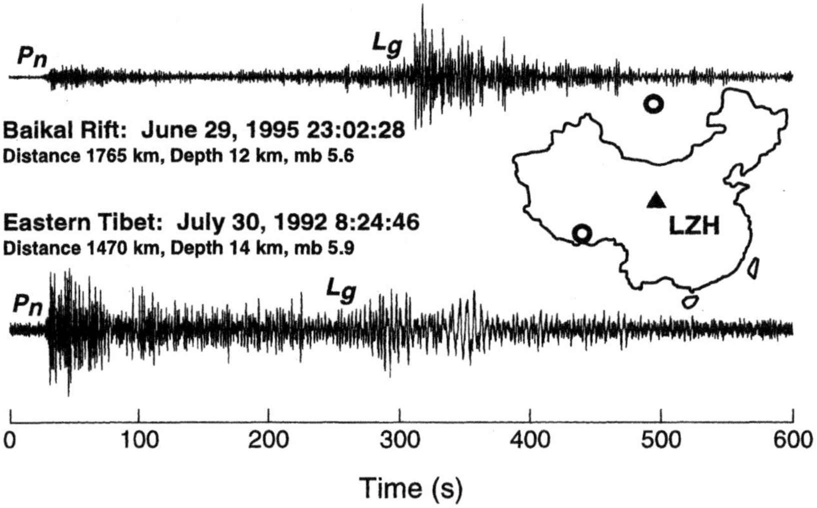

Figure 1
Sample, vertical component, broadband seismograms from the Baikal rift (top) and from eastern Tibet (bottom), recorded at LZH. The seismograms have been high-pass filtered at 0.5 Hz and aligned on their P_n arrival times. The L_g arrival is marked at its expected position in the Tibetan event record. Event (circles) and station (triangle) locations are indicated on the map.

epicenter. However, because we use only earthquakes to predict path effects, shallow path effects, or free-surface effects such as R_g to L_g scattering are largely ignored. Thus, after path correction, the discriminant ratio for an explosion or an exceptionally shallow earthquake will retain the shallow path or near-surface effect and may stand out on a discrimination plot for that reason. This depth effect is desirable if and when it occurs, because depth is an effective discriminant and depths are not generally well known for crustal sources.

Interpolation has been shown to be an effective path correction method, in terms of reducing variance in amplitude data (TAYLOR and HARTSE, 1998; PHILLIPS *et al.*, 1998; PHILLIPS, 1999; RODGERS *et al.*, 1999). These studies demonstrate that interpolated regional phase amplitude surfaces correlate well with regional geology, especially at lower frequencies (1 Hz); thus the results reflect path, rather than coherent source radiation or depth effects. Although earlier work was performed with simple, moving window smoothers, the non-stationary Bayesian kriging interpolator described by SCHULTZ *et al.* (1998) is more appropriate for monitoring work. Kriging provides error estimates and continuously differentiable surfaces, while the Bayesian modifications control extrapolation of the surface and errors into regions devoid of calibration events. In particular, the interpolated values go to zero and the interpolated errors go to the specified background error in such regions. This allows the inclusion of detailed background models by kriging the model residuals and merging the results with a map of the model predictions and errors. Background models provide the ability to extrapolate with added confidence and are the focus of this study. The resulting surface can be used as a lookup table to quickly correct for path effects during routine monitoring.

Another effective path correction method that has been tested extensively is the waveguide or path-parameter method. This method relies on correlations of amplitude data with readily available physical data such as topography, basin thickness or Moho depth to predict path effects (e.g., ZHANG and LAY, 1994; FAN and LAY, 1998; HARTSE *et al.*, 1998). The idea behind this method is that even though the topography, to choose one, may not be the physical cause of the path-effect variations, it may be correlated with the phenomena that are responsible. For example, elevated or rough topography could correlate with high intrinsic attenuation, both characteristic of an active tectonic region. The waveguide method is less effective than interpolation in reducing variance in amplitude data (PHILLIPS, 1999; RODGERS *et al.*, 1999), but is better able to predict data where no calibration events exist. PHILLIPS (1999) demonstrated the use of a waveguide model as a background model by applying Bayesian kriging interpolation to the waveguide model residuals; yielding a correction surface that matched data well in active regions and extrapolated more effectively into inactive regions.

Tomographic images of the attenuation of regional phases and coda can also be used to predict path effects (e.g., SINGH and HERRMANN, 1988; CAMPILLO, 1987; XIE and MITCHELL, 1990; CAMPILLO *et al.*, 1993; MITCHELL *et al.*, 1997; SANDVOL *et al.*,

1998). Tomography complements interpolation methods because images can be well resolved in areas where no calibration events occur. In addition, the physical basis of tomography increases our confidence in the derived path corrections. In this study we test tomography techniques using regional phase amplitude-ratio data, with emphasis on employing results as a background model, with or without additional kriged corrections, for use in an automatic monitoring scheme.

In initial tests we applied tomography techniques to 1 Hz P_g/L_g amplitude-ratio data. The use of ratio data decreases the influence of a possibly error prone, teleseismically determined m_b that is used to correct for source size and scaling prior to inversion. Of course, using ratios will lead to difficulties in interpreting results because two phases are involved. In addition, monitoring would require a correction surface for every ratio that might be desired, which is a large number if we consider all bands and phases that can be observed. For these reasons we support the use of single-phase correction surfaces for routine monitoring and plan to test the applicability of single-phase tomography to monitoring in future work.

The 1-Hz band we have chosen is useful for technique development in that it offers the most data and coverage. The 1-Hz ratios can improve the performance of multivariate discriminants (TAYLOR and HARTSE, 1997); however they generally perform poorly on their own (WALTER *et al.*, 1995; TAYLOR, 1996; HARTSE *et al.*, 1997). Results in this band can still be useful for magnitude estimation or for defining geophysically distinct regions that will be beneficial for assessing the progress of calibration efforts. As we move to higher frequencies, our hope is that path corrections will lower the frequency threshold at which source types are observed to separate (3–4 Hz), allowing the inclusion of smaller and more distant events that might be of poor signal-to-noise in the most effective discriminant bands (6–8 Hz).

Below we will outline data reduction procedures for stations in China and surrounding regions, and describe the tomographic method we applied. We will show relative attenuation results that correlate well with geological province as well as spatially coherent, unmodeled data that suggest new avenues of development.

Data

We used the Preliminary Determination of Epicenters catalog to guide the collection of regional earthquake data, archived at the Incorporated Research Institutions for Seismology, Data Management Center. For this study, we used data from earthquakes occurring between 1986 and 1996 recorded by 13 stations from China and surrounding areas: Enshi (ENH), Lhasa (LSA), Lanzhou (LZH), Urumchi (WMQ) and Xian (XAN) in China; as well as Boravoye (BRVK), Kurchatov (KURK) and Makanchi (MAK), Kazakhstan; Ala-Archa (AAK), Kirghizstan; Talaya (TLY) and Zalesovo (ZAL), Russia; Nilore (NIL) Pakistan; and Ulan Bator (ULN), Mongolia.

To process the data we first determined arrival times of the regional phases manually. After correcting for the instrument, RMS amplitudes were obtained for a number of pass bands between 0.5 and 8 Hz, using time-domain filtering techniques, aligning windows on picked arrivals if they existed or on preset group velocities if arrivals were not picked (HARTSE *et al.*, 1997). We used vertical component P_g/L_g ratios in the band 0.75 to 1.5 Hz in this study. The preset L_g group-velocity window was 3.6 to 3.0 km/s and the P_g group-velocity window was 6.2 to 5.2 km/s. Requiring paths to lie within a box defined by longitudes 70° to 110° and latitudes 25° to 55°, as well as distances between 300 and 2500 km, event depths less than 50 km and signal-to-noise (pre-*Pn*) greater than 2 for both phases, reduced the data set to 1488 measurements. A plot of the ray-path distribution indicates good coverage over most of the region (Fig. 2).

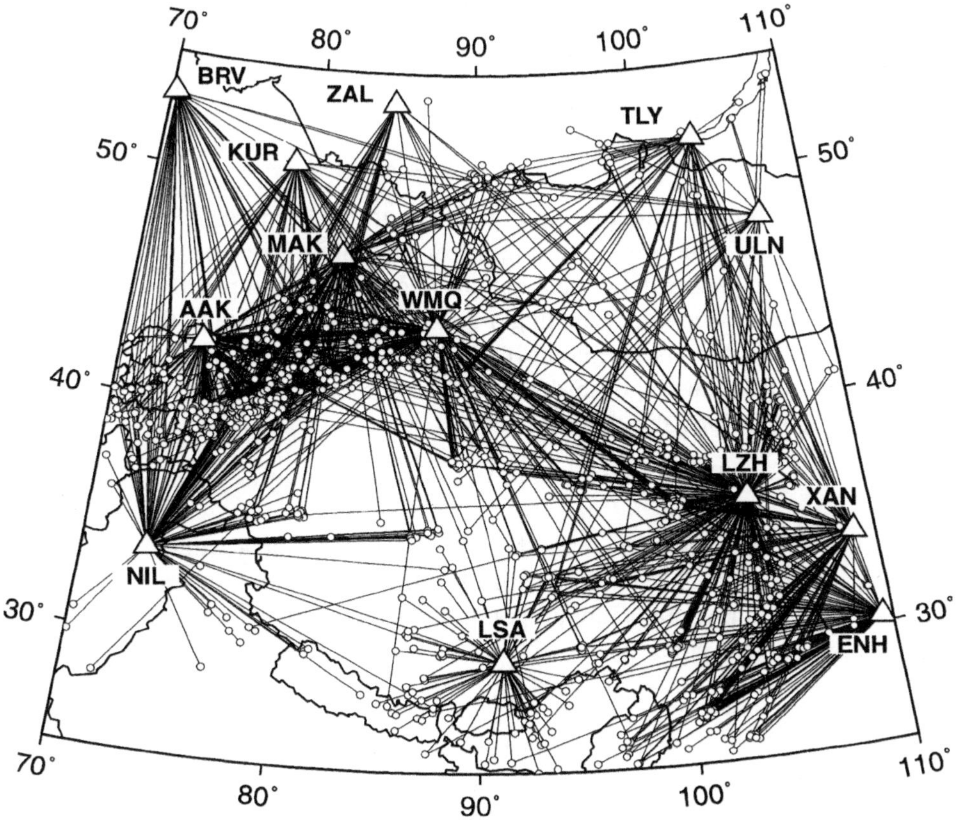

Figure 2
Great-circle ray paths representing data used in tomographic inversion. Stations are indicated by triangles and events by circles.

Tomography Method

Using tomographic techniques we solve for resolvable combinations of relative attenuation, site and spreading effects on regional phase amplitudes or amplitude ratios. For a homogeneous velocity model we express amplitude for a given frequency band,

$$a_{ijk} = c_k f(m_b^i) s_{jk} x_{ij}^{-\beta_k} \mathrm{e}^{-\alpha_k x_{ij}} \ ,$$

where the i, j, k indices represent source, site and phase type, respectively, c_k is a source generation term, f is source scaling, taken as a known function of m_b and frequency (CONG *et al.*, 1996; equations (6) and (10)), s_{jk} is site effect, x_{ij} is distance, β_k is the spreading coefficient and α_k is the spatial attenuation coefficient. We define a source-corrected amplitude,

$$d_{ijk} = a_{ijk}/f(m_b^i) \ .$$

Taking logarithms,

$$\log_{10} d_{ijk} = \log_{10} c_k + \log_{10} s_{jk} - \beta_k \log_{10} x_{ij} - \alpha_k x_{ij} \log_{10} \mathrm{e} \ .$$

Substituting capitals for $\log_{10}$ yields,

$$D_{ijk} = C_k + S_{jk} - \beta_k X_{ij} - \alpha_k x_{ij} \log_{10} \mathrm{e} \ .$$

We now discretize attenuation into an array of evenly spaced, latitude–longitude centered ellipses of constant α_k, setting major and minor axes to $\sqrt{2/2}$ of the local grid spacing so ellipses centered on a box of four adjacent grid points meet at a single point in the center (Fig. 3). Assuming a great-circle path, we take the chord length across each ellipse it intersects and normalize such that the sum of normalized chords becomes the total path length in degrees. Thus $\alpha_k x_{ij}$ becomes a sum of products of discretized α_k and normalized ellipse chords,

$$D_{ijk} = C_k + S_{jk} - \beta_k X_{ij} - \log_{10} \mathrm{e} \sum_m \alpha_{km} \delta x_{ijm} \ , \qquad (1)$$

where the new index, m, represents the discretization of the attenuation model and the δx_{ijm} are the normalized ellipse-chord lengths. After taking phase ratios, relative generation, site, spreading and attenuation terms replace those of equation (1) (dropping the phase index, k),

$$\Delta D_{ij} = \Delta C + \Delta S_j - \Delta\beta X_{ij} - \log_{10} \mathrm{e} \sum_m \Delta\alpha_m \delta x_{ijm} \ . \qquad (2)$$

This relies on our homogeneous model assumption that requires both rays to take the same, great-circle path. Note that the effect of the source-scaling corrections, $f(m_b^i)$, cancel or are reduced, depending on the frequency bands and event size. Tomographic inversion can be based on these linear equations once we rationalize singularities or tradeoffs between unknown parameters.

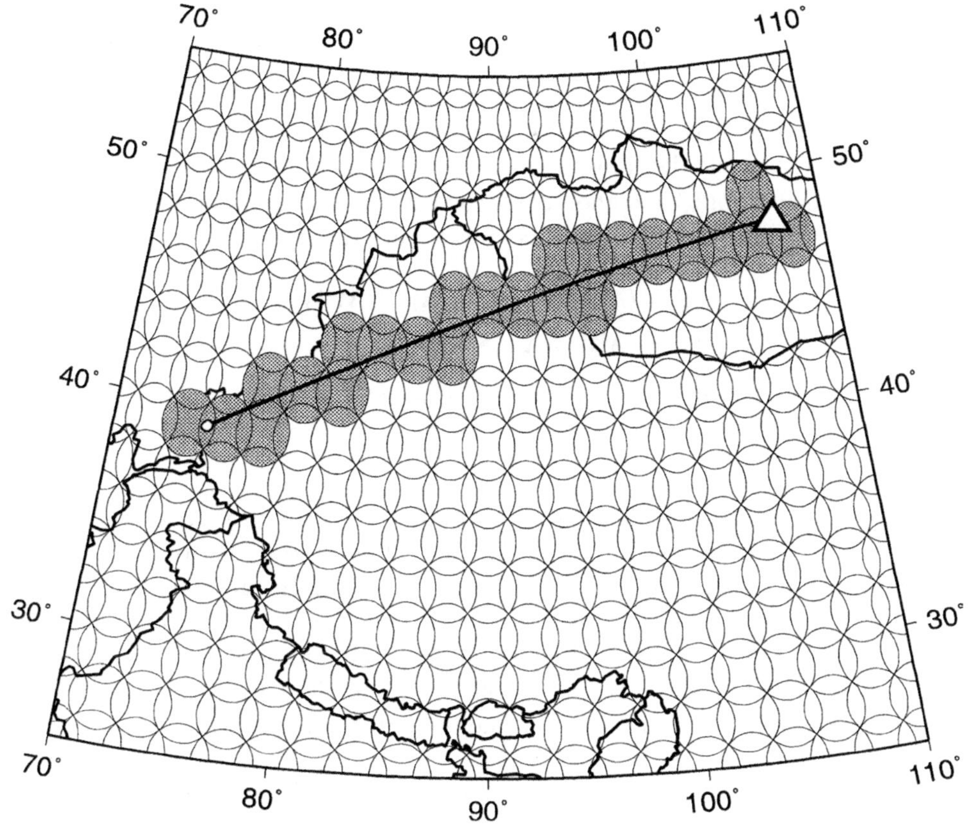

Figure 3
Array of ellipses used to parameterize attenuation. A 2° spacing is shown.

We use standard, least-squares techniques to solve equations (1) or (2), which can be written in matrix form,

$$\mathbf{Ax} = \mathbf{d} \ ,$$

where $\mathbf{d}$ is the data vector, $\mathbf{x}$ is the vector of unknowns and $\mathbf{A}$ represents the model equations. We apply first difference regularization to reduce the influence of noise on the results and to eliminate singularities due to unresolved combinations of model parameters. First difference regularization adds equations that constrain adjacent model parameters (in the case of attenuation terms) to be similar, weighted by a regularization parameter, which balances the importance of the constraint relative to constraints provided by the data. The larger the regularization parameter, the smoother the resulting image. Second difference regularization is often used in tomographic studies, although the first difference is more powerful in resolving singularities (e.g., PHILLIPS and FEHLER, 1991). The least-squares solution is written

$$\mathbf{x} = (\mathbf{A}^t\mathbf{W}^t\mathbf{W}\mathbf{A} + \mathbf{S}^t\Lambda^2\mathbf{S})^{-1}\mathbf{A}^t\mathbf{W}^t\mathbf{d} \ , \qquad (3)$$

where **S** represents the regularization equations, constraining certain model parameters to be similar, Λ is a diagonal matrix containing regularization weights, **W** is a diagonal matrix representing weights based on estimated data error and superscript t represents the transpose. In practice, we use one regularization weight for attenuation terms (λ_α) and another for site terms (λ_s). In addition, we assume data error to be uniform. Resolution and covariance are calculated via

$$\mathbf{R} = (\mathbf{A}^t\mathbf{W}^t\mathbf{W}\mathbf{A} + \mathbf{S}^t\Lambda^2\mathbf{S})^{-1}\mathbf{A}^t\mathbf{W}^t\mathbf{W}\mathbf{A},$$
$$\mathbf{C} = (\mathbf{A}^t\mathbf{W}^t\mathbf{W}\mathbf{A} + \mathbf{S}^t\Lambda^2\mathbf{S})^{-1} \ .$$

We can also estimate the variance of the predicted value along a new path,

$$\sigma_p^2 = \mathbf{a}^t\mathbf{C}\mathbf{a} \ ,$$

where **a** is a column vector representing equation (1) or (2) for that path, similar to a row of **A** (e.g., DRAPER and SMITH, 1981). This will provide errors for the amplitude correction surface generated from the tomographic image. In addition, the prediction error can be included when applying Bayesian kriging to tomography residuals, if we wish to combine tomography and interpolation results into one correction surface. The prediction error may also be useful for combining predictions of different background models, such as waveguide and tomography models, into one correction surface.

The tomographic calculations have been implemented using a Jacobi conjugate gradient solver (ITPACK). Coarse grid calculations were confirmed with a standard L-U decomposition code. We added switches that eliminate or combine unknown terms in equations (1) and (2) to avoid singularities. For example, the generation term (ΔC) must be combined with the site terms (ΔS_j) because neither can be resolved uniquely. More subtle tradeoffs can be evaluated by examining the resolution matrix.

We use residual variance to evaluate goodness of fit and model improvement. The trace of the resolution matrix is taken as the number of free parameters in the model (TARANTOLA, 1987), which is needed to calculate variance. Variance reduction will be quoted relative to a power-law fit with distance, which is the traditional correction for amplitude ratios in discrimination studies (e.g., HARTSE *et al.*, 1997).

Results

To parameterize the inversion we examined residual variance for a range of discretization intervals and regularization parameters. For this study, 1° and 2° attenuation term grids allowed similar fits to data, while a 5° grid was too coarse to fit the data well (Fig. 4). Results employing a grid spacing of 2° will be presented

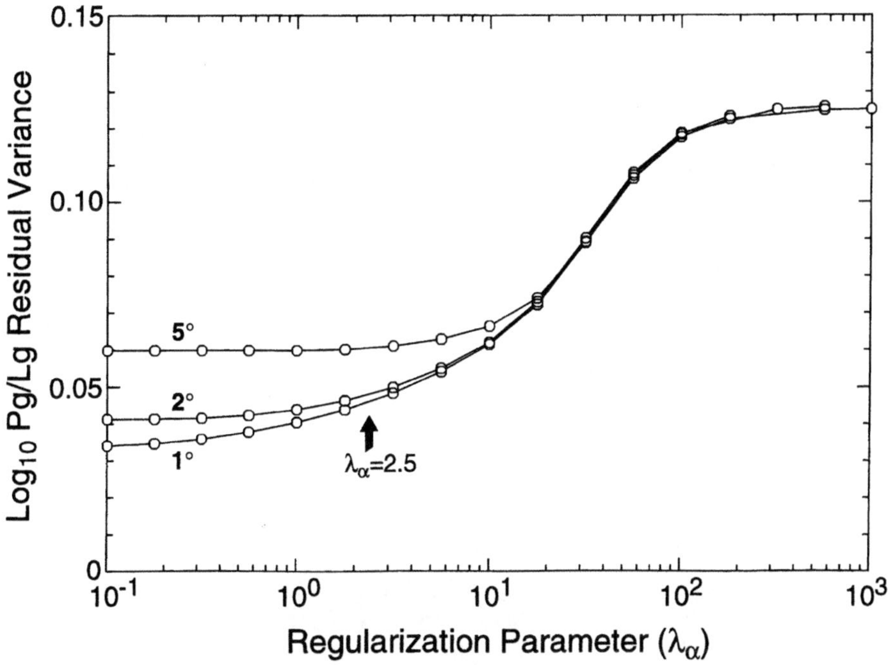

Figure 4
$\mathrm{Log}_{10}P_g/L_g$ residual variance versus attenuation regularization parameter. Results for 1°, 2° and 5° grid spacings are shown as marked. A regularization parameter of 2.5 was used in this study.

here. Regularization parameters must be chosen large enough to control the effects of noise, yet lie below the point where excessive smoothing causes the residual variance to increase rapidly (e.g., PHILLIPS and FEHLER, 1991). Our choice of 2.5 for attenuation regularization (Fig. 4) allowed 58 free parameters in the 2° model, as estimated from the trace of the attenuation portion of the resolution matrix. This number of free parameters is similar to that obtainable with an unregularized 5° grid; however the finer grid allows more spatial freedom to fit the data. Holding attenuation regularization fixed, the site term regularization parameter was determined to be 0.5, following the same procedure. This regularization constrains poorly resolved site terms to be similar to well resolved site terms and helps to avoid tradeoffs between site and near-station attenuation terms.

After removing a power-law fit with distance, we found the standard deviation of the residuals to be 0.36, representing the initial scatter in the ratio data. Allowing attenuation, spreading and source generation terms free produced a residual standard deviation of 0.23, or a variance reduction of 59% relative to the power-law fit. Replacing the source generation term with individual site terms resulted in only slight improvement and a variance reduction just over 60%. We plot relative P_g/L_g attenuation, $\Delta\alpha_m$, in color (Fig. 5). As the equations are set up, positive $\Delta\alpha_m$

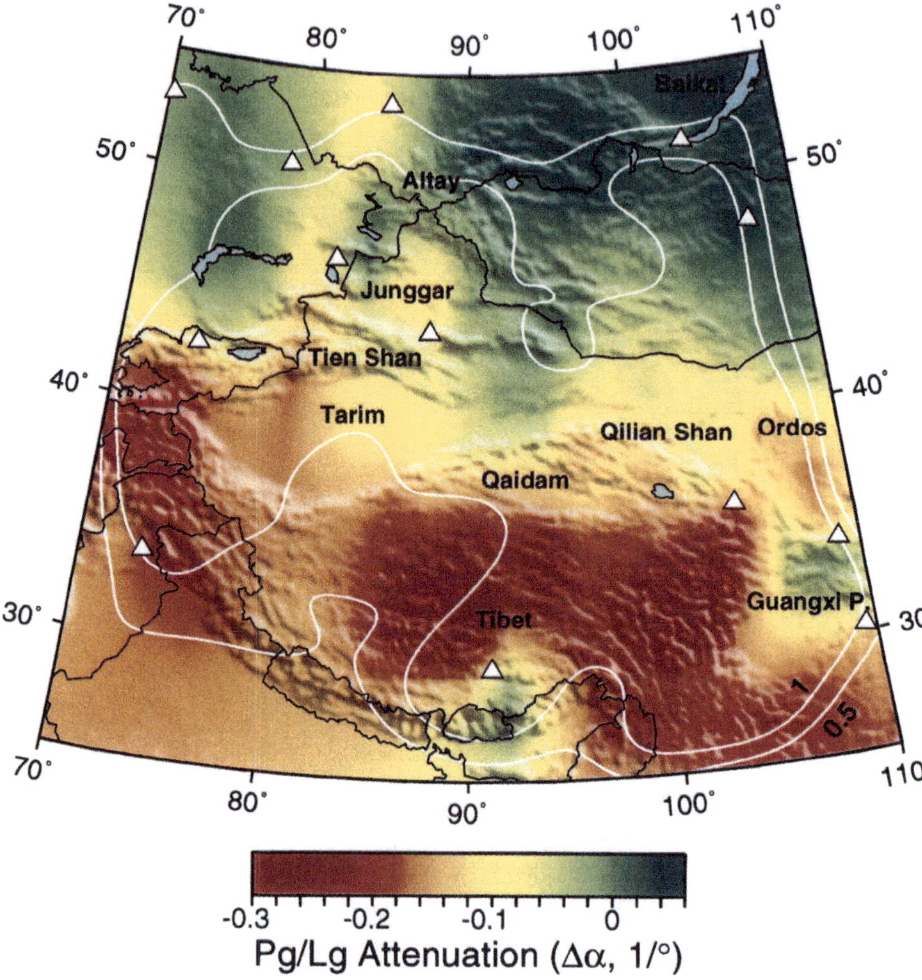

Figure 5
Relative attenuation from tomographic inversion of 1 Hz, P_g/L_g ratios. Green represents high relative attenuation (P_g attenuation high; L_g, low); red, *vice versa.* Colors are superposed on a topography gradient map. White contours indicate resolution as discussed in the text. Annotations mark geographical and geological features mentioned in the text. Triangles represent stations that provided data for this study.

(green) indicate P_g attenuates faster than L_g, while negative $\Delta\alpha_m$ (red) indicate the opposite. The tomography results are superposed on a topography map for interpretation purposes. Well-resolved portions of the image are indicated by contours of the diagonal component of the resolution, normalized by the ratio of the number of grid points to free parameters in the attenuation model. The normalization adjusts for regularization effects, which reduce resolution by averaging adjacent model parameters.

We found that P_g/L_g attenuation correlates well with geological province. The lowest values (low P_g, high L_g attenuation) fall within the Tibetan Plateau. This low is bounded sharply to the north by the Qaidam Basin, and to the east by the Guangxi platform. Intermediate to low values can be seen in the Qilian Shan and western Tian Shan ranges, as well as the Tarim, Ordos and Junggar Basins. Note the sharp transition between the Junggar Basin area and the surrounding mountain ranges. High P_g/L_g attenuation is found in the stable platform areas to the north and in the Guangxi platform to the east. Higher values are also found for trans-Himalaya paths to Lhasa (LSA), similar to patterns noted in spectral Q studies of regional phases recorded at LSA (REESE *et al.*, 1998). We expected higher values in the vicinity of station ZAL in Russia; however, the weak L_g observed for paths through the Tian Shan and border ranges, relative to other paths through the northern area, is responsible for the low pattern.

Relative site terms ranged from −0.8 to −0.3 for these data (Fig. 6), most clustering about −0.5. The mean can be considered the relative source generation term, which is incorporated into the site terms in this inversion. Including these terms improves the fit to data only slightly, yet adds definition to the attenuation image as site effects are more appropriately accounted for. The similarity between most well resolved, site terms may be an advantage of applying tomography to ratio data. However, two outlier site terms (KURK and LSA) appear to reflect the attenuation from nearby areas, even though these terms are reasonably well resolved.

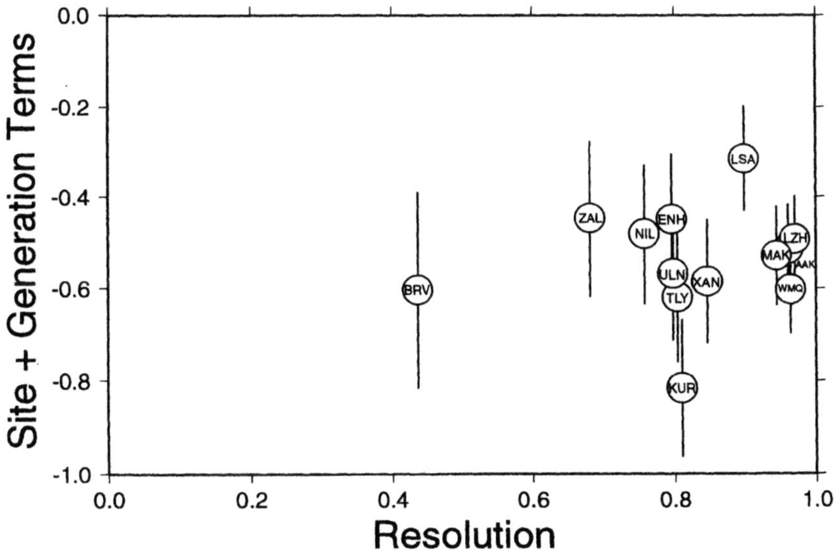

Figure 6
Relative site terms versus their resolution. Station names are indicated. Error bars represent twice the standard error of the estimate.

The relative spreading coefficient, $\Delta\beta$, was estimated to be 0.54 ± 0.11 (standard error). Again, the positive value indicates P_g spreads more quickly than L_g. Synthetic calculations generated using the inversion result and including similar levels of Gaussian noise resulted in correct recovery of the results, with some tradeoff evident between spreading and the range of attenuation terms. Based on 100 runs with different realizations of noise, the mean and standard deviation of the recovered $\Delta\beta$ was 0.50 ± 0.08.

Discussion

The tomography image is similar to results from previous Eurasian path effect studies. If we interpret the variation of P_g/L_g attenuation primarily to be the result of L_g effects (PHILLIPS *et al.*, 1998), we can compare with studies of L_g and L_g coda. RAPINE *et al.* (1997) prepared a qualitative map of L_g propagation in China, based on regional data from many stations. Most of China is described as efficient for L_g, while the Tibetan Plateau is described as inefficient. The boundary between the propagation regions matches our tomographic result reasonably well. We fail to see an attenuating zone for L_g along the Tibetan Plateau edges, as suggested by MCNAMARA *et al.* (1996); however, our signal-to-noise criteria have omitted most blocked L_g paths from the study and the regularization constraints act to smooth the influence of the few that remain. MITCHELL *et al.* (1997) produced a tomographic map of 1 Hz, L_g coda Q for Eurasia. This map is roughly consistent with ours, except for Mongolia, where we see efficient L_g (high P_g/L_g attenuation) and the L_g coda Q predicts inefficient L_g. However, the L_g coda-Q maps are of significantly larger scale and lower resolution than those presented here. Perhaps the best correspondence, in terms of spatial resolution, is with results of a local earthquake, crustal coda-Q study (JIN and AKI, 1988). The coda-Q measurements are unevenly distributed through China; however, where coverage is dense, such as along the eastern boundary of the Tibetan Plateau, the patterns are very similar to ours. This points out the value to discrimination research in performing local earthquake coda-Q studies in regions of dense station coverage, with care taken to limit the coda lapse times so only crustal materials are sampled (PHILLIPS *et al.*, 1988).

The distribution of data that are inconsistent with the tomography result can help us understand how well the underlying assumptions have been satisfied. We first examine residuals (data minus prediction) as a function of source–receiver distance by taking means within equally spaced bins in the logarithmic domain (Fig. 7). Recall that the relative spreading coefficient, $\Delta\beta$, was determined to be 0.54 ± 0.11 in our inversion. This indicates that P_g spreads faster than L_g so that ratios decrease with distance. In this discussion we refer to spreading loosely, intending apparent spreading, including possible measurement effects rather than pure geometrical spreading. Our $\Delta\beta$ of 0.54 can be thought to correspond to the zero-residual line in

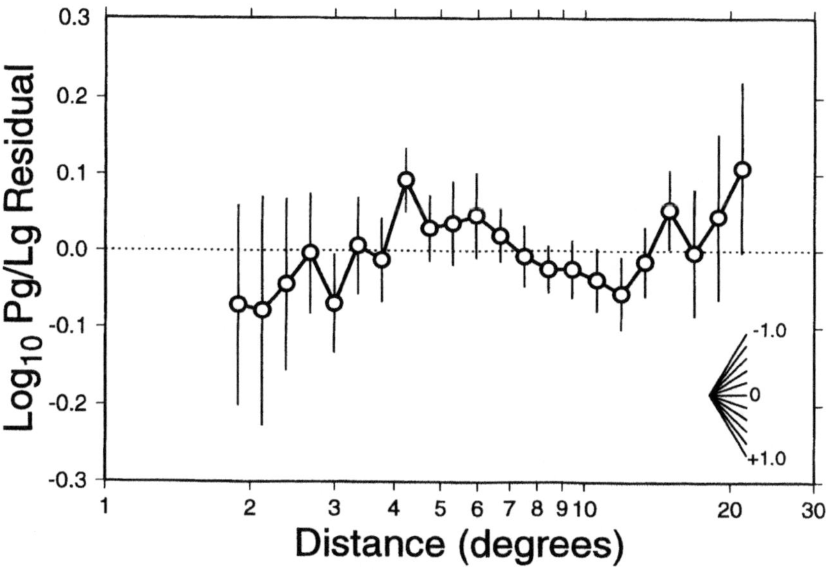

Figure 7
$\text{Log}_{10}P_g/L_g$ residuals (data minus prediction) binned by distance. Error bars represent twice the standard error of the mean. Spreading values relative to zero slope are indicated in the lower right corner.

Figure 7. We observe three ranges of similar trend in the residuals: up to 500 km, where spreading is 0.3 less than $\Delta\beta$; 500 to 1200 km, where spreading is 0.3 greater than $\Delta\beta$; and beyond 1200 km, where spreading is 0.4 less than $\Delta\beta$. Note the slopes read off the plot are opposite in sign to the change in spreading value because of the sign conventions in the inversion equations. Error bars show the middle trend is the most significant. These trends can result from a number of factors including: evolution from direct to coda energy; focusing of mantle P affecting P coda, thus P_g; tradeoffs with attenuation; and window-length, measurement effects. We further note that residuals are low for short distances and high for long distances, implying an underestimation of $\Delta\beta$ due to the influence of data from the extremes on the fitting. If we assume L_g spreading is 0.83 (NUTTLI, 1973; CAMPILLO *et al.*, 1984), our $\Delta\beta$ of 0.54 or greater implies a P_g spreading coefficient of 1.37 or greater, reasonably consistent with a numerical modeling estimate of 1.5 (CAMPILLO *et al.*, 1984). Clearly, further effort is warranted to understand distance trends in these data. Evaluation of various effects may be easier when single-phase data are inverted rather than ratios.

To examine the geographical distribution of inconsistent data we interpolate tomography residuals, one station at a time. Interpolation is performed with a Bayesian kriging algorithm (SCHULTZ *et al.*, 1998), using a data correlation function appropriate for L_g amplitudes in central China (ANDERSON and PHILLIPS, 1998). For all stations we see coherent geographical variations, indicating path effects that are inconsistent with the tomography result. For many stations, such as BRVK (Fig. 8),

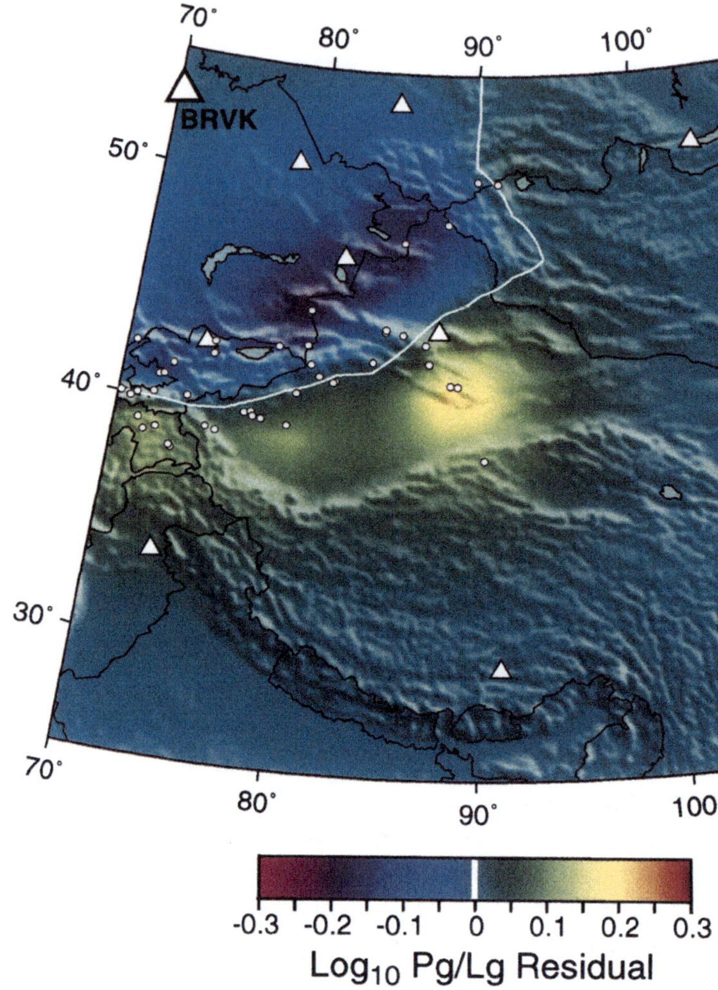

Figure 8
Spatial interpolation of $\log_{10}P_g/L_g$ residuals (data minus prediction) from station BRVK obtained by kriging. Warm colors represent underprediction, cold colors overprediction. Underpredicted data seek lower relative attenuation (warmer colors) along their paths through the tomography image (Fig. 5), overpredicted data, the opposite. Colors are superposed on a topography gradient map. A white line represents the zero contour. Triangles represent stations and small circles show locations of events recorded at BRVK that provided useable data to the study.

we see residual patterns that follow boundaries between geological regions. For BRVK a sudden increase in P_g/L_g ratios south of the Tian Shan may indicate a long, thin zone of high L_g attenuation, or a blockage. The high attenuation may be a structural or scattering, rather than an intrinsic, effect (e.g., BAUMGARDT, 1990). The blockage may have greater effect on ray paths traveling across the structure (to

BRVK, for example) than on ray paths traveling along the structure. Subparallel ray paths would tend to wash out the high attenuation zone during tomographic inversion. Our smoothing and coarse parameterization contribute to this effect.

The inconsistent data patterns suggest future modification of the tomography methods. We can explore the relaxation of regularization constraints in well resolved areas. We can also include anisotropic effects into the model parameterization, as has been done for velocity tomography (HEARN, 1996), and which may help resolve blockages. Lateral refraction and multipathing will also appear as inconsistent data; we may have to rely on numerical modeling to evaluate the importance of these effects.

More general improvement could come from the addition of constraints from censored (poor signal-to-noise) data into the tomography, especially for blocked L_g paths, as well as realistic data covariance reflecting correlation between data from similar paths. Furthermore, better magnitude estimates will reduce noise introduced by source corrections based on teleseismic m_b and facilitate the use of phase amplitude data, rather than ratios. The magnitudes could be improved by accounting for censoring effects on teleseismic amplitudes (e.g., MCLAUGHLIN, 1988), or by using regional envelopes or coda (AKI, 1980; MAYEDA, 1993). Tomography of phase amplitudes, as opposed to ratios, will be simpler to interpret and implement into production schemes. Comparison with ratio tomography results will indicate how well source effects have been accounted for.

The tomographic results can be used to correct for path effects and reduce scatter in discrimination studies. For a grid surrounding each station we can predict amplitudes or amplitude ratios, along with their errors. These surfaces can be used during the course of routine processing. In addition, the tomography can be employed as a background model, and corrections calculated by applying Bayesian kriging to the residuals (Fig. 8) to form the final correction surface. The kriging can allow for the tomography prediction errors in creating the correction surface. The final result will reflect the tomography model in areas where events rarely occur and an interpolated version of the data in areas where events do occur.

Conclusions

We have applied tomographic techniques to regional phase, amplitude-ratio data to solve for resolvable combinations of relative attenuation, source generation, site and spreading effects. The main benefit of using ratio data is the reduced importance of source corrections which introduce noise.

We obtained a variance reduction of 60% for 1488 P_g/L_g ratios in the 0.75–1.5 Hz band for a 30° by 40° box, covering western China and surrounding areas. Relative P_g/L_g attenuation correlated with geological province and compared well with previous studies of L_g, L_g coda and S coda of varying geographical resolution. Regions such as western Tibet were poorly resolved because blocked L_g paths were excluded from the study; the inclusion of the censored data constraints should improve the final model. Relative spreading was estimated to be 0.54 ± 0.11, consistent with literature values for P_g and L_g spreading. Interpolation of residuals showed coherent spatial patterns, indicating significant amounts of unmodeled path effect. Some unmodeled path variation can be attributed to long, thin zones of high attenuation, or blockage that are smoothed over in the tomographic image. Relaxing regularization constraints in well-resolved regions, and including anisotropy, may allow these zones to be modeled more effectively.

The tomography results can be used to predict correction factors and errors for source discrimination studies. In addition, residuals can be interpolated and added to corrections for a given station to combine the extrapolation power of tomography with the better fit to data of the interpolation methods.

Acknowledgments

Chris Young and Jim Hipp provided the Bayesian kriging code as well as needed guidance. We also thank Dan McNamara and an anonymous reviewer for their suggestions. Seismic data were retrieved from the IRIS DMC, GTOPO30 topography from the USGS. GMT and SAC software were used extensively. This work was performed under the auspices of the US Department of Energy by Los Alamos National Laboratory under Contract W-7405-ENG-36.

References

Aki, K. (1980), *Physical Basis for the Duration Magnitude and Recommended Practice for Coda Magnitude Determination*, Proc. 17th Assemb. ESC, Budapest, 73–77.

Anderson, K. K., and Phillips, W.S. (1998), *A Simple Procedure for Kriging Left-censored Data*, EOS Trans. Am. Geophys. U. *79*, F555.

Baumgardt, D. (1990), *Investigation of Teleseismic L_g Blockage and Scattering Using Regional Arrays*, Bull. Seismol. Soc. Am. *80*, 2261–2281.

Campillo, M., Bouchon, M., and Massinon, B. (1984), *Theoretical Study of the Excitation, Spectral Characteristics and Geometrical Attenuation of Regional Seismic Phases*, Bull. Seismol. Soc. Am. *74*, 79–90.

Campillo, M. (1987), *L_g Wave Propagation in a Laterally Varying Crust and the Distribution of the Apparent Quality Factor in Central France*, J. Geophys. Res. *92*, 12604–12614.

Campillo, M., Feignier, B., Bouchon, M., and Bethoux, N. (1993), *Attenuation of Crustal Waves across the Alpine Range*, J. Geophys. Res. *98*, 1987–1996.

Cong, L. L., Xie, J. K., and Mitchell, B. J. (1996), *Excitation and Propagation of L_g from Earthquakes in Central Asia with Implication for Explosion/Earthquake Discrimination*, J. Geophys. Res. *101*, 27779–27789.

Draper, N. R., and Smith, H., *Applied Regression Analysis, Second Edition* (John Wiley & Sons, New York 1981).

Fan, G., and Lay, T. (1998), *Statistical Analysis of Irregular Waveguide Influences on Regional Seismic Discriminants in China*, Bull. Seismol. Soc. Am. *88*, 74–88.

HARTSE, H. E., TAYLOR, S. R., PHILLIPS, W. S., and RANDALL, G. E. (1997), *A Preliminary Study of Regional Seismic Discrimination in Central Asia with Emphasis on Western China*, Bull. Seismol. Soc. Am. *87*, 551–568.

HARTSE, H. E., FLORES, R. A., and JOHNSON, P. A. (1998), *Correcting Regional Seismic Discriminants for Path Effects in Western China*, Bull. Seismol. Soc. Am. *88*, 596–608.

HEARN, T. (1996), *Anisotropic Pn Tomography in the Western United States*, J. Geophys. Res. *101*, 8403–8414.

JIN, A., and AKI, K. (1988), *Spatial and Temporal Correlation between Coda Q and Seismicity in China*, Bull. Seismol. Soc. Am. *78*, 741–769.

MCNAMARA, D. E., OWENS, T. J., and WALTER, W. R. (1996), *Propagation Characteristics of L_g across the Tibetan Plateau*, Bull. Seismol. Soc. Am. *86*, 457–469.

MAYEDA, K. (1993), $m_{bL_g\mathrm{coda}}$*: A Stable Single Station Estimator of Magnitude*, Bull. Seismol. Soc. Am. *83*, 851–861.

MCLAUGHLIN, K. L. (1988), *Maximum-likelihood Event Magnitude Estimation with Bootstrapping for Uncertainty Estimation*, Bull. Seismol. Soc. Am. *76*, 855–862.

MITCHELL, B. J., PAN, Y., XIE, J., and CONG, L. (1997), *L_g Coda Q Variation across Eurasia and its Relation to Crustal Evolution*, J. Geophys. Res. *102*, 22,767–22,779.

NUTTLI, O. W. (1973), *Seismic Wave Attenuation and Magnitude Relations for Eastern North America*, J. Geophys. Res. *78*, 876–855.

PHILLIPS, W. S., LEE, W. H. K., and NEWBERRY, J. T. (1988), *Spatial Variation of Crustal Coda Q in California*, Pure appl. geophys. *128*, 251–260.

PHILLIPS, W. S., and FEHLER, M. C. (1991), *Traveltime Tomography: A Comparison of Popular Methods*, Geophysics *56*, 1639–1649.

PHILLIPS, W. S., RANDALL, G. E., and TAYLOR, S. R. (1998), *Path Correction Using Interpolated Amplitude Residuals: An Example from Central China*, Geophys. Res. Lett. *25*, 2729–2732.

PHILLIPS, W. S. (1999), *Empirical Path Corrections for Regional Phase Amplitudes*, Bull. Seismol. Soc. Am. *89*, 384–393.

RAPINE, R. R., NI, J. F., and HEARN, T. M. (1997), *Regional Wave Propagation in China and its Surrounding Regions*, Bull Seismol. Soc. Am. *87*, 1622–1636.

REESE, C. C., RAPINE, R. R., and NI, J. F. (1998), *Lateral Variation of P_n and L_g Attenuation at the CDSN Station LSA*, Bull. Seismol. Soc. Am. *88*, 325–330.

RODGERS, A. J., WALTER, W. R., SCHULTZ, C., MYERS, S., and LAY, T. (1999), *A Comparison of Methodologies for Representing Path Effects on Regional P/S Discriminants*, Bull. Seismol. Soc. Am. *89*, 394–408.

RUZAIKIN, A. I., NERSESOV, I. L., KHALTURIN, V. I., and MOLNAR, P. (1977), *Propagation of L_g and Lateral Variations in Crustal Structure in Asia*, J. Geophys. Res. *82*, 307–316.

SANDVOL, E., SEBER, D., BARAZANGI, M., MOHAMAD, R., TURKELLI, N., GURBUZ, C., ZOR, E., and GOK, R. (1998), *Seismological Research in the Middle East*, Proceedings, 20th Ann. Seis. Res. Symp. Santa Fe, 745–753.

SCHULTZ, C., MYERS, S., HIPP, J., and YOUNG, C. (1998), *Nonstationary Bayesian Kriging: Application of Spatial Corrections to Improve Seismic Detection, Location and Identification*, Bull. Seismol. Soc. Am. *88*, 1275–1288.

SINGH, S., and HERRMANN, R. B. (1988), *Regionalization of Crustal Coda-Q in the Continental United States*, J. Geophys. Res. *88*, 527–538.

TARANTOLA, A. *Inverse Problem Theory* (Elsevier, Amsterdam 1987).

TAYLOR, S. R., DENNY, M. D., VERGINO, E. S., and GLASER, R. E. (1989), *Regional Discrimination between NTS Explosions and Western U.S. Earthquakes*, Bull. Seismol. Soc. Am. *79*, 1142–1176.

TAYLOR, S. R. (1996), *Analysis of High-frequency Pg/Lg Ratios from NTS Explosions and Western U.S. Earthquakes*, Bull. Seismol. Soc. Am. *86*, 1042–1053.

TAYLOR, S. R., and HARTSE, H. E. (1997), *An Evaluation of Generalized Likelihood Ratio Outlier Detection to Identification of Seismic Events in Western China*, Bull. Seismol. Soc. Am. *87*, 824–831.

TAYLOR, S. R., and HARTSE, H. E. (1998), *A Procedure for Estimation of Source and Propagation Amplitude Corrections for Regional Seismic Discriminants*, J. Geophys. Res. *103*, 2781–2789.

WALTER, W. R., MAYEDA, K. M., and PATTON, H. J. (1995), *Phase and Spectral Ratio Discrimination between NTS Earthquakes and Explosions. Part I: Empirical Observations*, Bull. Seismol. Soc. Am. *85*, 1050–1067.

XIE, J. K., and MITCHELL, B. J. (1990), *A Back-projection Method for Imaging Large-scale Lateral Variations of L_g Coda Q with Application to Continental Africa*, Geophys. J. Int. *100*, 161–181.

ZHANG, T. R., and LAY, T. (1994), *Analysis of Short-period Regional Phase Path Effects Associated with Topography in Eurasia*, Bull. Seismol. Soc. Am. *84*, 119–132.

(Received August 24, 1999, revised February 10, 2000, accepted April 24, 2000)

Pure appl. geophys. 158 (2001) 1207–1250
0033–4553/01/071207–44 $ 1.50 + 0.20/0

Pure and Applied Geophysics

Sedimentary Basins and the Blockage of *Lg* Wave Propagation in the Continents

DOUGLAS R. BAUMGARDT[1]

Abstract—The phenomenon of "*Lg* blockage," where *Lg* is strongly attenuated by crustal heterogeneities, poses a serious problem to CTBT monitoring because *Lg* is an important seismic phase for discrimination. This paper examines blockage in three continental regions where the *Lg* blockages may be caused by large, enclosed sedimentary basins along the propagation path. The Barents Sea Basin blocks *Lg* propagation across the Barents Sea from the Russian nuclear test sites at Novaya Zemlya to Scandinavian stations. Also, "early *Lg*" waves are observed in *Sn* codas on NORSAR, NORESS, and ARCESS recordings of Novaya Zemlya explosions where direct *Lg* is blocked. Early *Lg* waves may have resulted from *Sn*-to-*Lg* mode conversion at the contact between the Barents Basin and the Kola Peninsula. The Northern and Southern Caspian Sea Basins also block *Lg* waves from PNEs and earthquakes, perhaps due to thick, low-velocity, low-Q sediments replacing the granitic layer rocks in the crust. *Lg* blockage has also been observed in the Western Mediterranean/Levantine Basin due to low-Q sediments and crustal thinning. A "basin capture" model is proposed to explain *Lg* blockage in sedimentary basins. In this model, shear waves that reverberate in the crust and constitute the *Lg* wave train are captured, delayed, and attenuated by thick, low-velocity sediments that replace the "granitic" layer rocks of the upper crust along part of the propagation path. *Sn* waves, which propagate below the basin, would not be blocked and in fact, the blocked *Lg* waves may be diverted downward into *Sn* waves by the low velocity sediments in the basin.

Key words: *Lg* blockage, sedimentary basins, nuclear explosions, earthquakes, discrimination.

Introduction

The *Lg* phase has assumed critical importance in monitoring the Comprehensive Test-Ban Treaty (CTBT), especially for event identification. Recent studies in seismic discrimination have shown that explosions and earthquakes may be separable, based on the earthquakes having higher *Lg* excitation relative to P than explosions (BENNETT *et al.*, 1995; BAUMGARDT and YOUNG, 1990; HARTSE *et al.*, 1997; KIM *et al.*, 1997). The proper use of *Lg* amplitude measurements for seismic source characterization requires that propagation path effects on *Lg* amplitudes be understood. MURPHY *et al.* (1997), in a study of Peaceful Nuclear Explosions

[1] ENSCO, Inc., 5400 Port Royal Rd., Springfield, VA 22151, USA. Fax: 703-321-4609, E-mail: doug@ensco.com

(PNEs) in the Former Soviet Union (FSU), have observed uncharacteristically small *P*/*Lg* amplitude ratios for nuclear explosions, presumably caused by unexplained effects of heterogeneous crustal structure on regional phase propagation. Other studies have shown that laterally heterogeneous crustal structure can cause *Lg* "blockage" which makes measurements of *P*/*Lg* amplitude ratios impossible (BAUMGARDT, 1990).

The regional seismic phase identified as *Lg* is usually, but not always, the most prominent phase on regional seismograms. It has usually been interpreted as the superposition of higher-mode Rayleigh waves that primarily propagate in the crust (KNOPOFF *et al.*, 1973) and, alternatively, as the superposition of supercritically reflected shear waves in the continental crust (CAMPILLO *et al.*, 1985; KENNETT, 1986). *Lg* propagation has only been observed in the continents and does not appear to cross ocean basins or paths that contain 100 to 200 km of oceanic crust (EWING *et al.*, 1957).

The term "*Lg* blockage" refers to the sudden disappearance of the phase along a particular propagation path that crosses a geologic or tectonic structure boundary or lateral heterogeneity. Although *Lg* propagates efficiently across continental regions, *Lg* blockages have been observed in several continental areas. Explanations for the cause of *Lg* blockage include (1) Drastic variations in crustal thickness and topography, (2) scattering from tectonic boundaries, (3) strong anelastic attenuation in upper-crustal sediments, and (4) lack of an upper-crustal "granitic layer."

RUZAIKIN *et al.* (1997) proposed crustal thickness variations to explain the inefficient propagation of *Lg* in the Tibetan Plateau. The region is known to have unusually thick (up to 70 km) crust due to the convergent type tectonics of the region. KENNETT *et al.* (1985) also suggested that *Lg* propagation across the Norwegian Sea might be blocked by a region of crustal thinning beneath a graben structure. KENNETT (1986) has shown with ray-tracing modeling that *Lg* blockage can be caused by sudden crustal thinning, as might occur at a continent/ocean boundary (e.g., SHAPIRO *et al.*, 1996). The second explanation has been invoked by a number of investigators who have used the efficiency of *Lg* propagation as a method for mapping tectonic boundaries (e.g., KADINSKY-CADE *et al.*, 1981; NI and BARAZANGI, 1983). BAUMGARDT (1985, 1990) has provided evidence for *Lg* scattering in the Ural Mountains, a collisional suture zone, which partially blocks the propagation of *Lg* waves from the Semipalatinsk test site in eastern Kazakh to the ARCESS array. MITCHELL and HWANG (1987) have suggested that the attenuation of *Lg* may be strongly affected by low *Q* sediments, which are known to be present in the Barents Basin (CHAN and MITCHELL, 1985). MCNAMARA *et al.* (1996) attribute *Lg* blockage in Tibet to low *Q* and perhaps the unusually thick crust beneath the Tibetan Plateau. The "missing granitic layer" explanation originates from the idea that such a layer is essential to provide the waveguide for *Lg* propagation, a phase only observed in continental regions. PIWINSKII (1981) has explained the poor propagation of *Lg* in the Caspian Sea region of the Soviet Union in terms of an implied "missing granitic layer" that has also been reported in the Soviet geophysical literature for the Caspian Sea sedimentary basin. Theoretical studies of CAO and MUIREHEAD (1993) and JIH (1995) show how anomalous crustal structure associated with substantial crustal thinning can cause partial *Lg* blockage. ZHANG and LAY (1995) argue that the extremely thin oceanic crust is primarily responsible for the inefficient propagation of *Lg* in the oceans.

This paper revisits this issue by reviewing data that reveals *Lg* blockages in three continental regions: the Barents Sea near the FSU test site of Novaya Zemlya, the Russian/Iranian border region, including both the northern and southern Caspian Sea, and in the Eastern Mediterranean/Levant region in the Middle East. The method used to identify the cause of *Lg* blockage is to make crustal cross sections for the blocked and unblocked paths, and to look for differences between the two paths. These crustal cross sections are derived from available databases of elevation and depth to the crystalline basement (sediment depths), and inferred and measured crustal thicknesses. This method assumes that regional phases propagate along great circle paths from source to receiver. This analysis reveals that the deep sedimentary basins in these regions may account for the *Lg* blockages. In addition to presenting the data examples of *Lg* blockage in the sedimentary basins, a mechanism to explain *Lg* blockages in sedimentary basins, called "basin capture," is presented.

The Barents Sea Sedimentary Basin

NORSAR Observations

The characteristics of regional waveforms recorded at NORSAR from several nuclear explosions have been analyzed. These include two Soviet peaceful nuclear explosions (PNEs) on the Kola Peninsula and in western Russia near the White Sea and five explosions at Novaya Zemlya. The source parameters for these explosions are given in Table 1, and the locations of the explosions and great circle paths from

Table 1

Parameters for events studied at NORSAR

Date	Time	Latitude	Longitude	m_b	Region
09/04/72	07:00:3.6	67.69	33.45	4.6	Kola Peninsula PNE
08/10/78	07:59:57.7	73.31	54.70	5.9	Novaya Zemlya
10/11/80	07:09:57.2	73.36	54.82	5.7	Novaya Zemlya
08/18/83	16:00:58.6	73.38	54.87	5.9	Novaya Zemlya
09/25/83	13:09:57.9	73.35	54.39	5.8	Novaya Zemlya
10/25/84	06:29:57.7	73.37	54.96	5.9	Novaya Zemlya
07/18/85	21:14:57.4	65.97	40.86	5.0	White Sea/ Archangel'sk PNE

the explosions to NORSAR are shown on the map in Figure 1(A). The paths from the two PNEs are entirely continental whereas the path from Novaya Zemlya to

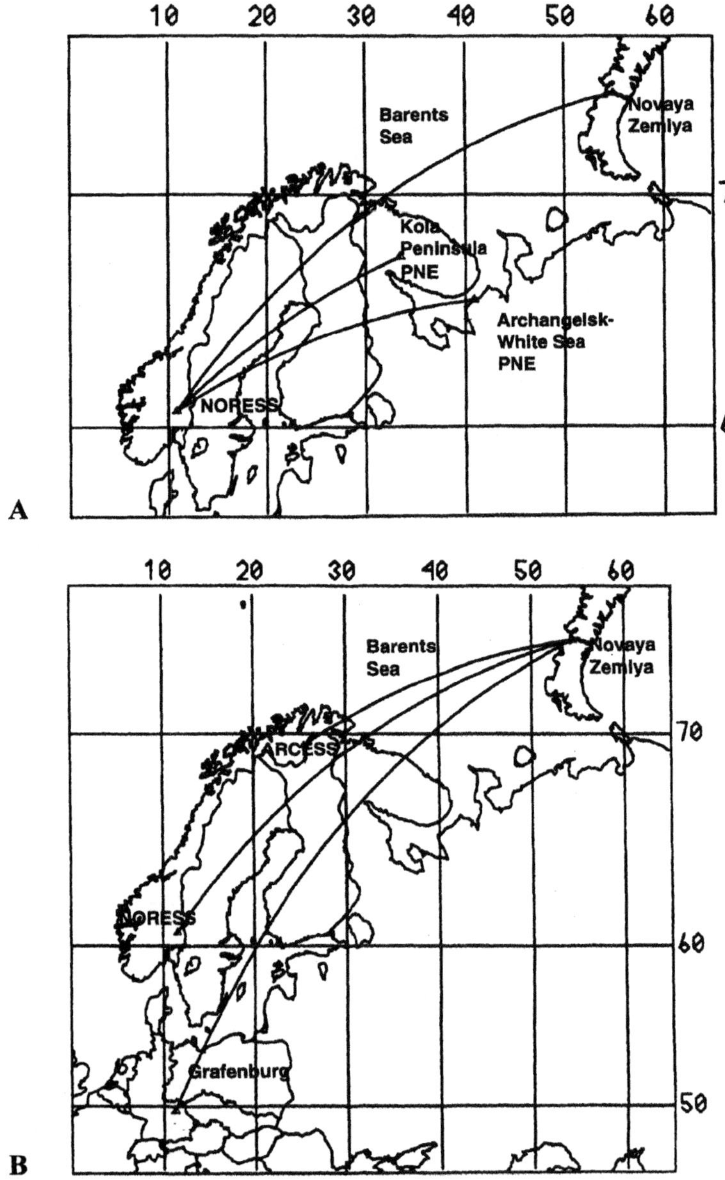

Figure 1
(A) Map showing the locations of the Kola and White Sea/Archangel'sk PNEs and the great-circle propagation paths from the Novaya Zemlya nuclear test site to the NORSAR array. (B) Map showing the great-circle propagation paths from the Novaya Zemlya test site to the ARCESS, NORESS, and Grafenburg arrays.

NORSAR has a segment that includes the Barents Sea. Thus, comparison of the *Lg* waves from the PNEs with those from Novaya Zemlya will show directly the effects of the Barents segment on the *Lg* propagation efficiency. The Novaya Zemlya explosions selected for this study registered magnitudes less than 6.0 and were not clipped at the times when *Sn* and *Lg* are expected.

Figure 2 shows a record section of the waveforms recorded at the N01A0 channel of NORSAR for the two PNEs and one of the Novaya Zemlya explosions. Each waveform has been filtered in the 0.6 to 3.0 Hz band. The arrivals identified as *Pn*, *Sn*, and *Lg* are marked on the waveforms. The two PNEs produced waveforms that resemble typical regional waveforms, with relatively sharp *Pn* onsets, an emergent arrival corresponding to the *Sn* onset, and strong *Lg* arrivals. The Novaya Zemlya explosion produced a strong *P* signal at NORSAR, an emergent *Sn* onset similar to the one from the PNEs, but no corresponding *Lg* arrival. It should be noted that the *Pn* signal from Novaya Zemlya is clipped, therefore the actual amplitude of *Pn* relative to *Sn* is actually much greater than that implied in Figure 2.

Figure 2 clearly shows that *Lg* has been greatly reduced relative to *P* for the Novaya Zemlya explosion, compared with those of the nearby PNEs. Although some

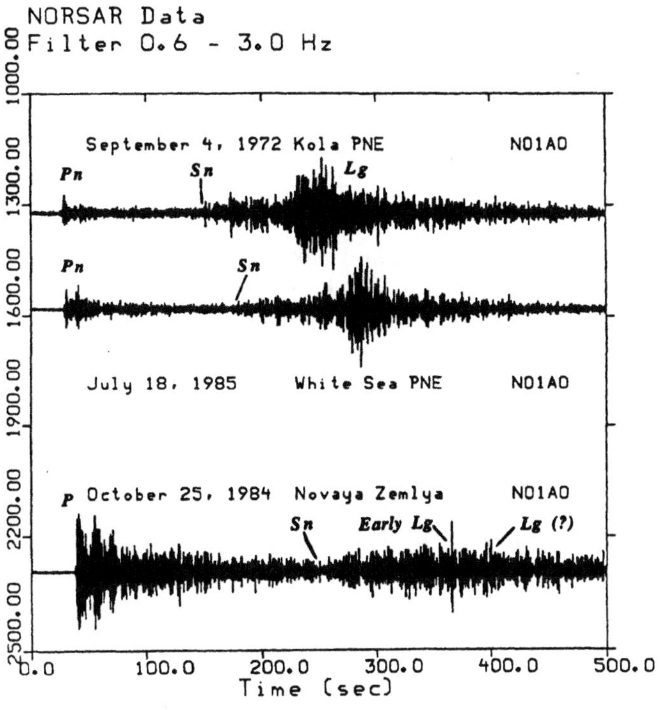

Figure 2

Record section of waveforms, recorded at the N01A0 element of the NORSAR array, from the Kola and White Sea/Archangel'sk PNEs and a Novaya Zemlya nuclear explosion. The numbers on the vertical axis denote the epicentral distance in km.

attenuation in *Lg* would be expected due to the increased distance of Novaya Zemlya from NORSAR compared with the PNEs, the lack of *Lg* from Novaya Zemlya cannot be explained by anelastic attenuation effects alone.

One notable feature of the Novaya Zemlya recording is a strong arrival that appears in the *Sn* coda well ahead of the expected 3.5 km/sec *Lg* arrival. This arrival corresponds in time to the "early *Lg*" phase discussed by BAUMGARDT (1990), that was also observed for recent Novaya Zemlya explosions recorded at NORESS. This phase was interpreted as being an *Lg* phase because its phase velocity, measured by the frequency-wavenumber (FK) method, was less than 4.0 km/sec, the value expected for *Lg*. However, its group arrival time is too early to be the *Lg* that propagates directly from Novaya Zemlya with group velocity of 3.5 km/sec.

This phase is interpreted to be an *Lg* phase produced by mode conversions of *Sn* at the interface of the Barents Sea and the Kola Peninsula, similar to those observed at coastlines in other regions for offshore events recorded on land (e.g., ISACKS and STEPHENS, 1975; BARAZANGI, 1977; CHINN *et al.*, 1980; SEBER *et al.*, 1993). The point at which the direct *Sn* from Novaya Zemlya intersects the Kola Peninsula is at a distance of about 13° from NORSAR, as shown in Figure 3(A). If *Lg* originates at that point, due to mode conversion from *Sn*, the converted *Lg* would lag the *Sn* phase by about 110 seconds, using the travel-time curve of Eurasia of GUPTA *et al.* (1980), shown in Figure 3(B). This is about the time delay between the *Sn* onset and the *Sn*-coda arrival in Figure 2.

Figure 4 shows a plot of the incoherent beams of all five Novaya Zemlya explosions recorded at NORSAR. Incoherent beams are array-stacked log-rms envelopes computed from the NORSAR vertical-component seismograms. Each trace begins after the *Pn* onset time and has been shifted vertically by one log-rms unit for display purposes, consequently the vertical axis only gives relative log-rms amplitudes. Note that the envelope traces begin after the maximum amplitude of the first arrival *P*, which is clipped on most channels, and the horizontal dashed lines show the mean noise level over two minutes ahead of the *P* onset. Each waveform was filtered from 0.6 to 3.0 Hz before the log-rms stack traces were computed.

The first solid vertical line indicates the inferred onset of *Sn*, and the vertical dashed line marks the expected arrival time of the presumed *Sn*-to-*Lg* conversion at 110 seconds after the *Sn* phase. At this time a peak is evident on the beam with log-rms amplitude above that of the maximum *Sn* amplitude. The second solid vertical line marks the inferred arrival time of the direct *Lg* phase, which should have a group velocity of 3.5 km/sec. This arrival seems to be manifested on each trace by a flattening of the coda level after the early-*Lg* onset. The signal-to-noise ratio at the time of the expected on-time *Lg* is quite large, although the *Lg* amplitude is much less than the *Sn* amplitude.

In summary, the NORSAR data establish that the propagation of the *Lg* phase from Novaya Zemlya to Norway is partially blocked, and that the shear

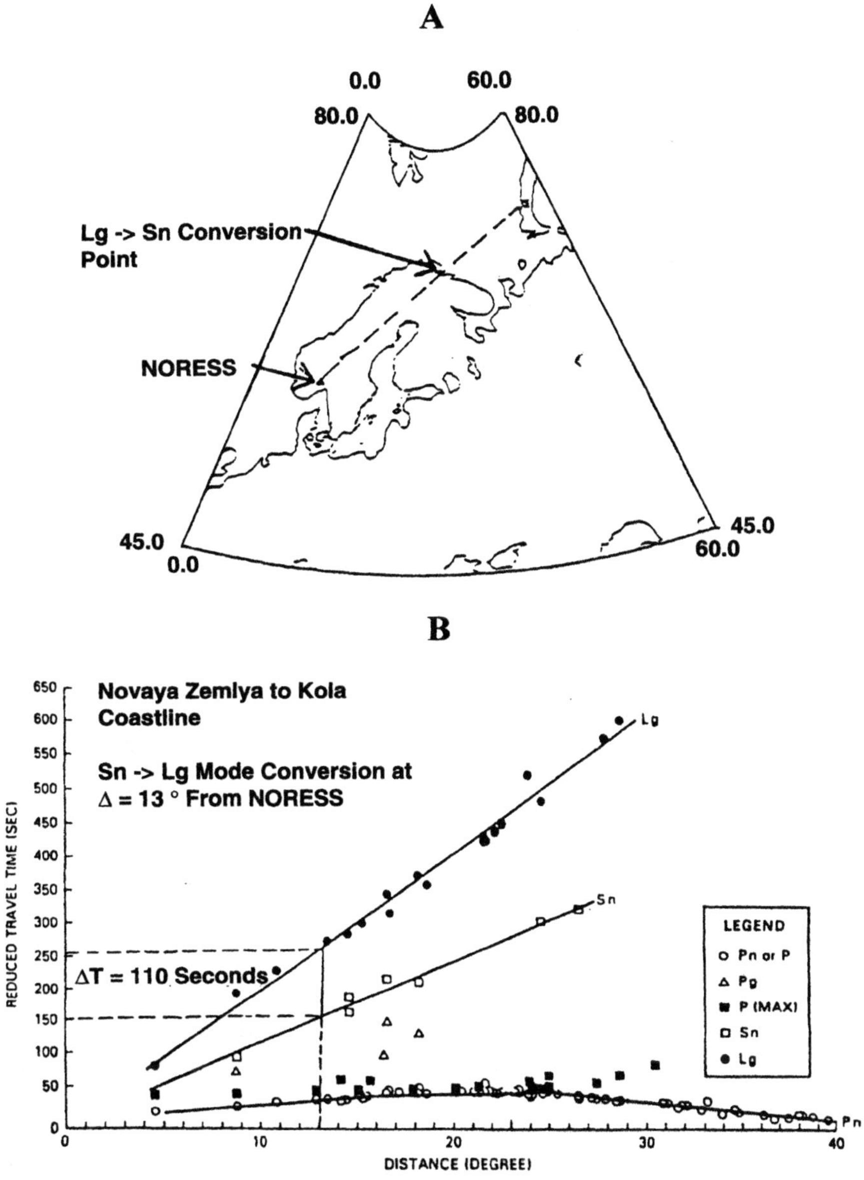

Figure 3

(A) Map showing the path from Novaya Zemlya to NORSAR. The distance from the edge of the Kola Peninsula to NORSAR is indicated. (B) Travel-time curve, after GUPTA *et al.* (1980), showing the distance and the *Sn* to *Lg* time difference of 110 seconds for a presumed *Sn*-to-*Lg* mode conversion at the coast.

phases from Novaya Zemlya are dominated by the *Sn* phase and an *Sn*-to-*Lg* mode conversion at the Barents Sea–Kola Peninsula interface. A direct *Lg* can also be inferred on the incoherent-beam traces, which is more apparent on the

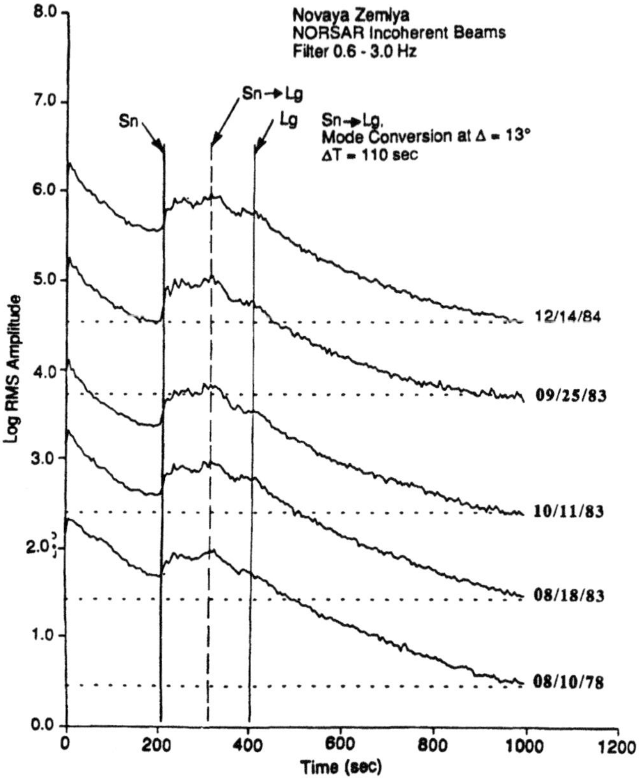

Figure 4
Incoherent beam plots at NORSAR, starting just after the maximum *Pn* amplitude, for Novaya Zemlya explosions. Plots have been shifted for display purposes. Dashed lines indicate the average noise level ahead of the *Pn* phase. The assumed times for the *Sn* and presumed *Sn*-to-*Lg* mode conversion at the Kola Peninsula coast are indicated.

NORSAR beams than on the NORESS beams studied by BAUMGARDT (1990). Although the *Sn* coda at the expected time for *Lg* does have high signal-to-noise ratio, this phase has apparently been considerably attenuated along the propagation path.

ARCESS, NORESS, and Grafenberg Array Observations

Figure 1(B) shows the propagation paths for a "reversed profile" of the one shown in Figure 1(A) that incorporates the broadband Grafenberg array and the regional arrays, NORESS and ARCESS. The event parameters of the Novaya Zemlya explosions that have occurred since ARCESS began recording data are presented in Table 2.

The vertical-component waveforms for the three explosions recorded at the ARCESS array center element (ARA0) are shown in Figure 5. The phase marked "early *Lg*" has an apparent group velocity of about 4.3 km/sec, which is considerably faster than the expected 3.5 km/sec velocity of direct *Lg*. There is virtually no energy at the 3.5 km/sec time.

The results of incoherent beam analysis for the December 4, 1988 event are shown in Figure 6 as a record section of logged rms incoherent beams for the ARCESS and

Table 2

Event parameters for Novaya Zemlya events recorded at NORESS, ARCESS, and Grafenburg

Date	Time	Latitude	Longitude	m_b
05/07/88	22:49:58.0	73.35	54.47	5.6
12/04/88	05:19:53.6	73.49	54.18	5.7
10/24/90	14:57:54.7	42.86	54.66	5.4

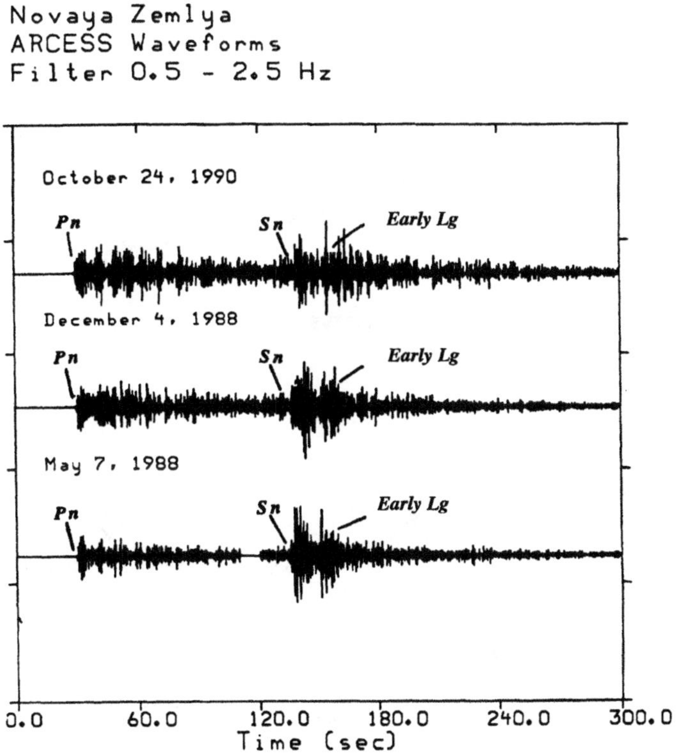

Figure 5
ARCESS recordings of recent Novaya Zemlya explosions. The times of *Pn*, *Sn*, and the inferred early *Lg* are indicated.

NORESS vertical components. The interesting features visible in the ARCESS recordings of the Novaya Zemlya explosions are that the *Sn* waves have clear onsets and amplitudes which exceed the *P* wave amplitudes, and the amplitudes increase 10 to 15 seconds into the *Sn* coda. This later coda energy is the "early *Lg*" arrival observed at NORSAR and NORESS. Since NORESS and NORSAR are collatated, the "early *Lg*" phase can be seen in the short-period data from both arrays. The relative arrival time between the secondary pulse at ARCESS and the "early *Lg*" at NORESS is consistent with an *Lg* group velocity of 3.5 km/sec.

Figure 7 shows a plot of the Grafenburg incoherent beams for the Novaya Zemlya explosions. A 0.6 to 3.0 Hz bandpass filter was applied to each waveform prior to computing the incoherent beams. The instrument response of the Grafenburg sensors, described by HARJES and SEIDL (1978), does not significantly differ from that of the Norwegian arrays in the narrow 0.6 to 3.0 Hz passband.

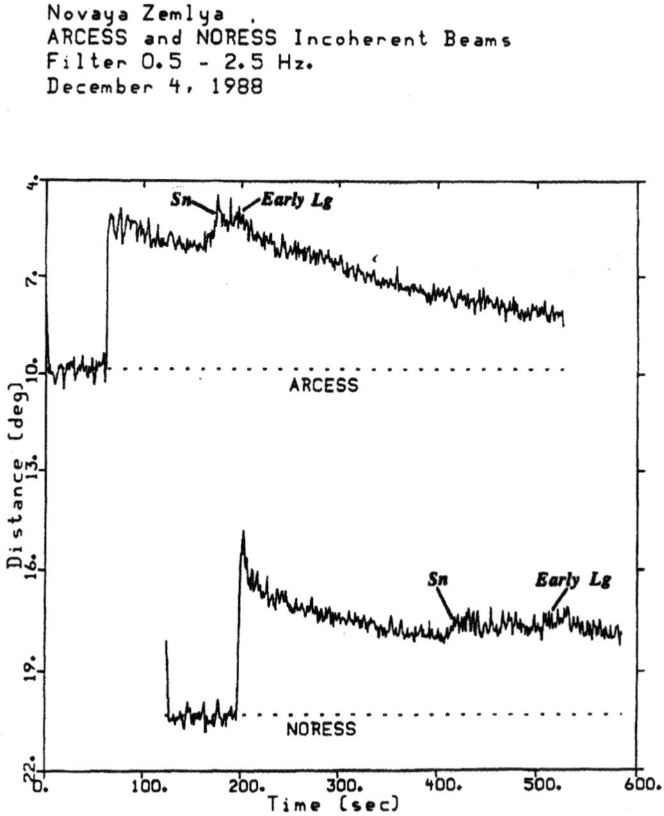

Figure 6
Record section of the incoherent beams for the December 4, 1988 Novaya Zemlya explosion recorded at ARCESS and NORESS.

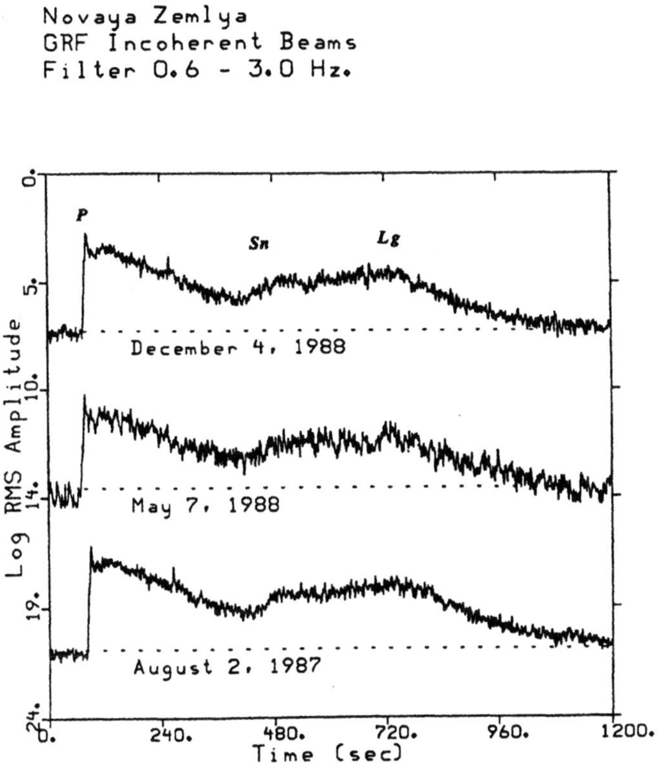

Figure 7
Incoherent beams for the Novaya Zemlya explosions recorded at the Grafenburg array.

Figure 7 shows that *Lg* is larger than *Sn* at Grafenburg for the Novaya Zemlya events, although *Lg* is emergent. The beginning of the *Sn* energy is also quite emergent with no clear onset. The *Sn* coda also appears to gradually increase with time to the peak *Lg* amplitude which indicates that the *Lg* energy may be superimposed on the scattered energy of the *Sn* coda. The expected time for the arrival of direct *Lg* for the three incoherent beams is indicated in the top plot. An *Lg* onset is most visible on the incoherent beam of the May 7, 1988 incoherent beam, whereas on the other beams the *Sn* coda appears to be continuous through the *Lg* time. Thus, although *Lg* is blocked to the Norwegian arrays such that the on-time *Lg* has energy less than the maximum *Sn* energy, more *Lg* energy does appear to get through to the Grafenburg array at approximately the time expected for direct *Lg*.

Comparison of Kola PNE (NORSAR) and Novaya Zemlya Nuclear Explosion (ARCESS)

Figure 8 compares the great-circle propagation paths for *Lg* from two different events, recorded at different times and at different arrays. The two great circle propagation paths are a similar distance. However, the Barents Sea constitutes most of the propagation path from Novaya Zemlya to ARCESS. The path from the Kola Peninsula to NORSAR does cross the northern tip or the Gulf of Bothnia, but most of the path is continental.

Figure 9 compares the ARCESS (top) and NORSAR (bottom) waveforms for the two events in the bandpass filter from 0.6 to 3.0 Hz. This filter reduces low-frequency noise and allows comparison NORSAR and ARCESS seismograms even though they have different recording instrumentation. The ARCESS recording has no apparent *Lg* at the expected arrival time for the 3.5 km/sec *Lg*. However, a strong *Lg* was recorded from the Kola Peninsula PNE.

Geological Explanation for the Barents Sea Lg Blockage

To illustrate how these geological heterogeneities may block *Lg* propagation across the Barents Basin, crustal cross sections for the great-circle propagation paths from Novaya Zemlya to the Norwegian arrays and the Grafenburg array have been plotted in Figures 10(A), (B), and (C). These plots were produced using information available from the Cornell Geographic Information System (GIS) database for Eurasia (FIELDING *et al.*, 1992) and the Profile Maker GIS access process at the Cornell website (http://atlas.geo.cornell.edu). The top plot shows the elevations obtained for the great-circle path from Novaya Zemlya through the receiver. The topography comes from the GTOPO30 database of the EROS Data Center (EROS, 1996). The horizontal axis indicates the distance from the array along the path, with the array located on the left edge of the plot. The approximate location of the Novaya Zemlya explosion site is marked with an asterisks. The bottom plots give the inferred crustal cross section of depth to basement and Moho depth. The current database at Cornell for basement and Moho depths in Eurasia derives from

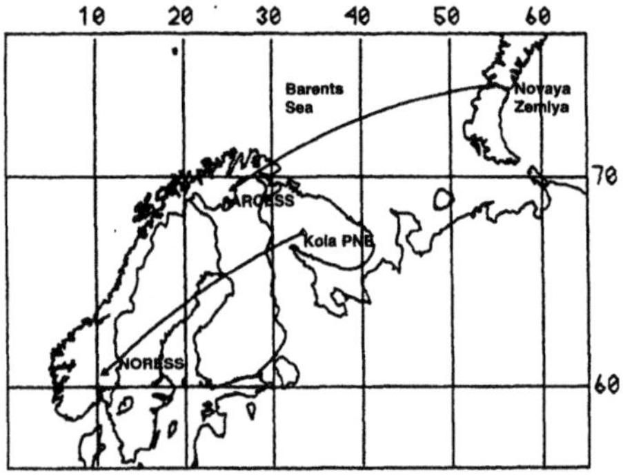

Figure 8
Map showing great-circle paths from the Novaya Zemlya test site to ARCESS and the Kola PNE to NORSAR.

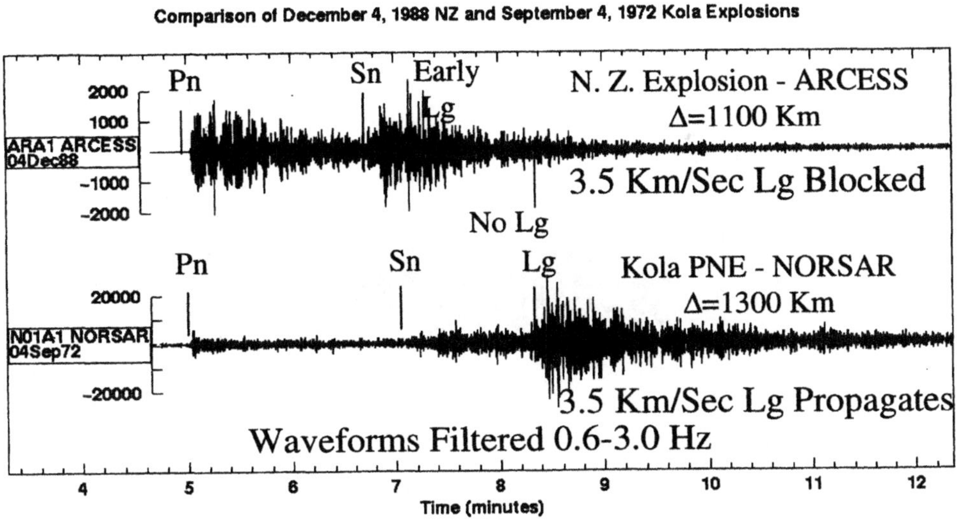

Figure 9
Comparison of the ARCESS ARA1 recording of a Novaya Zemlya explosion and the NORSAR NO1A1 recording of the Kola PNE with 0.6 to 3.0 prefilter.

digitization of maps obtained from the Institute of Physics of the Earth in Moscow, Russia (KUNIN *et al.*, 1987, 1988). The labels on the figures were obtained from tectonic maps of ALVERSON *et al.* (1967) and CLARKE and RACHLIN (1990).

In the case of the paths to ARCESS, NORESS, and Grafenburg in Figures 10(A), (B), and (C), respectively, the elevation falls quickly to the south of the Novaya Zemlya elevations, up to 6000 m, into the Barents Basin, which has a gradual slope to its deepest depth of below 300 m beneath sea level. For the path to Grafenburg in Figure 10(C), the depth and width of the Barents Sea is not quite as large as it is for the ARCESS and NORESS paths. For all three paths a very sharp increase in elevation marks the interface between the Southern Barents Basin and the Kola Peninsula.

For all three paths sediment thicknesses reaching 20 km are indicated in the same part of the structure where the Moho rises by about 5 to 10 km. This essentially reflects the general view of the basin portrayed in the Soviet geological literature, according to CLARKE and RACHLIN (1990), that the terrigenous sediments replace the granitic layer present in other parts of the path, specifically beneath Novaya Zemlya and the Kola Peninsula. Based on seismic profiles of the basin, the sediment velocities range from 3.9 to 5.5 km/sec in the sediments in contrast to the 6.0 to 6.5 km/sec velocities in the granitic layer. This would correspond to a shear-wave velocity contrast of about 2.2 to 3.17 km/sec for the sediments and 3.46 to 3.75 km/sec for the granitic rocks. This contrast in velocities would produce a strong impedance contrast at the boundaries of the basin and thus seismic energy

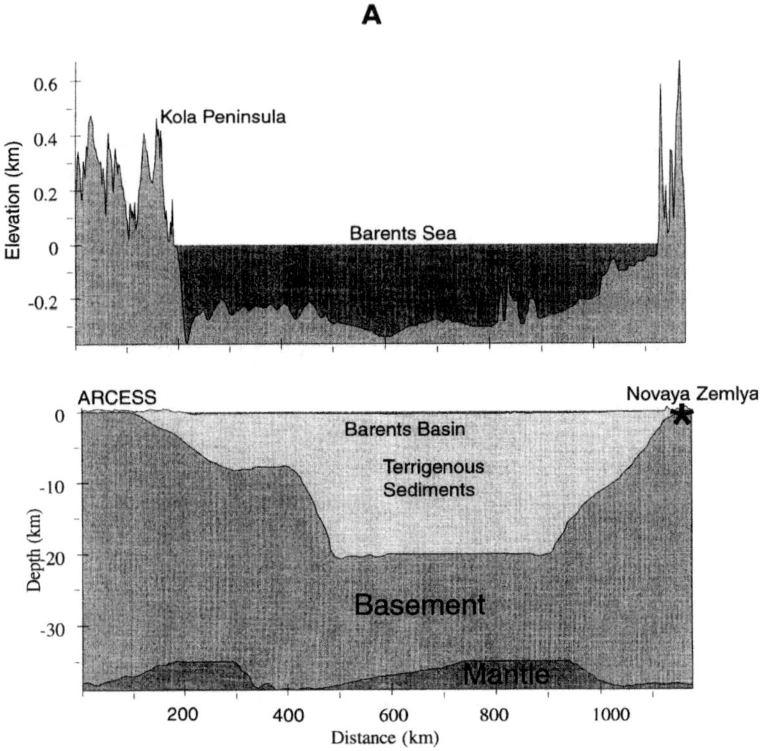

Figure 10A

that propagates into the basin may be captured and attenuated by repeated reverberations within the basin.

Figure 10(A) indicates that the distance of ARCESS from the Barents–Kola interface is about 250 km. We argued earlier that the *Sn* phase might convert to *Lg* at this interface, producing the "early *Lg*" at ARCESS. Assuming a group velocity of 4.5 km/sec for *Sn* and 3.5 km/sec for *Lg*, this would imply that the *Sn*-to-*Lg* conversion should arrive about 15 seconds after the onset of *Sn*. This is consistent with the time interval of "early *Lg*" observed at ARCESS in Figures 5 and 6. Figure 10(B) shows that the distance of this interface from NORESS is approximately 1400 km, or about 13°, which implies that the mode converted *Lg* should lag *Sn* by 110 seconds, as shown in Figure 3. This is again consistent with the observations at both NORESS and NORSAR.

The cross section from Novaya Zemlya to Grafenburg in Figure 10(C) reveals crustal thinning under the Barents Sea due to a rise in the Moho depth of about 5 km, a feature not as pronounced along the ARCESS and NORSAR paths. The rest of the path to Grafenburg crosses the elevated regions of the Kola Peninsula,

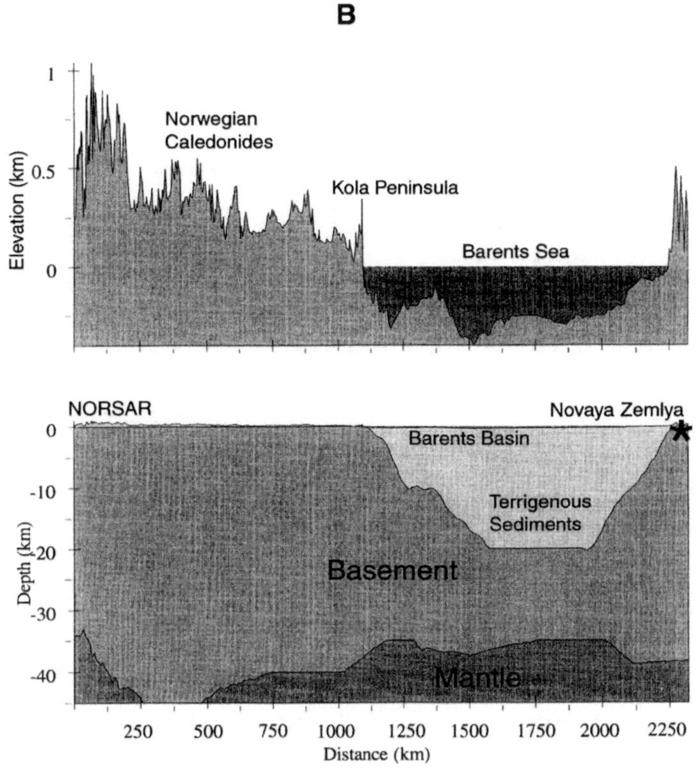

Figure 10B

Scandinavia, and the Baltic Sea, which also includes a sedimentary basin with sediment depths reaching 10 km. Moho topography varies by about ± 5 km about the average of about 35 km, although a sharp Moho dip appears at approximately 1350 km distance.

Figure 11 presents a comparison of the crustal part of the propagation paths for the two paths shown in Figure 8. The top plot shows the Barents Sea blocked-*Lg* path, with the associated sedimentary basin with sediment depths of some 20 km, and the bottom plot shows the unblocked-*Lg* path. Obviously, the sedimentary basin is the main feature of the crust that differs between these two paths and would appear to be the primary cause of *Lg* blockage across the Barents Sea.

It has been proposed that *Lg* blockage may be caused by high anelastic attenuation of *Lg* in basins with low-*Q* sediments (MITCHELL and HWANG, 1987). To check this possibility, *Lg Q* cross sections, obtained from the database of MITCHELL *et al.* (1997), are plotted in Figure 12. These *Q* values, derived by the coda method, stem from tomographic analysis of many paths in Eurasia, and the two cross sections in Figure 12 for blocked (top) and unblocked (bottom) paths were taken from the

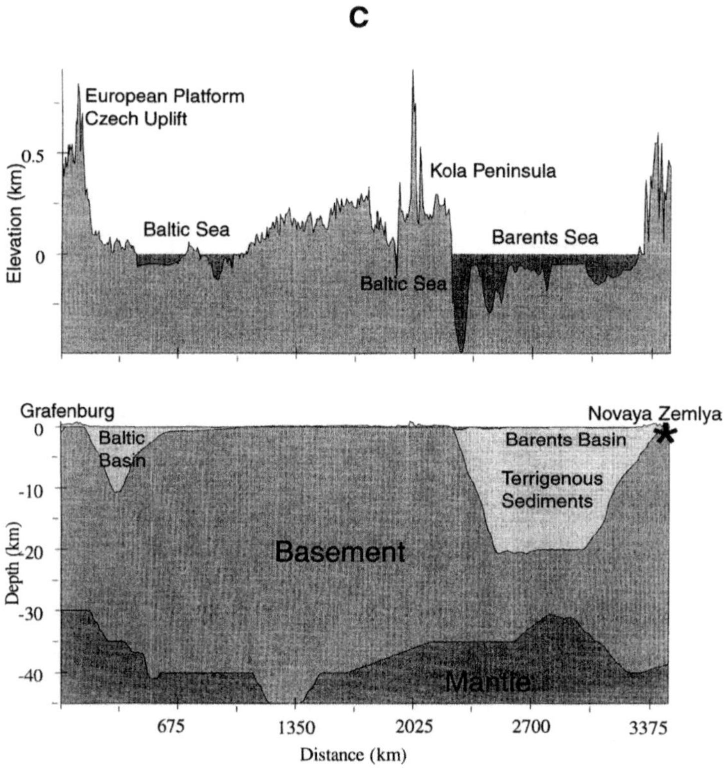

Figure 10
(A) Geological cross sections of the propagation path from Novaya Zemlya ARCESS. Top plot is a cross section of topography on an expanded vertical scale, and the bottom plot indicates the elevation, depth to basement, and depth to Moho on the same vertical scale, inferred from the Cornell Website server Profile Maker. (B) Crustal cross section from Novaya Zemlya to NORSAR. (C) Crustal cross section from Novaya Zemlya to the Grafenberg Array.

tomographic images. Both plots show that the average Q in both regions is high, with averages of 917 and 875 for the blocked and unblocked paths, respectively. Thus the data set does not reveal low Q in the Barents Sea Basin. However we cannot rule it out because the tomographic maps may not be accurate for this region, since the Barents Sea is near the edge of the overall Eurasian map region and the values in this region may reflect extrapolation.

Based on this analysis of crustal cross sections for the *Lg* propagation paths from Novaya Zemlya to the ARCESS, NORESS, and Grafenburg arrays, we conclude that the main cause of the *Lg* blockage in the Barents Sea, for the paths to the Norwegian stations, is the presence of thick sediment accumulations in the sedimentary basin beneath the southern part of the Barents Sea. Moho depth variations do not appear to be significant over most of these paths. Elevation changes are quite sharp but are actually quite small when compared to the overall thickness of

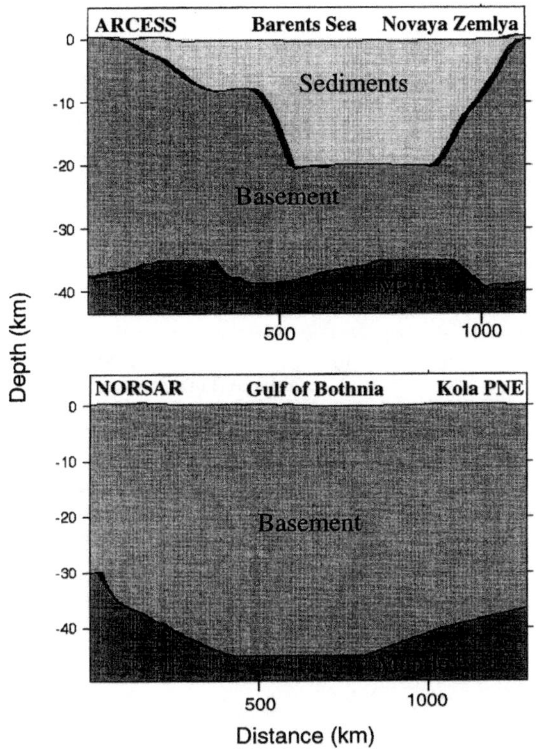

Figure 11
Comparison of crustal cross sections from Novaya Zemlya to ARCESS (top) and from the Kola PNE to NORSAR (bottom).

the crust. The same can be said for the presence of water in the basins, which is only a few hundred meters on average. Since the wavelength of *Lg* is on the order of a few kilometers at frequencies near 1 Hz, the variation in elevation and water depth is decidedly less than a wavelength. Since sediment depth variations are on the order of several wavelengths, this would appear to be a more likely cause of *Lg* blockage in this region.

Lg is better observed at Grafenburg than at either ARCESS and NORESS, in spite of the considerably longer propagation path to Grafenburg. Moreover, the Grafenburg incoherent beams contain significant energy in the codas of both *P* and *Sn*. These coda waves may be generated by scattering of the crustal guided phases, primarily *Sn* and *Lg*, from the geological and topographic heterogeneities along the path. For example, Figure 10(C) displays several topographic and geologic heterogeneities along the propagation path from Novaya Zemlya to the Grafenburg array that may be scattering sources, including the Kola Peninsula, the Barents and Baltic Basins, and the contacts between the Barents Basin and Kola Peninsula and between the Baltic Basin and European Platform and Czech Uplift. Both shear modes, *Sn* and

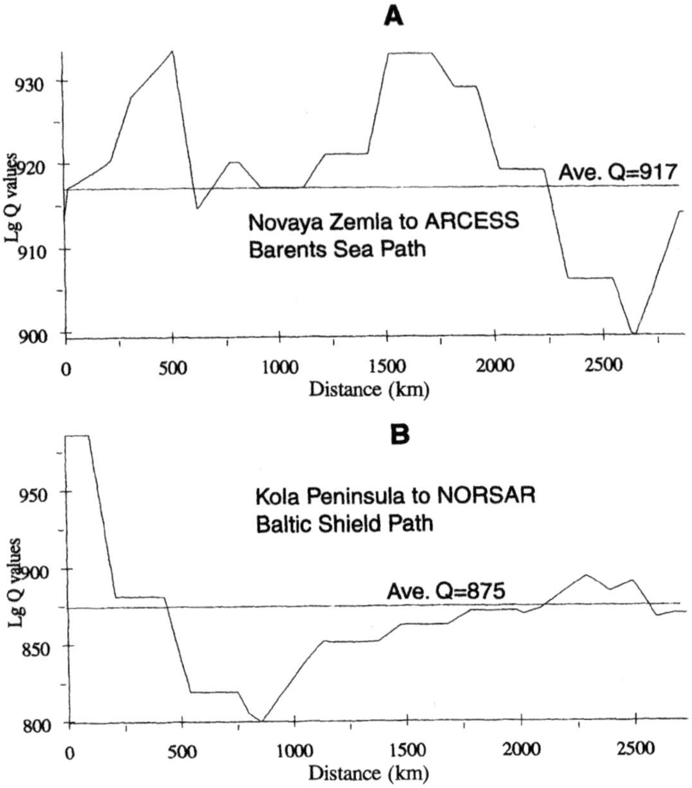

Figure 12
Cross sections through the *Lg Q* tomographic map for Eurasia (MITCHELL *et al.*, 1997). (A) Barents Sea path. (B) Baltic Shield path.

Lg, can interconvert to each other as well as forward scatter into *Pn* and *P* modes, thus building up the pre-*Sn* coda. Figure 7 shows that the *Sn* coda builds up after the onset of *Sn* and that "early *Lg*" energy may be produced by cumulative scattering from the heterogeneities along the propagation path.

Southern Urals and Caspian Sea Basins

We now consider the long-range ($\Delta > 2000$ km) propagation of *Lg* across sedimentary basins in Eurasia to NORSAR. An earlier study (BAUMGARDT, 1985) has revealed partial blockage of *Lg* across the Ural mountains, although the precise cause of the blockage was considered to be scattering of *Lg* in the crust beneath the Urals. Moreover, *Lg* has been known for some time to propagate inefficiently across the Caspian Sea region (e.g., KADINSKY-CADE *et al.*, 1981; PIWINSKII, 1981;

RODGERS *et al.*, 1997). Two sets of data have been examined that probe this region. First, historical PNEs in the FSU, recorded at the NORSAR array, have been studied with paths that cross the Southern Urals and the northern part of the Caspian Sea. Second, a new data set of seismic events recorded at the Iranian Long Period Array (ILPA) has paths that cross the southern part of the Caspian Sea.

NORSAR Observations – Southern Urals and Northern Caspian Sea

Figure 13 shows the great circle paths from four PNEs on either side of the Ural Mountains and the Northern Caspian region to NORSAR. The source parameters of the events are listed in Table 3. The two Southern Urals PNEs, 10/22/71 and 11/24/72, occurred on opposite sides of the Ural Mountains. The Astrakhan event of 10/08/80 and the East Caspian Sea event off 12/23/70 occurred on opposite sides of the Northern Caspian Sea. The paths from these events to NORSAR cross over several

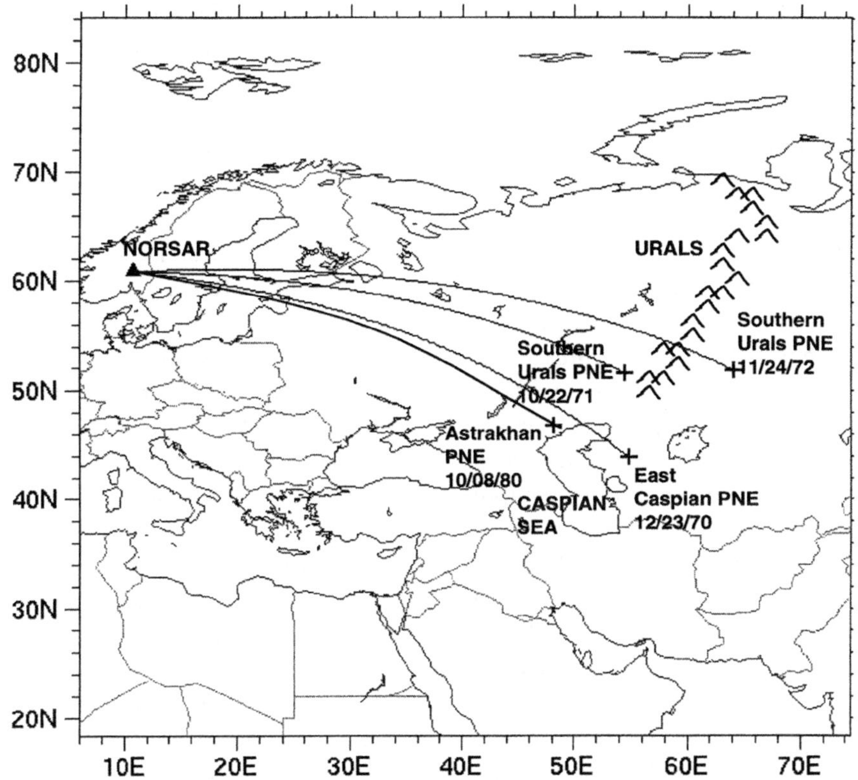

Figure 13
Great circle paths to NORSAR from PNEs in the southern Ural Mountains, Astrakhan, and east of the Caspian Sea.

Table 3

Event parameters for PNEs near Northern Caspian Sea recorded at NORSAR

Date	Time	Latitude	Longitude	m_b	Region
12/23/70	07:00:57.3	48.83	54.85	6.6	East of Caspian Sea
10/22/71	05:00:00.0	51.56	54.53	5.2	Southern Urals
11/24/72	09:59:57.8	54.84	64.15	5.2	Southern Urals
10/08/80	06:00:00.0	46.70	48.21	5.2	Astrakhan

heterogeneous structures in the source region but share common paths nearer the receiver.

The incoherent beam at NORSAR starts five seconds after the *P* onset time, as described by BAUMGARDT (1985). The reason for this is that these events occur at the epicentral distance, 25 to 31°, where upper mantle triplications can significantly affect the *P*-wave amplitudes, and the relative amplitudes of *P* and *Lg* may relate more to upper mantle effects on *P* rather than to crustal structure effects on *Lg*. In order to focus more on possible blockages of *Lg*, the *P* amplitude variations are ignored by starting the windows five seconds into the coda and comparing the *Lg* amplitudes with *P*-coda levels.

Furthermore, as was done in the earlier study of BAUMGARDT (1985), the coda incoherent beam envelopes were computed from the smaller NORSAR array configuration that was set up after 1 October 1976. This then ensures that the same sensors were used in the analysis of the events before and after 1976. The windows used to calculate the incoherent beams begin 5 seconds after the *P*-onset time on each trace. These times were measured by hand with an accuracy of about 0.3 to 0.5 seconds. Thus, these beams have, in affect, been "steered" to the velocity and azimuth of the *P* wave. Only the vertical component waveforms were used in computing the envelopes, and the waveforms were all bandpass-filtered from 0.6 to 3.0 Hz prior to calculating the beams. The average noise levels in the prior 50 seconds of noise were also computed and plotted as horizontal lines on the incoherent beam plots.

Figure 14(A) compares the NORSAR incoherent beams for the 10/22/71 Southern Urals event and the 10/08/80 Astrakhan event. The two events have the same magnitudes ($m_b = 5.2$) and epicentral distances (about 26°) from NORSAR. The beams are plotted on the same absolute amplitude scale, with log-rms amplitude units, and the horizontal line indicates the average noise level for both events. The early *P*-coda levels and the secondary *PP* phase nearly overlap with the same amplitude. The phase labeled *PP* may also consist of other long-range coda phases, such as $P_{410}P$ and "whispering gallery" phases (WG and WG_{fs}), that have been identified by MOROZOVA *et al.* (1998) on long-range Quartz profiles across the Urals, and correspond to the large-distance propagation of the *Pn* phase. However,

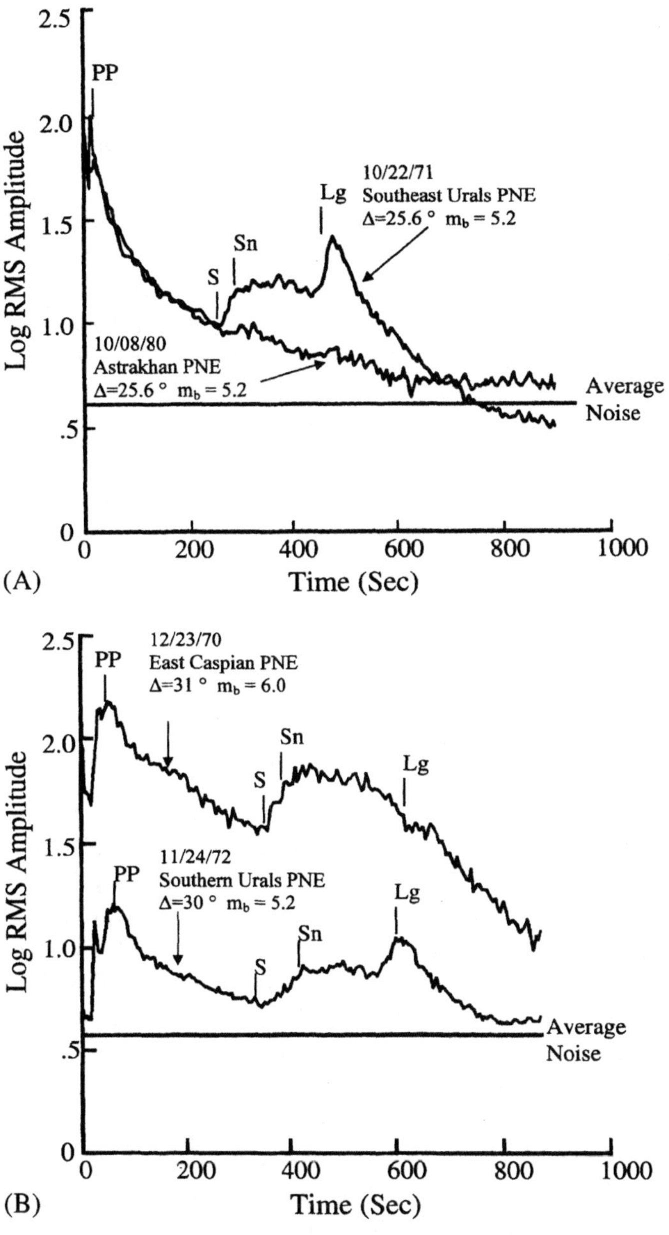

Figure 14
(A) NORSAR incoherent beams of the Astrakhan PNE and the Southern Urals PNE, located west of the Urals, beginning just after the peak *P* amplitude. (B) Comparison of NORSAR incoherent beams from the East Caspian PNE and the Southern Urals PNE located east of the Urals.

beginning at the *S* onset time, the beams are very different. The Southern Urals event has a very strong *Sn* and *Lg* phase whereas these phases are absent for the Astrakhan event.

Figure 14(B) compares the 11/24/72 Southern Urals event with the East Caspian PNE of 12/23/79. These events are also about the same distance, 30° and 31°, however the magnitudes are very different which is why the East Caspian PNE, with $m_b = 6.6$, has a much higher amplitude in the entire coda than the Southern Urals PNE, with $m_b = 5.2$. Aside from the amplitude differences, the codas of these two events have similar features and decay slope from *PP* to the expected *S*-arrival time. However, the East Caspian PNE does not have a strong direct *Lg* whereas the 11/24/72 Southern Urals PNE has an *Lg* that is larger than the *S* and *Sn* phases.

Thus, these comparisons suggest that *Lg* may be blocked from the Astrakhan and East Caspian events, both located in the Northern Caspian region, although *Lg* propagates efficiently from the two Southern Urals PNEs at comparable distance. Also, *Sn* is missing for the Astrakhan event but propagates efficiently from the other events to NORSAR. Figure 13 illustrates that the two Southern Urals PNEs occurred on opposite sides of the Urals relative to NORSAR. However, *Lg* does not appear to be blocked from the 11/24/72 event whose *Lg* propagation path crosses the Urals. Both events had equally strong *Lg* with amplitudes greater than the preceding *S* and *Sn* phases.

The South Ural and Pri-Caspian Sedimentary Basin Structures and Lg Blockage

To search for possible causes of the various blockages, we compare the four crustal cross sections for these events, plotted in Figures 15(A), (B), (C), and (D), that were derived from the Cornell Eurasia elevation, basement-depth, and Moho-depth databases using Profile Maker, as discussed earlier. The region northwest of the Caspian Sea includes a large sedimentary basin called the Pri-Caspian (Piwinskii, 1981) that appears in the cross sections for the Astrakhan and East Caspian events in Figures 15(B) and 15(D), respectively. This basin is similar to the Barents Basin in that the post-Paleozoic sediments reach extraordinary depths in excess of 20 km. Also, the basin contains extensive evaporite deposits in the form of salt domes that are exposed at the surface. The East Caspian event occurred on the extreme east end of the basin and thus *Lg* would propagate across this very deep basin before entering the Voronezh Uplift on the Russian Plateau. The Astrakhan event occurred in the basin, north of the Caspian Sea, and the *Lg* propagation path did not cross the water segment. This indicates that the North Caspian or Pri-Caspian Depression, not the shallow Caspian Sea itself, blocks the *Lg* propagation across the region.

The missing *Sn* as well as *Lg* in the case of the Astrakhan explosion may not be entirely explainable as a blockage effect. Astrakhan explosions are known to have been detonated in salt (Murphy *et al.*, 1997), and the salt source medium may be fairly homogeneous. Thus, these detonations in salt may not have excited strong

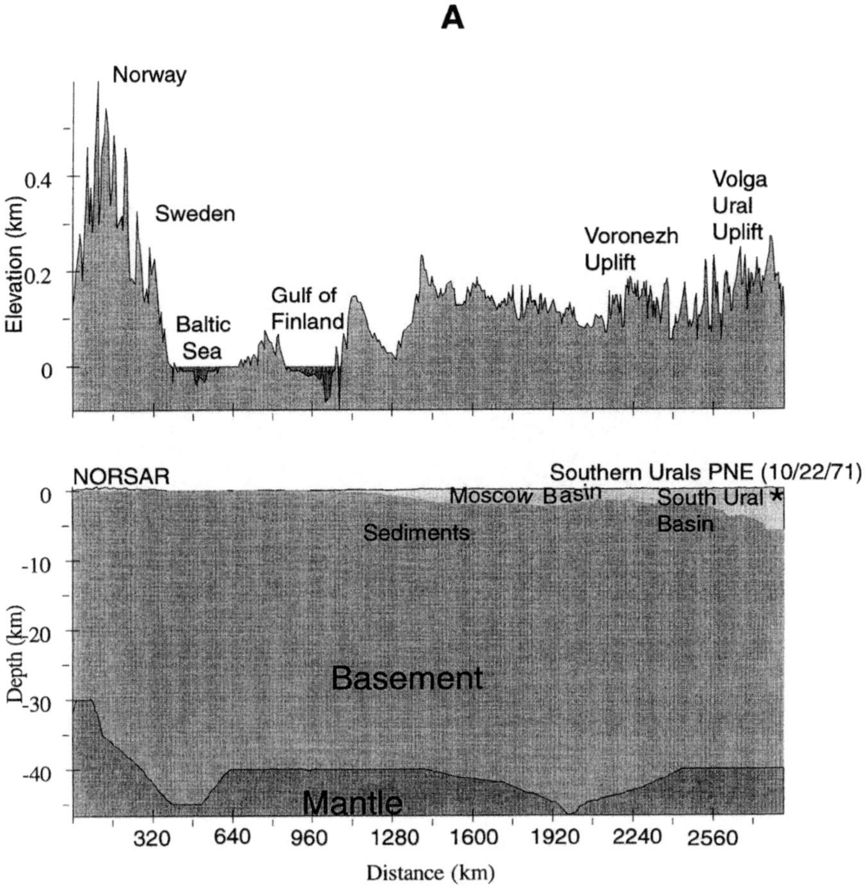

Figure 15A

shear waves. MURPHY *et al.* (1997) also reported low *Sn* amplitudes relative to *Pn* for Astrakhan explosions recorded at the Russian station at Borovoye.

The two Southern Urals PNEs were also detonated in a sedimentary basin called the South Ural Basin, which is evident in the crustal cross sections in Figures 15(A) and (C). However, this basin does not block the *Lg*, and in fact, *Lg* is strongly excited from both these sources. One possible explanation for why *Lg* is not blocked here, and is blocked by the Caspian Basin, may relate to where the PNE occurs relative to the basin, the depth of sediments and width of the basin. If the event occurs outside the basin and *Lg* must cross the entire basin, as in the case of the two Caspian events, there may be more *Lg* blockage than if the event occurs in the basin itself.

The Caspian Basin is considerably larger, both in width and depth of sediments, than the Southern Urals Basin. Also, in the case of the Caspian Basin, there is a sharp topographic interface between the basin and the adjacent Voronezh Uplift,

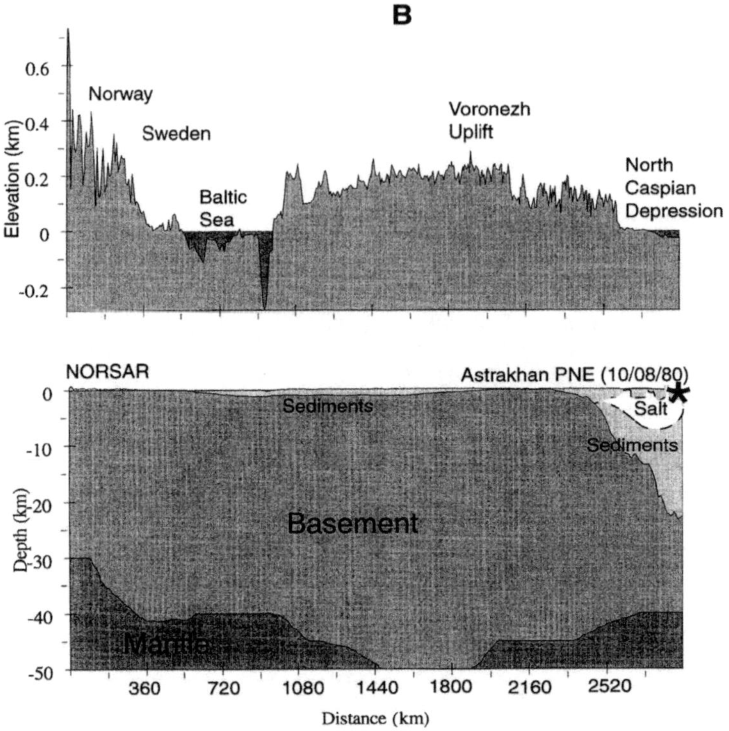

Figure 15B

similar to the Barents Sea–Kola interface, which may significantly backscatter *Lg* waves. This interface could also be a source of *Sn*-to-*Lg* scattering that produces the complex *Sn* coda observed from the East Caspian event. The "bumps" in the *Sn* coda in Figure 14(B) may actually be "early *Lg*" waves produced by this conversion. As regards the Southern Urals PNE located to the west of the Urals, shown in Figure 15(A), the South Ural Basin actually underlies an elevated region and there is no sharp vertical interface between the Volga-Ural Uplift and the edge of the basin that could backscatter *Lg*. However, Figure 15(C) illustrates that the path from the 11/24/72 Southern Urals PNE must cross both the South Ural Basin and the Moscow Basin as well as the Ural Mountains and yet none of these features appear to block *Lg* propagation from this event.

Thus we are left with the question of why the Pri-Caspian Basin blocks *Lg* propagation whereas the South Urals Basin does not? As discussed above, the difference in size and depth of sediments in the two basins, with the Pri-Caspian Basin being substantially larger with thicker sediments than the South Ural Basin, may be part of the explanation. Another may be that the sediments in the Pri-Caspian Basin have lower *Q* than those in the South Ural Basin. To check this possibility we examine *Lg Q* cross sections for the four paths, taken from the

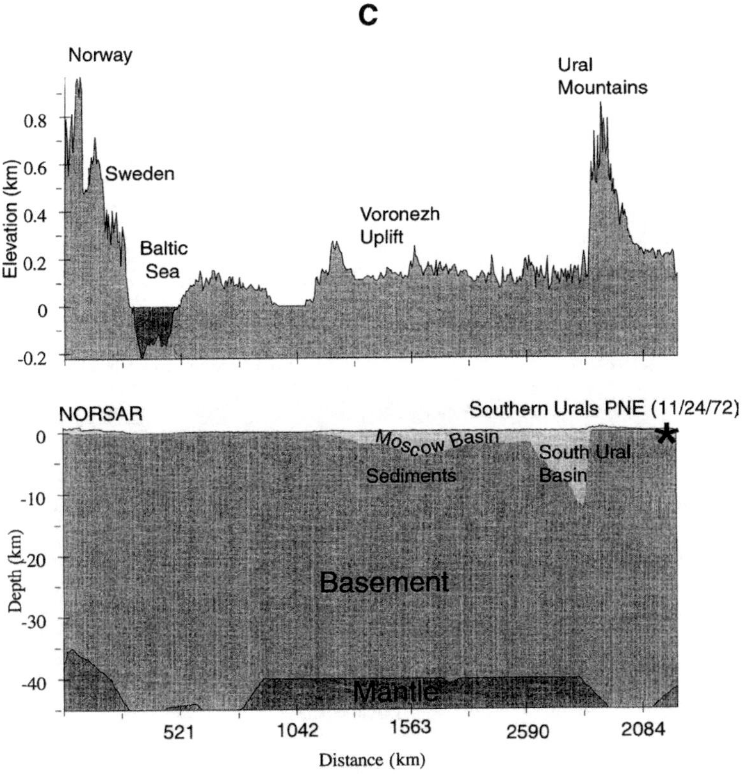

Figure 15C

Eurasian *Lg Q* tomographic map of MITCHELL *et al.* (1997). Figures 16(A) and (B) show the *Lg Q* cross sections for the Southern Urals PNEs and Figures (C) and D show the values for the Astrakhan and East Caspian PNEs, respectively. *Lg Q* does decrease along the path from NORSAR to both basins, although the drop in *Lg Q* is far greater for the Caspian paths than for the Urals paths. The average *Q* is the Pri-Caspian Basin is less than 400, significantly less than the average value of about 700 in the South Ural Basin. Consequently the combination of the low *Lg Q* as well as differences in morphology of the basins might explain why the Pri-Caspian Basin blocks *Lg* although the South Urals Basin does not.

Iranian Long-period Array (ILPA) Observations – Central and Southern Caspian Sea

We now consider *Lg* propagation across the central and southern parts of the Caspian Sea region using a new data set that has recently been made available, consisting of seismic events in the Southern Caspian Sea region recorded at the

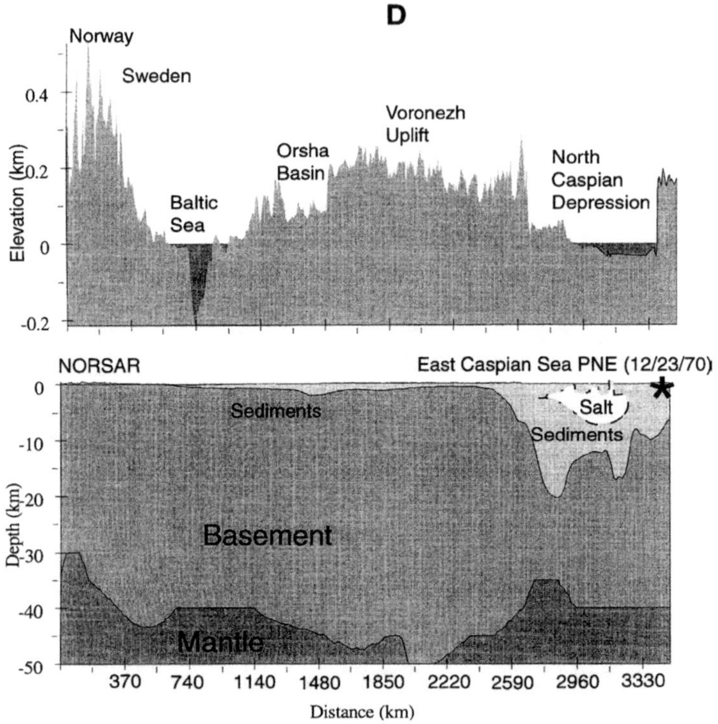

Figure 15
Crustal cross sections from NORSAR to (A) 10/22/71 Southern Urals PNE, (B) Astrakhan PNE, (C) 11/24/72 Southern URALS PNE, and (D) East Caspian PNE.

Iranian Long-period Array (ILPA) (GRANT *et al.*, 1996). The event parameters are given in Table 4. The body-wave magnitudes of the events range from 4.0 to 6.0. Many of the events have determined depths that range from very shallow to subcrustal (greater than 40 km). One event, on 08/13/78, has an estimated depth of 127.9 km which is in the mantle. Note that most of events have reported depths of 33 km, which is the NEIC constrained depth when depth cannot be determined.

The KS36000 seismometers of ILPA were primarily designed for long-period seismic recording. However, ILPA also recorded in the short-period band to 10 Hz Nyquist (TEXAS INSTRUMENTS, 1977). This region was previously studied by SIKHARULIDZHE (1964), as reported by SHISHKEVISH (1979), who observed no propagation of *Lg* for paths to Russian stations across the central and southern parts of this region. RODGERS *et al.* (1997) has also studied this ILPA data set, along with other data, and reported *Lg* blockage in the Caspian Sea and other regions of the Middle East. KADINSKY-CADE *et al.* (1981) and RODGERS *et al.* (1997) attribute *Lg* blockage in this region to the presence of deep oceanic crust.

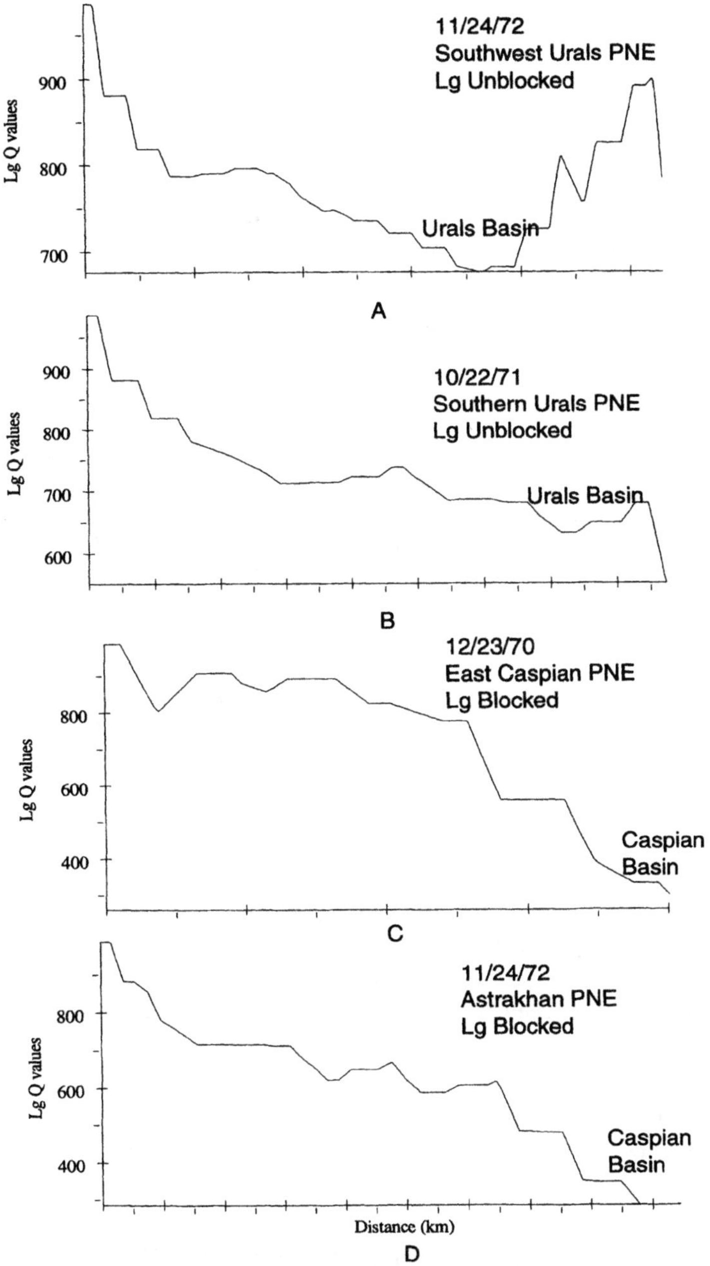

Figure 16

Lg Q cross sections to NORSAR from (A) 11/24/72 Southern Urals PNE, (B) 10/22/71 Southern Urals PNE, (C) East Caspian PNE, and (D) Astrakhan PNE.

Table 4

NEIS event parameters for events near the Southern Caspian Sea recorded at ILPA

Date	Time	Latitude	Longitude	Depth	m_b
07/20/78	10:20:25	38.664	55.243	33	4.3
07/31/78	16:37:15	40.371	52.914	33	4.3
08/13/78	17:20:25	39.328	41.072	127.9	4.3
08/15/78	09:04:22	41.249	43.989	7.8	4.7
08/20/78	02:30:03	42.292	49.982	33	4.0
08/22/78	22:48:11	41.936	43.87	3.9	4.8
08/26/78	13:10:18	36.898	54.333	33	–
08/29/78	00:19:17	41.231	49.526	0	4.4
09/03/78	00:21:17	44.451	38.012	33	5.7
09/07/78	09:19:52	38.058	58.571	33	–
09/21/78	11:08:49	38.062	38.645	30.6	4.6
10/17/78	16:45:14	39.675	41.735	33	–
11/03/78	18:54:07	42.501	45.259	33	4.4
11/04/78	15:22:20	37.713	48.946	36.8	6.0
11/16/78	05:46:09	43.499	46.623	33	3.9
11/23/78	15:24:39	44.171	39.337	25.4	4.4
11/25/78	08:57:25	39.901	44.069	10	4.4
12/04/78	03:12:38	38.068	37.43	37.3	5.0
12/14/78	05:08:23	45.494	37.36	10	4.7
12/18/78	07:59:56	47.783	48.136	33	5.9
01/01/79	04:25:39	37.638	50.176	33	4.0
05/12/79	16:19:55	41.419	48.155	33	4.2
06/08/79	17:46:10	37.074	55.616	3	4.4
06/15/79	21:15:04	36.117	54.521	33	–
08/02/79	22:42:10	40.727	52.016	67.7	4.3

Figure 17(A) shows locations of seismic events with *Lg* propagation paths to ILPA that cross the central and southern Caspian Basin. A record section of vertical component waveforms from the first ILPA channel (IR1) is plotted in Figure 18(A). The 18 December 1978 event, at a distance near 1400 km, is a nuclear explosion in the Azgir region north of the Caspian Sea. The other events appear to be earthquakes because they are located in the Caspian Sea and generally have larger *Sn* waves than *Pn* waves.

The waveforms in Figure 18(A) contain *Pn* and *Sn* arrivals. The dashed line indicates the expected arrival time of the *Lg*, however no clear arrivals are apparent at the indicated time of *Lg*. For these paths, *Lg* propagation has been blocked.

Figure 17(B) shows paths for other events recorded at ILPA, but which do not cross the Caspian Sea. Figure 18(B) shows the record section of the ILPA (IR1) recordings of these events. These waveforms are distinctly different than those in Figure 18(A) in that they contain significant energy at the expected arrival time of *Lg*. Also, these events produced considerably stronger *Pg* and coda waves than the events in Figure 18(A). The greater complexity of these waveforms may be due to the

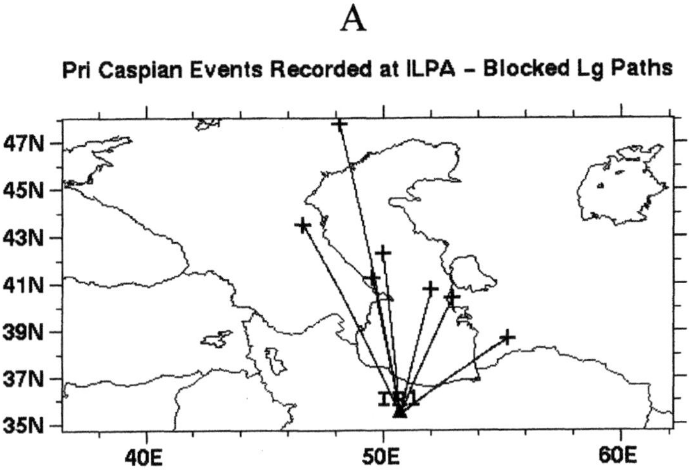

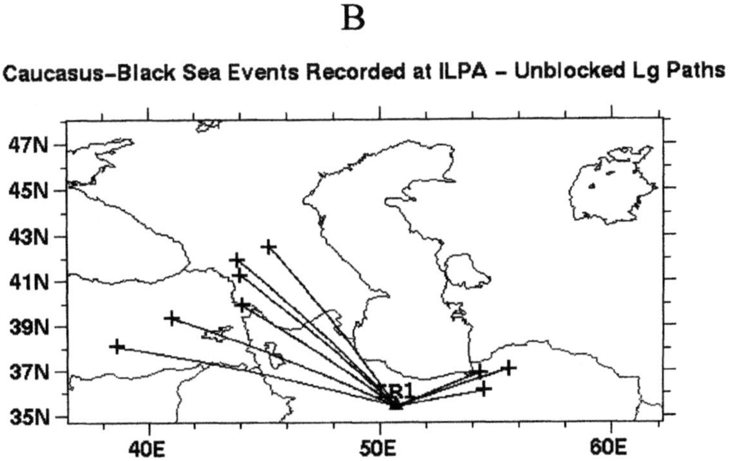

Figure 17

(A) Map showing great-circle paths from earthquakes to the ILPA array that appear to have *Lg* blockage. (B) Map showing great-circle paths from earthquakes to the ILPA array where *Lg* is not blocked.

increased complexity of the propagation paths, since many of the paths cross the Caucasus and the thrust belts of northern Iran.

Variations in source parameters, especially magnitude and depth, must also be considered as a possible cause of *Lg* amplitude variations. A detailed systematic study of possible effects of source parameters on *Lg* amplitudes has not yet been carried out. However, a cursory examination of Table 4 does not reveal strong evidence that extreme values of source parameters are a major cause of the lack of *Lg* for the events in Figure 18(A). For example, deeper events might be expected to

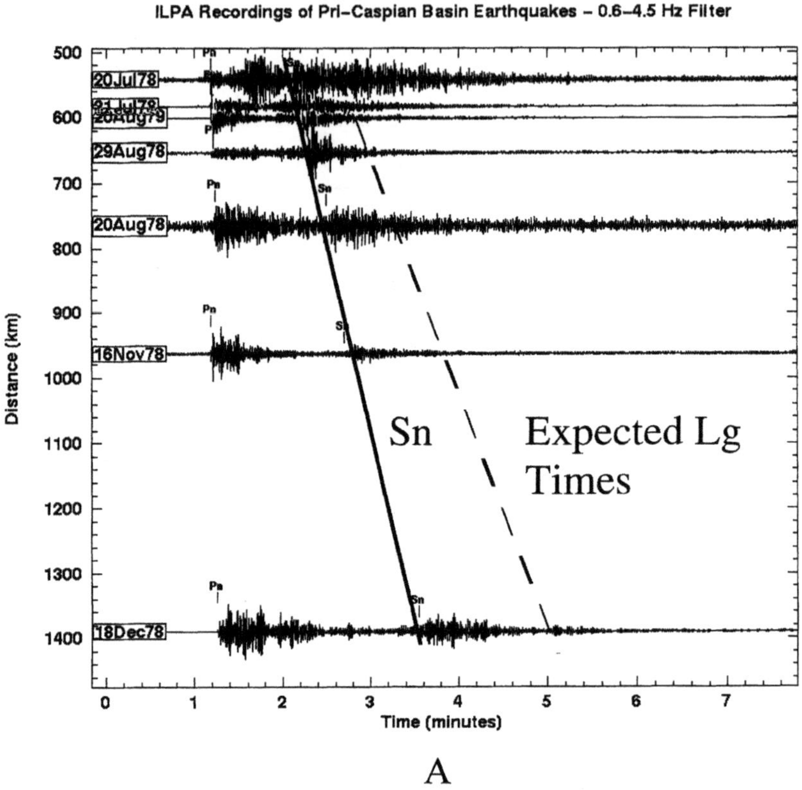

A

Figure 18A

underexcite *Lg* whereas shallower events would produce larger *Lg* amplitudes. The subcrustal event of 8/20/79, with depth of 67.7 km, does fall in the "blocked *Lg*" group in Figure 18(A). However, the 8/13/78 event with the mantle depth of 127.9 km produced a large *Lg* arrival in Figure 18(B), although admittedly in a very complex *P* coda. Moreover, the 8/29/78 event had the shallowest reported depth, 0 km, yet produced no *Lg*, as shown in Figure 8(A), but did produce a very large *Sn*.

South Caspian Sedimentary Basin Structure and Lg Blockage

Figure 19 gives a comparative analysis of waveforms and crustal structure for two selected paths that had blocked and unblocked *Lg*. The locations of the events and their great-circle paths to ILPA are shown in the map in Figure 19(A); their crustal cross sections are shown in (B), and (C) compares the waveforms of the two events. The West Caspian earthquake (16 November, 1978) is the event with the blocked *Lg*, and the South Caucasus earthquake (22 August, 1978) produced a strong *Lg*. Note that the blocked path from the West Caspian earthquake just skirts

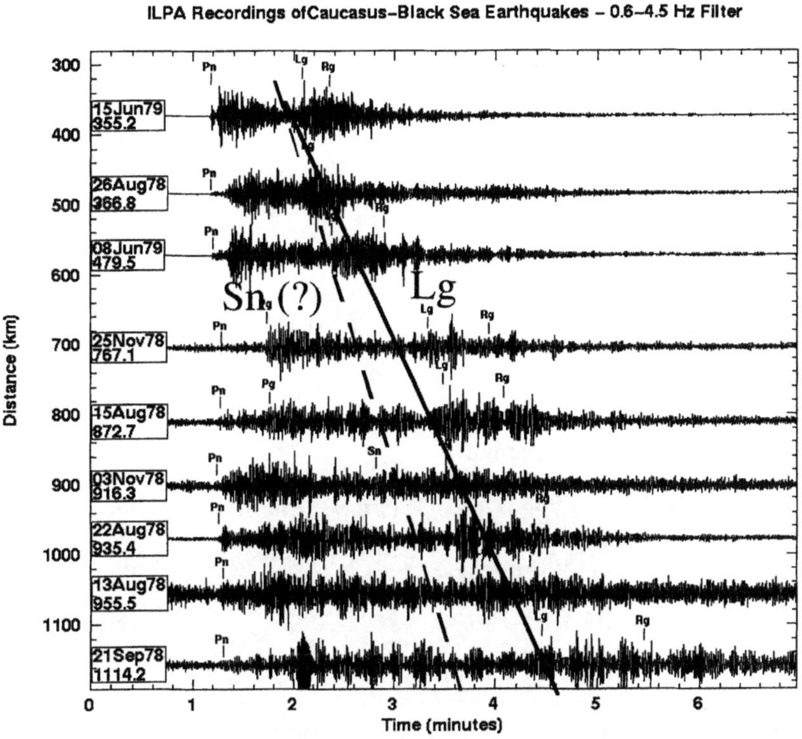

B

Figure 18

(A) Record section, aligned to *Pn* and corresponding to the paths in Figure 17(A), of waveforms recorded at the ILPA IR1 sensor where *Lg* arrivals phases appear to have been blocked. (B) Record section, aligned to *Pn* and corresponding to the events in Figure 13(B), of waveforms recorded at IR1 where *Lg* phases are observed.

the western edge of the southern Caspian Basin that is an enclosed sedimentary basin, with sediments reaching 12-km thickness in this region. The unblocked path misses the sedimentary basin.

The crustal structure of the deep sedimentary basins in the Caspian has been explained as a remnant of a vastly larger marginal sea that was buried during the Mesozoic collision of the Arabian promontory and the Eurasian margin (e.g., ZONENSHAIN and LE PICHON, 1986). As revealed by Russian DSS studies, the sediments in the basins have low velocity (about 3.5 km/sec) which overlie the basement rocks of considerably higher velocity (6.5 km/sec), the latter associated with the oceanic crust of the buried (Tethys) ocean. The basement rocks on either side of the basin appear to have normal granitic velocities (6.0 km/sec) (BERBERIAN, 1983). This interpretation is shown in the top cross section for the blocked path in Figure 19(B).

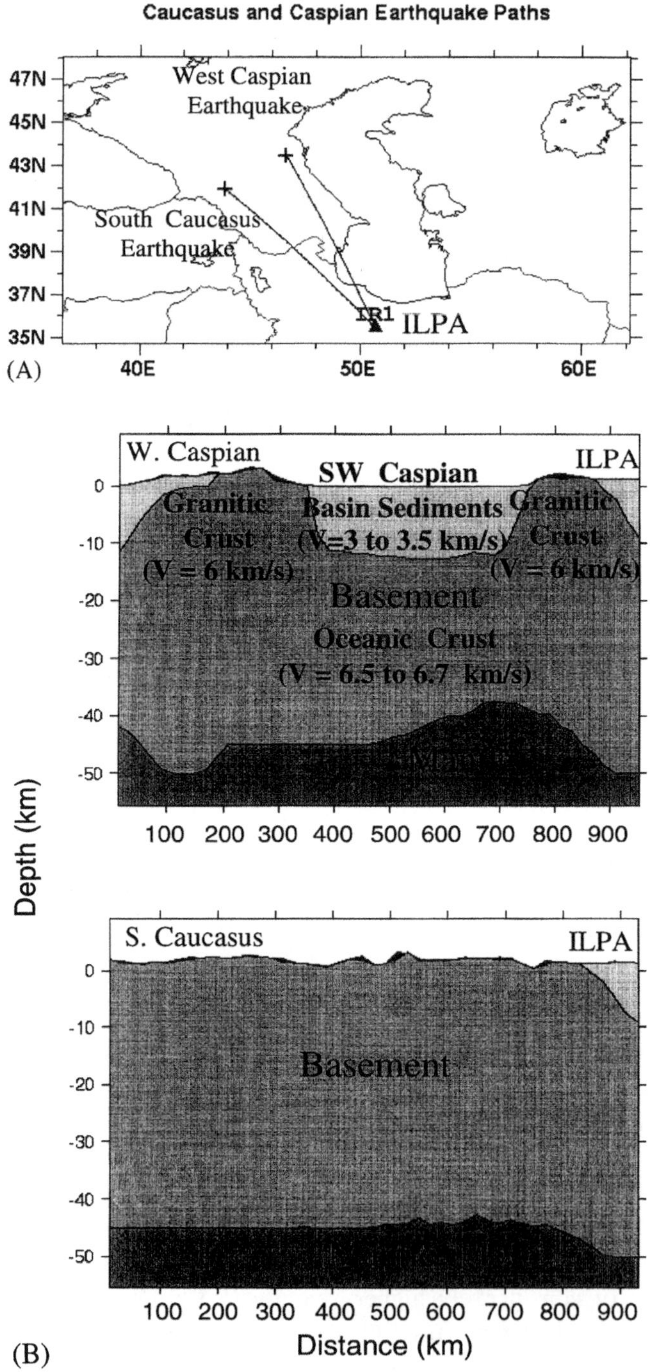
Caucasus and Caspian Earthquake Paths
West Caspian
Earthquake
South Caucasus
Earthquake
IR1
ILPA
47N
45N
43N
41N
39N
37N
35N
40E
50E
60E
(A)
W. Caspian
SW Caspian
ILPA
Granitic
Crust
(V = 6 km/s)
Basin Sediments
(V=3 to 3.5 km/s)
Granitic
Crust
(V = 6 km/s)
Basement
Oceanic Crust
(V = 6.5 to 6.7 km/s)
Depth (km)
S. Caucasus
ILPA
Basement
0
-10
-20
-30
-40
-50
100 200 300 400 500 600 700 800 900
Distance (km)
(B)

Figure 19A, B

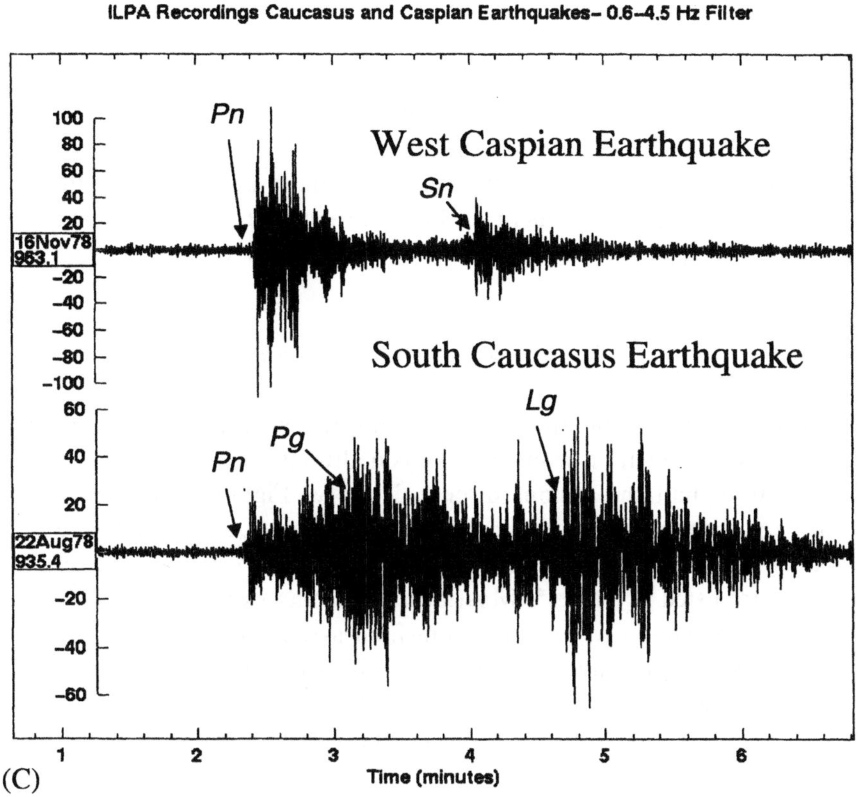

(C)

Figure 19
(A) Propagation paths from events west of the Caspian Sea with blocked and unblocked paths. (B) Crustal cross sections for the unblocked (blocked) and unblocked (bottom) paths. (C) Comparison of the ILPA IR1 recordings of the blocked (top) and unblocked waveforms.

The structure is the same as the Barents and North Caspian Basins in that the upper-crustal heterogeneity caused by the sedimentary basins may cause the *Lg* blockage across these regions. Evidently, whether or not there is oceanic crust beneath the sedimentary basin is immaterial to the *Lg* blockage. ZHANG and LAY (1995) modeled *Lg* propagation through various ocean-basin earth models and showed that the primary cause of poor *Lg* propagation across the oceanic crust is the thinness of the crust itself, not the composition. They demonstrate that a typical 6-km thick oceanic crust cannot accommodate a sufficient number of modes that can combine to produce *Lg*, as is possible in a 15-km thick continental crust. Even for mixed oceanic/continental crust, with the requisite 100 km of oceanic crust for *Lg* blockage, it is the thinness of the oceanic segment that blocks the *Lg*.

In the case at hand of the Southern Caspian Basin, even if there are subducted oceanic crustal rocks of the Tethys present beneath the basin, Figure 19(B) shows that the crust has uniform thickness on the order of 45 km. However, it would appear that the large velocity contrast between the sediments of the contained sedimentary basin and the surrounding crust, whether continental or oceanic, could trap the *Lg* waves in the basin and may explain the blockage of *Lg* across the southern Caspian Sea. Moreover, *Lg Q* may be low in these sediments as well which can cause the anomalous attenuation of *Lg* waves that cross the basin. Thus, this analysis suggests that it is the low velocity and low *Q* sediments in the basin, rather than the high velocity oceanic rocks in the crust under the basin, that are responsible for *Lg* blockage.

Mediterranean/Levantine Basin

We show one final region, the Eastern Mediterranean/Levantine Basin, where *Lg* blockage is associated with a sedimentary basin although with considerably different crustal structure than the regions discussed above. Figure 20 presents a map of propagation paths for earthquakes in Syria, Jordan, Israel, and Lebanon recorded at the Mednet station Kottyama, Egypt (KEG). The source parameters of the earthquakes are listed in Table 5. Most of these events were located along a region defined by the Jordan-Dead Sea transform fault system that separates the Arabian and Sinai plates. Tectonically, the Dead Sea transform system has evolved since mid-Cenozoic time as a result of the breakup of the Arabian plate from the African plate (AMBRASEYS and BARAZANGI, 1989). The system includes the pull-apart basins of the

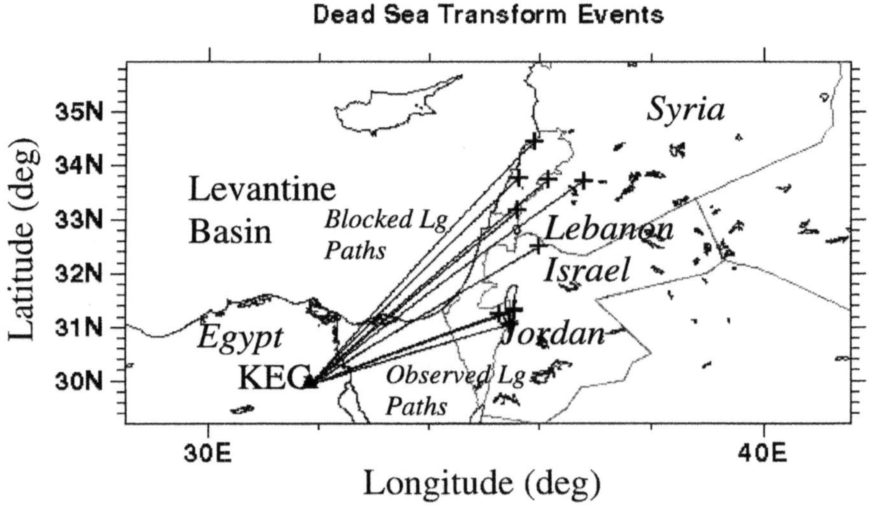

Figure 20
Map showing the propagation paths from events along the Dead Sea Transform to the Egyptian Mednet station at Kottyama (KEG).

Table 5

NEIS event parameters for events on the Dead Sea Fault recorded at KEG

Date	Time	Latitude	Longitude	Depth	Magnitude
06/23/91	13:02:22	31.312	35.557	10	–
09/28/91	00:43:07	31.048	35.464	10	3.9 (M_L)
10/18/91	16:48:50	33.702	36.792	10	4.9 (M_b)
07/29/92	05:30:43	32.489	36.007	10	3.3 (M_L)
03/09/92	19:54:27	34.454	35.925	5	4.2 (M_b)
09/07/92	02:57:52	31.298	35.551	10	3.7 (M_L)
04/24/93	03:45:14	33.154	35.615	24	3.6 (M_L)
07/30/93	11:46:20	33.727	36.173	33	3.7 (M_L)
11/12/93	20:34:54	33.758	35.628	10	4.2 (M_L)

Sea of Galilee and the Dead Sea and extends as far north to the Ghab and Yammouneh faults in Syria and south into the Gulf of Aqaba.

The crustal structure in the region is known to be quite variable because it was formed by the convergence of a number of displaced terranes (BEN-AVRAHAM and GINZBURG, 1990). The eastern Mediterranean Sea is underlain by the quaternary Levantine Sedimentary Basin. Seismic refraction surveys of the basin have inferred seismic velocities of 1.5 to 2.1 km/sec in the sediments in the eastern part of the basin, extending to depths of 8 to 10 km depth underlain by velocities of 4.1 to 4.5 km/sec. The adjoining Galilee-Lebanon, Samaria, and Negev regions have very different crustal structures, perhaps thinner crust, and the seismic velocities in the upper crust range from 3.7 to 6.2 km/sec. As shown in Figure 20, all the events occurred east of the Levantine Basin, however the great-circle propagation paths from the northern events in Syria and Lebanon to KEG cross the eastern part of the Levantine Basin.

A record section of these events, with vertical component waveforms filtered from 0.6 to 6 Hz, is shown in Figure 21. The most notable observation in this plot is the apparent blockage of *Lg* between 388 km and 488 km distance from KEG.

Figure 22 shows a comparison crustal cross sections for the unblocked and blocked paths. The blocked paths are associated with the paths that cross the Levantine Basin. However, Figure 22 shows that the Moho is elevated by as much as 15 km beneath the basin, and water depths approach 4 km.

The record section in Figure 21 also reveals that *Sn* is non-existent for the shorter distance where *Lg* is observed, but is strong at the greater distances where *Lg* is blocked. Also, *Pg* apparently disappears when *Lg* does, and the *Pn* is more impulsive for the paths where *Sn* is strongest.

These results are consistent with the observation in the Barents Sea and Caspian Sea basins. The *Pg*, *Lg* blockage is associated with paths which cross the sedimentary basin. However, the Levantine Basin has the additional features of 15 km of crustal thinning and 4 km water depths, both of which approach the wavelength of *Lg*.

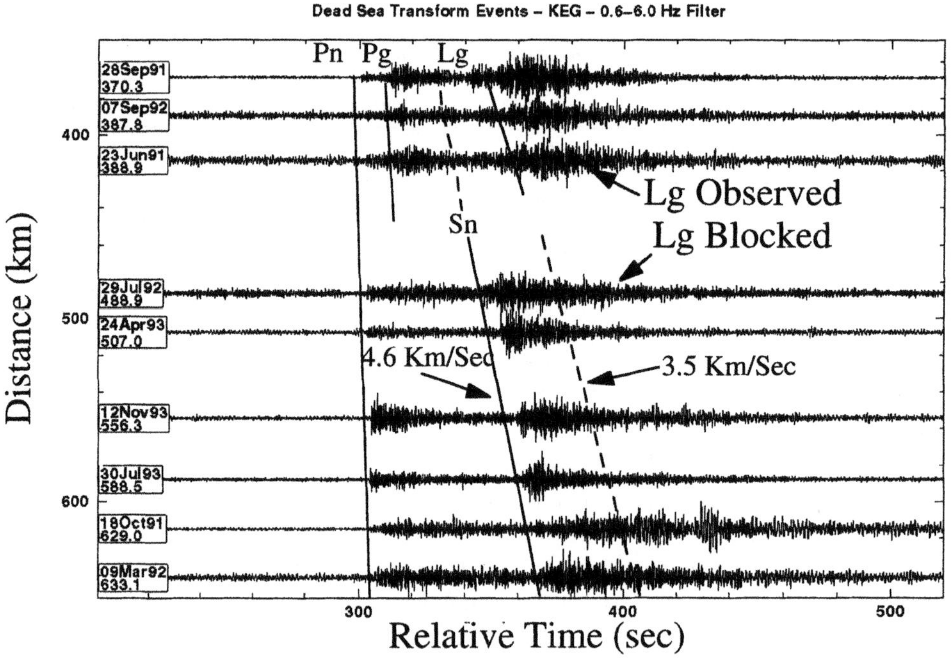

Figure 21
Record section of KEG waveforms, prefiltered from 0.6 to 6 Hz, from the Dead Sea Transform events. Expected times for *Pn*, *Pg*, *Sn*, and *Lg* are indicated.

Basin Capture Model for Lg Blockage

The examples presented above strongly suggest that sedimentary basins block *Lg* waves. The modeling study of *Lg* blockage by CAO and MUIREHEAD (1993) demonstrated that a sedimentary basin structure can partially block *Lg*. However, the crustal structure also included a sudden decrease in the Moho depth and a crustal pinch out. Sudden crustal thinning was also considered by KENNETT (1986) and proposed as a possible explanation of *Lg* blockage. Such structures might explain the *Lg* blockage in the Levantine Basin where the crustal pinch out is very prominent in Figure 22. However, this model would not explain *Lg* blockages in the Barents and Caspian Sea Basins, where Moho depths either do not vary or become greater and the pinch out structures are not apparent.

Also, low *Q* for *Lg* has also been proposed as a cause of *Lg* blockage, for example, across Tibet (MCNAMARA *et al.*, 1996). The tomographic *Lg-Q* maps of MITCHELL *et al.* (1997), and the cross sections shown in this paper from those maps, do suggest that low *Lg Q* in Northern and Southern Caspian Basins may explain why *Lg* is blocked in these regions, but does not appear to be blocked by the Urals Basin which has higher *Q*.

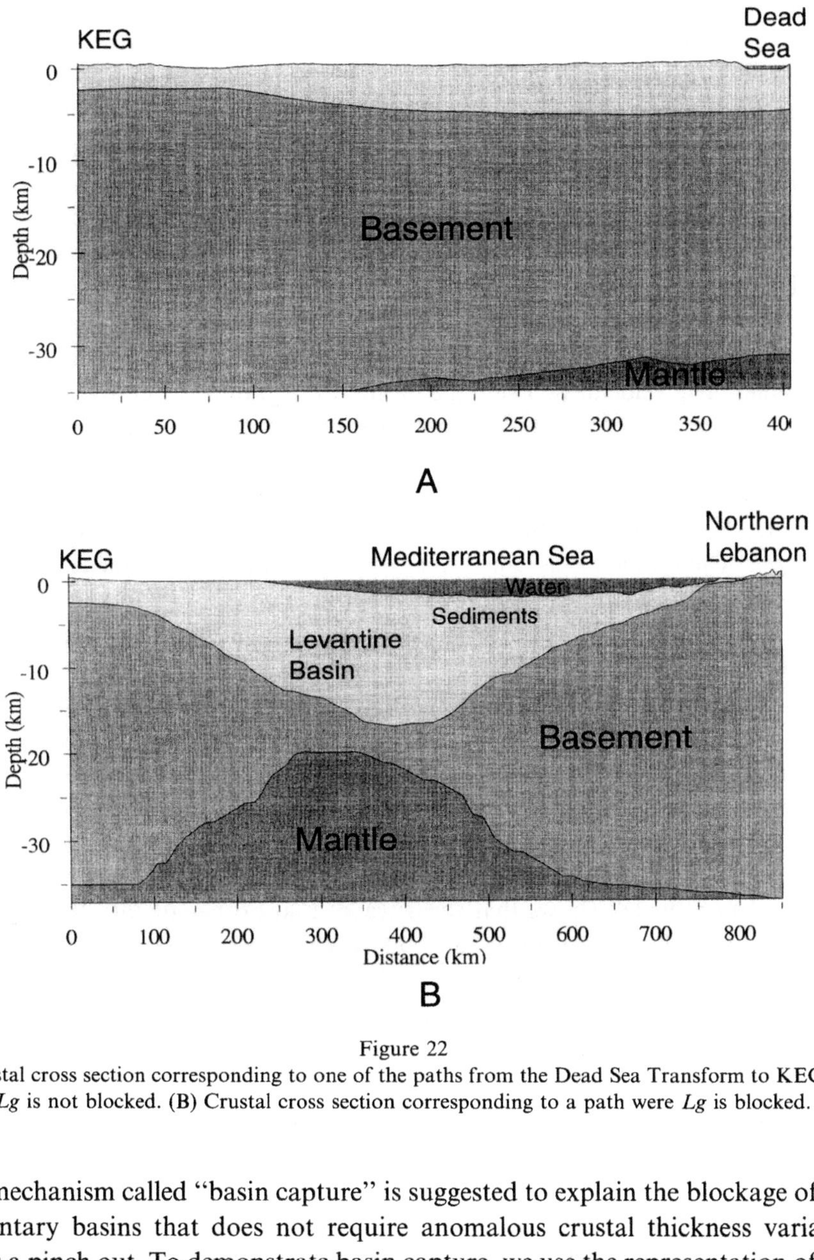

Figure 22
(A) Crustal cross section corresponding to one of the paths from the Dead Sea Transform to KEG where *Lg* is not blocked. (B) Crustal cross section corresponding to a path were *Lg* is blocked.

A mechanism called "basin capture" is suggested to explain the blockage of *Lg* in sedimentary basins that does not require anomalous crustal thickness variations, such as a pinch out. To demonstrate basin capture, we use the representation of *Lg* as the superposition of shear waves, multiply-reflected in the upper crust. The ray-tracing method of KENNETT (1986) has been used to model the interaction of *Lg* with structural boundaries by tracing *S*-type rays through the two-dimensional structure. The phase velocity of the wave is the horizontal component of shear-wave velocity in the upper layer of the velocity model, which is controlled by the shear-wave velocity in the layer and the takeoff angle of the ray from the source. The phase velocity is assumed to be about 4.5 to 5.0 km/sec for *Sn* and 4.0 to 4.5 for *Lg*. The group velocity is the distance that the wave traverses divided by the travel time. Group velocities are about 4.5 km/sec for *Sn* and 3.5 km/sec for *Lg* waves.

A sedimentary basin like the Barents Basin is modeled as an inverted trapezoidal structure whose top width is 5000 km and bottom width is 2000 km. The basin is assumed to be filled with sediments reaching a maximum depth of 15 km. The shear-wave velocities inside the basin are assumed to be 2.6 km/sec at the top which gradually increases to about 3.46 km/sec at the bottom and sides of the basin. These velocities correspond to compressional velocities of 4.5 km/sec in the sediments and 6.0 km/sec at the sides of the basin. The velocities in between are interpolated by cubic splines. The velocities outside the basin in the granitic layer are assumed to be 3.46 km/sec. The rest of the model is assumed to be a normal, continental structure, with two layers in the crust overlying the mantle.

The source is assumed to be located on the left side of the structure, approximately 500 km from the upper corner of the basin, and the rays travel from left to right. The receiver is assumed to be located somewhere on the right-hand side of the structure outside of the basin. This approximates the situation of the Novaya Zemlya to ARCESS path. The ray-tracing calculations were made using RAY81 (ÇERVENY *et al.*, 1977).

Figures 23(A) and (B) contrast the propagation of an *Lg*-type ray in a structure without (left) and with (right) the sedimentary basin, respectively. Starting with an input phase velocity of 4.9 km/sec, the output group velocity is 3.5 km/sec as expected for *Lg*. As shown, Figure 23(A) on the left, simple shear-wave reverberation in the "granitic" layer, or the upper crust, of the laterally homogeneous structure produces the combination of phase and group velocities expected for *Lg*. Figure 23(B) on the left shows that shear waves that propagate into the lower crust below 15 km, with phase velocity of 4.0 km/sec, would have a group velocity of 3.73 km/sec, which is faster than is expected for first arrival *Lg*. These reverberations may contribute to the later-arriving *Sn* coda. However, shear-wave reverberations in the granitic layer of the crust have phase and group velocities closer to those expected for the first arrival *Lg*.

The presence of the sedimentary basin in the structure severely disrupts the pattern of propagation of the reverberations in the granitic layer wherein the shear waves that compose the *Lg* wave train are effectively captured in the low-velocity sediments of the basin. One effect is that the waves that propagate into the basin are bent closer to vertical by the velocity contrast between the higher granitic velocities (about 6.0 km/sec) outside the basin and the slower, sedimentary velocities (3 to 4 km/sec) inside the basin. Another is that the rays become trapped in the upper layers of the structure. The waves are then guided by the slower velocities in the upper part of the basin structure until they arrive at the other side of the basin. After they submerge back into the homogeneous structure, they have been considerably

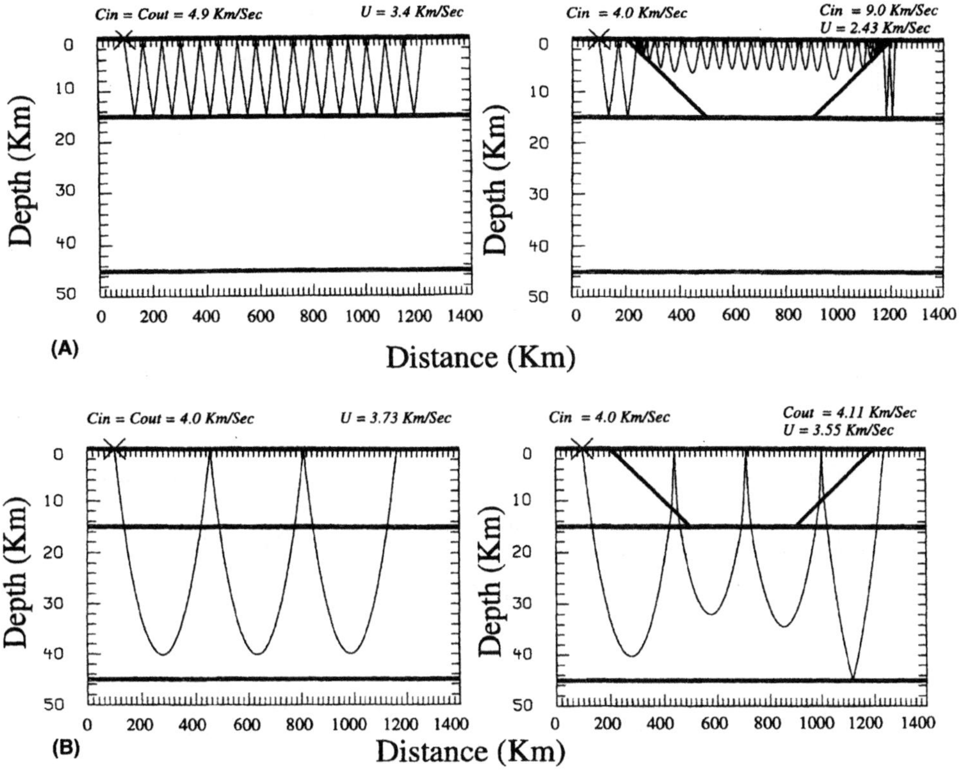

Figure 23A, B

slowed down. In this case the group velocity of the rays comprising *Lg* would only be 2.43 km/sec. Furthermore, the original phase velocity has been greatly altered because of the dipping interfaces between the basin structure and the surrounding homogeneous structure. For the model in Figure 23(**B**), the *S* wave reverberations, which start out with *Lg* phase velocities, have been converted to steeply dipping rays with phase velocities near 9 km/sec. These phases may become *Sn* type modes or convert to *P* waves. This mechanism might explain how *Lg* waves scatter into *P* and *Sn* coda waves.

Thus, the effect of the basin structure on *Lg* is that shear waves are captured in the upper layers and that direct *Lg* waves are essentially delayed and spread out over decidedly later times. Because of the repeated reverberations in the upper layers and the likely lower *Q* in these layers, the original *Lg* waves would be severely dissipated. Whether the *Lg* is completely dissipated or not depends on the dimensions of the basin through which the *Lg* waves must propagate. The most likely result would be that the *Lg* modes, instead of arriving as a distinct arrival, would be spread out into long coda with amplitudes which diminish with time. This example shows that it is lateral heterogeneity in the upper crust which is required to block the propagation of

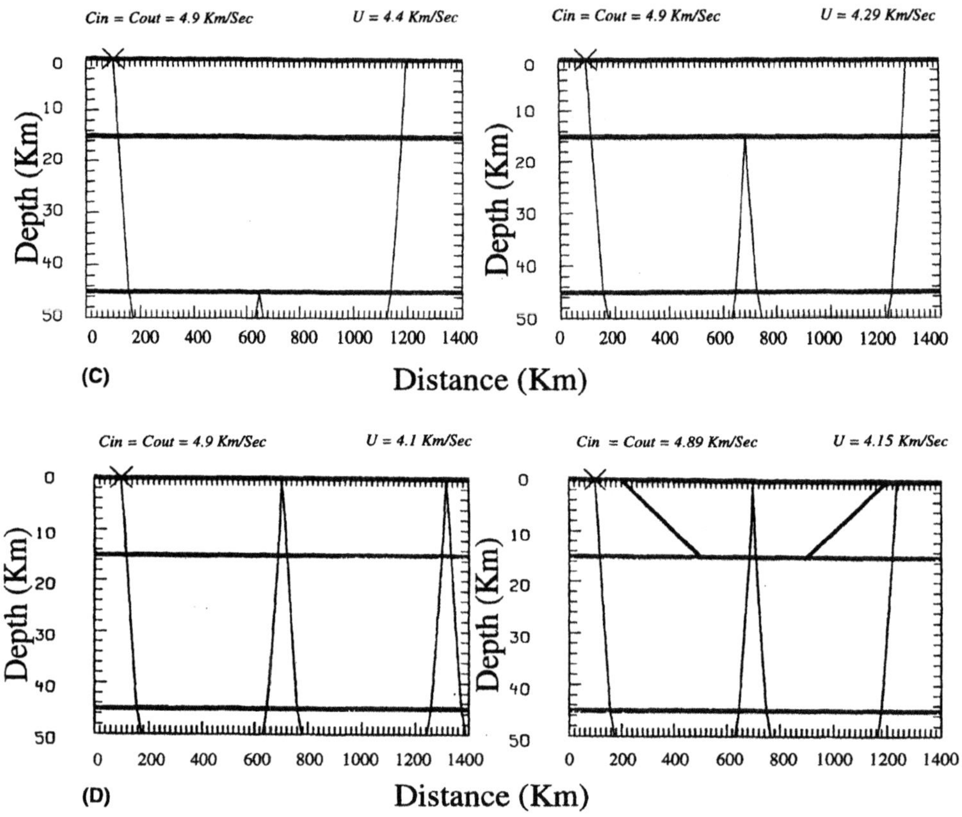

Figure 23
(A) Ray plots of shear-wave reflections in a crustal granitic layer that give propagation properties of *Lg*-type waves. Homogeneous path is shown on the left and effect of a sedimentary basin in the path is shown on the right. (B) Ray plots of shear waves refracting through the lower crust. Homogeneous path is shown on the left and the effect of a sedimentary basin is shown on the right. (C) Rays paths of diving shear waves in the mantle, corresponding to *Sn*-type waves, that propagate below any lateral heterogeneity in the upper crust. (D) *Sn*-type waves that pass twice through the upper crust. Effect of sedimentary basin is shown on the right.

Lg, and which effects in the lower crust or at the Moho would have little effect on *Lg*. Figure 23(B) on the right indicates that shear waves propagating through the entire crust would be delayed by the basin and produce group velocities of close to 3.55 km/sec. However, they are not as strongly affected as shear waves propagating in the upper crust, and the main effect of the basin would be to stretch out the *Sn* coda wave arrivals to later times.

Two examples of *Sn* type propagation are shown in Figures 23(C) and (D), where *Sn* is modeled as a diving wave in the lower crust and upper mantle. No attempt has been made to model decreased depth of the Moho, which has been observed in the Barents Basin. However, if *Sn* is primarily composed of diving waves in the upper mantle, they would not be greatly affected by such an effect since it is mainly controlled by the structure of the upper mantle, although Moho depth variations may change the group velocities of *Sn* modes which bounce off the bottom of the Moho. Also, the *Sn* wave would propagate below the sedimentary basin and be totally unaffected by it.

Figure 23(D) on the right depicts another type of *Sn* propagation which might be affected by the presence of the basin. Shear-wave reverberations which involve the entire crust would produce group velocities of about 4.15 km/sec which are on the lower end of the range expected for *Sn*. However, although this mode would propagate into the basin structure, because of the high-phase velocities of these *Sn* modes and steeper angles of incidence, they are not captured by the basin and essentially propagate through the structure with virtually no blockage effects.

Conclusions

The well-recognized problem of "*Lg* blockage" has been reexamined in terms of the effect of sedimentary basin structures. This study has shown that the three regions where strong *Lg* blockage has been observed, the Barents and Caspian Seas, and eastern Mediterranean regions, are associated with the presence of deep enclosed sedimentary basins where low-velocity sediments replace the granite velocity rocks in the upper crust.

Simple ray-tracing has been used to illustrate a "basin capture" model which explains *Lg* blockage by the sedimentary basin capturing and delaying the *Lg* modes and their amplitudes being reduced by repeated reverberations and scattering within the basin. The ray-tracing study only gives a first-order representation of the propagation of *Lg* through the sedimentary basin. Only the primary rays, i.e., shear waves that reverberate in the crustal granitic layer, were included in the ray-tracing. Actual *Lg* waves consist of many such rays, and undoubtedly some may follow paths that do not pass through the heterogeneous structure. A true modeling of *Lg* should include all possible rays that may show that some of the *Lg* energy can propagate through the heterogeneous structure. The incoherent beam analysis of the Novaya Zemlya events has, in fact, shown that the coda shapes flatten at the expected time of the onset of on-time *Lg* so that *Lg* is not completely blocked. However, the ray plots of the primary rays represent the paths traversed by most of the *Lg* energy, and they will clearly be diverted by the sedimentary basin along the path.

However, other explanations may also be invoked to explain *Lg* blockage, perhaps in combination with the basin capture idea. First, unusually low-*Q* associated with the sediments may also explain the strong attenuation, as suggested by MITCHELL and HWANG (1987) and MITCHELL *et al.* (1997). Low *Lg Q* has been observed in the Caspian Sea and eastern Mediterranean Sea regions. The fact that the *Lg Q* tomography map for the Barents Region does not reveal low *Q* may be an artifact of the inversion process and lack of resolution, as has been pointed out by MITCHELL *et al.* (1997). Second, the crustal pinch out blockage appears to be clearly present in the Levantine Basin.

Also, basin capture cannot completely explain all the observations at NORSAR from the Southern Urals, especially the lack of blockage of *Lg* across the South Urals Basins. The fact that these basins are small and *Lg Q* is high, might be part of the explanation. The lack of *Sn* and *Lg* from the Astrakhan event may be more of a source effect than a blockage effect, since this event was probably detonated in salt in the Pri-Caspian Basin. Source-parameter effects also cannot be completely ruled out for the earthquakes as a cause of the variations in *Lg* amplitude. However, we might expect source mechanism to be less important at high frequencies.

Finally, the question of the utility of *Lg* for event identification in the face of blockages must be addressed. Obviously, *P*/*Lg* amplitude-ratio discriminants cannot be used if *Lg* is blocked by the propagation path. Crustal cross sections, derived from GIS databases such as those at Cornell, may have some predictive value for *Lg* blockage. However, uncertainties in the parameters in these cross sections must be considered when making these predictions. This study has shown that when *Lg* is blocked, *Sn* sometimes appears enhanced. In fact, earthquakes can be identified on the basis of small *P*/*Sn* amplitude ratio when weak or no *Lg* is recorded. Often, however, when *Lg* propagates with high energy, *Sn* is not well observed. This apparent tradeoff, if it can be shown to hold in general, may be related to the way in which the total shear-wave budget from an earthquake is partitioned in the laterally heterogeneous crust. Future studies should address this apparent tradeoff, both experimentally and in models.

REFERENCES

ALVERSON, D. C., COX, D. P., WOLOSHIN, A. J., TERMAN, M. J., and WOO, C. C. (1967), *Atlas of Asia and Eastern Europe to Support Detection of Underground Nuclear Testing, Vol II., Tectonics*, prepared by the Department of the Interior, U.S. Geological Survey, for the Advanced research Projects Agency.

AMBRASEYS, N. N., and BARAZANGI, M. (1989), *The 1759 Earthquake in the Beak Valley: Implications for Earthquake Hazard Assessment in the Eastern Mediterranean Region*, J. Geophys. Res. *94*, 4007–4013.

BARAZANGI, M. J. (1977), *Relative Excitation of the Seismic Shear Waves Sn and Lg as a Function of Source Depth and their Propagation from Melanesia and Banda Arcs to Australia*, Annali di Geofisica *30*, 385–407.

BAUMGARDT, D. R. (1985), *Comparative Analysis of Teleseismic P Coda and Lg Waves from Underground Nuclear Explosions in Eurasia*, Bull. Seismol. Soc. Am. *80*, 1413–1433.

BAUMGARDT, D. R. (1990), *Investigation of Teleseismic Lg Blockage and Scattering Using Regional Arrays*, Bull. Seismol. Soc. Am. *80*, 2261–2281.

BAUMGARDT, D. R., and YOUNG, G. (1990), *Regional Seismic Waveform Discriminants and Case Based Event Identification Using Regional Arrays*, Bull. Seismol. Soc. Am. *80*, Part B, 1874–1892.

BEN-AVRAHAM, Z., and GINZBURG, A. (1990), *Displaced Terranes and Crustal Evolution of the Levante and the Eastern Mediterranean*, Tectonics *9*, 613–622.

BENNETT, T. J., BARKER, B. W., MARSHALL, M. E., and MURPHY, J. R. (1995), *Detection and Identification of Small Regional Seismic Events*, Final Report, PL-TR-95-2125, S-Scubed, La Jolla, CA.

BERBERIAN, M. (1983), *The Southern Caspian: A Compressional Depression Floored by a Trapped, Modified Oceanic Crust*, Can. J. Earth Sci. *69*, 163–183.

CAO, S., and MUIREHEAD, K. J. (1993), *Finite Difference Modeling of Lg Blockage*, Geophys. J. Int. *115*, 85–96.

CAMPILLO, M., PLANTET, J.-L., and BOUCHON, M. (1985), *Frequency-dependent Attenuation in the Crust Beneath Central France from Lg Waves: Data Analysis and Numerical Modeling*, Bull. Seismol. Soc. Am. *75*, 1395–1411.

ÇERVENY, V., MOLOTKOV, I. A., and PSENÇIK, I. (1977), *Ray Method in Seismology*, Univerzita Karlova, Praha.

CHAN, W. W., and MITCHELL, B. J. (1985), *Surface Wave Dispersion, Crustal Structure, and Sediment Thickness Variations across the Barents Shelf*, Geophys. J. R. Astr. Soc. *80*, 329–344.

CHINN, D. S., ISACKS, B. L., and BARAZANGI, M. (1980), *High-frequency Seismic Wave Propagation in Western South America along the Continental Margin, in the Nazca Plate and across the Altiplano*, Geophys. J. R. Astr. Soc. *60*, 209–244.

CLARKE, J. W., and RACHLIN, J. (1990), *Geology of the Barents Sea Structural Basin.*, US Geological Survey, Military Geology Project, Open-File Report, July, 1990.

EROS (Earth Resources Observation Systems) Data Center (EDC) (1996), *GTOPO30 (a Digital Elevation Model)*, United States Geological Survey (USGS).

EWING, M., JARDETSKY, W. S., and PRESS, F., *Elastic Waves in Layered Media* (McGraw-Hill, New York, 1957).

FIELDING, E. J., ISACKS, B. L., and BARAZANGI, M. (1992), *A Geological and Geophysical Information System for Eurasia*, Technical Report No. 2, Cornell University, Ithaca, New York.

GUPTA, I. N., BARKER, B. W., BURNETTI, J. A., and DER, Z. A. (1980), *A Study of Regional Phases from Earthquakes and Explosions in Western Russia*, Bull. Seismol. Soc. Am. *70*, 851–872.

GRANT, L., RYALL, F., HENSON, I., and RIVERS, W. (1996), *Ground truth database for seismic discrimination research*, Proc. of the 18th Annual Seismic Res. Symp. on Monitoring a Comprehensive Test-Ban Treaty, 4–6 September 1996, *PL-TR-96-2153*, Environmental Research Papers, no. 1195 (eds LEWKOWICZ, J. F., MCPHETRES, J. M., REITER, D. T.) pp. 997–1006.

HARJES, H. P., and SEIDL, D. (1978), *Digital Recording and Analysis of Broadband Seismic Data at the Grafenburg (GRF) Array*, J. Geophys. Res. *65*, 511–523.

HARTSE, H. E., TAYLOR, S. R., PHILLIPS, W. S., and RANDALL, G. E. (1997), *A Preliminary Study of Regional Seismic Discrimination in Central Asia with Emphasis on Western China*, Bull. Seismol. Soc. Am. *87*, 551–568.

ISACKS, B. L., and STEPHENS, C. (1975), *Conversion of Sn to Lg at a Continental Margin*, Bull. Seismol. Soc. Am. *65*, 224–235.

JIH, R. S. (1995), *Waveguide Effects of Large-scale Structural Variation, Anelastic Attenuation, and Random Heterogeneity on SV, Lg Propagation: A Finite-difference Modeling Study*, PL-TR-96-2016, Environmental Research Papers *1186*, Phillips Laboratory, Hanscom, AFB, MA.

KADINSKY-CADE, K., BARAZANGI, M., OLIVER, J., and ISACKS, B. (1981), *Lateral Variations of High-frequency Seismic Wave Propagation at Regional Distances across the Turkish and Iranian Plateaus*, J. Geophys. Res. *86*, 9377–9396.

KENNETT, B. L. N. (1986), *Lg Waves and Structural Boundaries*, Bull. Seismol. Soc. Am. *76*, 1133–1142.

KENNETT, B. L. N., GREGERSEN, S., MYKKELTVEIT, S., and NEWMARK, R. (1985), *Mapping of Crustal Heterogeneity in the North Sea Basin via the Propagation of Lg Waves*, Geophys. J. R. Astr. Soc. *83*, 299–306.

KIM, W. Y., AHARONIAN, V., LERNER-LAM, A. L., and RICHARDS, P. G. (1997), *Discrimination of Earthquakes and Explosions in Southern Russia Using Regional High-frequency Three-component Data from the IRIS/JSP Caucasus Network*, Bull. Seismol. Soc. Am. *87*, 569–588.

KNOPOFF, L., SCHWAB, F., and KAUSEL, E. (1973), *Interpretation of Lg*, Geophys. J. Roy. Astr. Soc. *33*, 389–404.

KUNIN, N., GONCHAROVA, N. V., IOGANSON, L. I., SEMENOVA, G. I., USENKO, S. V., and SHEIKH-ZADE, E. R. (1987), *Map of Depth to Seismic Basement*, Institute of Physics of the Earth, Moscow, Russia.

KUNIN, N., GONCHAROVA, N. V., SEMENOVA, G. I., USENKO, S. V., and SHEIKH-ZADE, E. R. (1988), *Map of Depth to Moho*, Institute of Physics of the Earth, Moscow, Russia.

MC NAMARA, D. E., OWENS, T. J. and WALTER W. R. (1996), *Propagation Characteristics of Lg across the Tibetan Plateau*, Bull. Seismol. Soc. Am. *86*, 457–469.

MITCHELL, B. J., and HWANG, H. J. (1987), *Effect of low-Q Sediments and Crustal Q on Lg Attenuation in the United States*, Bull. Seismol. Soc. Am. *77*, 1197–1210.

MITCHELL, B. J., PAN, Y., XIE, J., and CONG, L. (1997), *Lg Coda Q Variation across Eurasia and its Relation to Crustal Evolution*, J. Geophys. Res. *102*, 22767–22779.

MOROZOVA, I. B., MOROZOVA, E. A., and SMITHSON, S. B. (1998), *On the Nature of Teleseismic Pn Phase Observed in the Recordings from the Ultra-long Profile "Quartz", Russia*, Bull. Seismol. Soc. Am. *88*, 62–73.

MURPHY, J. R., SULTANOV, D. D., BARKER, B. W., KITOV, I.O., and MARSHALL, M. E. (1997), *Further Analyses of Regional Seismic Data Recorded from the Soviet PNE Program: Implications with Respect to CTBT Monitoring*, Final Report, PL-TR-97-2141, October, 1997.

NI, J., and BARAZANGI, M. (1983), *High-frequency Seismic Wave Propagation beneath the Indian Shield, Himalayan Arc, Tibetan Plateau and Surrounding Regions: High Uppermost Mantle Velocities and Efficient Sn Propagation beneath Tibet*, Geophys. J. R. Astr. Soc. *72*, 665–689.

PIWINSKII, A. J. (1981), *Deep Structure of the Earth's Crust and Upper Mantle in the USSR According to Geological, Geophysical and Seismological Data: Dneiper-Donetsk and Pri-Caspian Depressions*, UCID-19203, Lawrence Livermore Laboratory, Livermore, CA.

RODGERS, A. J., NI, J. F., and HEARN, T. M. (1997), *Propagation Characteristics of Short-period Sn and Lg in the Middle East*, Bull. Seismol. Soc. Am. *87*, 396–413.

RUZAIKIN, A., NERSESOV, I., KHALTURIN, V., and MOLNAR, P. (1977), *Propagation of Lg and Lateral Variations in Crustal Structure of Asia*, J. Geophys. Res. *82*, 307–316.

SEBER, D., BARAZANGI, M., TADILI, B. A., RAMDANI, M., IBENBRAHIM, A., SARI, D. B., and EL ALAMI, S. O. (1993), *Sn to Sg Conversion and Focusing along the Atlantic Margin, Morocco: Implications for Earthquake Hazard Evaluation*, Geophys. Research Letters, *20*, 1503–1506.

SHAPIRO, N., BETHOUX, N., CAPILLO, M., and PAUL, A. (1996), *Regional Seismic Phases across the Ligurian Sea: Lg Blockage and Oceanic Propagation*, Phys. Earth and Planetary Int. *93*, 257–268.

SHISHKEVISH, C. (1979), *Propagation of Lg Seismic Waves in the Soviet Union*, N-1014-ARPA, February 1979, Rand, Santa Monica, CA.

SIKHARULIDZHE, I. I. (1964), *The Nature of Lg and Rg and the Investigation of the Structure of the Earth's Crust*, AS Georgian SSR, Transactions of the Geophysical Institute *22*, 57–70.

TEXAS INSTRUMENTS (1977), *Iranian Long-period Array*, Final Report, Equipment Group, Texas Instruments, Inc., 8 April 1977.

ZHANG, T. R., and LAY, T. (1995), *Why the Lg Phase Does not Traverse Oceanic Crust*, Bull. Seismol. Soc. Am. *85*, 1665–1678.

ZONENSHAIN, L. P., and LE PICHON, X. (1986), *Deep Basins of the Black Sea and Caspian Sea as Remnants of Mesozoic Backarc Basins*, Tectonophys. *123*, 181–211.

(Received July 1, 1999, revised February 16, 2000, accepted April 24, 2000)

Pure appl. geophys. 158 (2001) 1251–1277
0033–4553/01/071251–27 $ 1.50 + 0.20/0

Pure and Applied Geophysics

A Hybrid BE-GS Method for Modeling Regional Wave Propagation

Li-Yun Fu[1,2] and Ru-Shan Wu[1]

Abstract—We present a hybrid boundary-element (BE) and generalized screen propagator (GSP) method for the 2-D SH problem to model the combined effects of arbitrarily irregular topography, large-scale crustal variation, and the associated small-scale heterogeneities on regional wave propagation. We develop a boundary connection technique to couple the wave fields calculated by the BE method with those of the GSP method. Its validity is tested by numerical experiments. For a long crustal waveguide, the relatively short sections with severe surface topography can be modeled by the time-consuming BE method to high frequencies, and the exterior field in the relatively weak heterogeneous media of large volume can be calculated by the GSP method. For the waveguide with severe topography, the BE method can be used section by section via the boundary connection technique to model the combined effects of rough topography and large-scale structural variation on *Lg* wave propagation at extended regional distances.

Numerical comparisons with independent methods showed that the hybrid method is relatively accurate for *Lg* simulation. We apply the hybrid method to *Lg* wave propagation in two real crustal waveguides in the Tibet region; one with *Lg* blockage and another without blockage. We found that the most characteristic effect from the irregular topography is the strong scattering by the topographic structures. The scattering by local irregular topographies leads to anomalous near-receive effects and tends to remove energy from the guided waves, which causes decay of amplitude and waveform distortion. It can be expected that rough surface topography and random heterogeneities with scale length close to the dominant wavelength will be very efficient in attenuating regional waves. The dramatic lateral variation of the topography-Moho large-scale structure combined with the small-scale rough topography and random heterogeneities could be the cause of *Lg* anomalous attenuation and blockage observed in this region. More quantitative assessment of the topographic effects must be conducted in the future.

Key words: Hybrid modeling, boundary element method, generalized screen method, wave-field connection technique, regional wave propagation, *Lg* phase, topographic effects.

Introduction

Regional phases, generated by shallow events, have attracted wide attention in recent years because of their potential uses in both crustal structure determination and yield estimation of underground explosions. The most prominent regional phase is the *Lg* phase (PRESS and EWING, 1952), observed from about 150 km up to several thousand kilometers in the frequency range 0–25 Hz. Numerous studies on

[1] Institute of Tectonics, University of California at Santa Cruz, Santa Cruz, California, USA.
[2] Present address: CSIRO Petroleum, PO Box 1130, Bentley, WA6102, Australia.

Lg propagation in different geological and geophysical environments (e.g., WETMILLER, 1974; KENNETT et al., 1985; BAUMGARDT, 1990) have shown that path changes in complex structures and heterogeneities have a critical role in anomalous propagation of *Lg*. Numerical simulations can provide insight into the path effects on the pattern of phase and amplitude changes of *Lg* across continental margins (e.g., KENNETT, 1986; CAMPILLO, 1987; MAUPIN, 1989).

A variety of numerical methods exist for regional wave propagation simulation, each with its own strengths and limitations. The full-waveform numerical methods, allowing all conversions and all orders of multiple-scatterings, include the finite-difference (FD) method (e.g., BOORE, 1970; FRANKEL and CLAYTON, 1986; CAO and MUIRHEAD, 1993; JIH, 1996) and pseudo-spectral method (e.g., SCHATZMAN, 1996; FURUMURA and KENNETT, 1997) for arbitrarily heterogeneous media, discrete wavenumber method (BOUCHON, 1982; CAMPILLO and BOUCHON, 1985; BOUCHON and COUTANT, 1994) and global generalized reflection/transmission matrices method (CHEN, 1995) for stratified media with irregular interfaces, and coupled mode method (KENNETT, 1984, 1998; MAUPIN and KENNETT, 1987) for stratified and moderately heterogeneous media. To reduce computation time, some alternative and flexible approaches have been developed for *Lg* simulation, for example, ray diagrams (KENNETT, 1986; BOSTOCK and KENNETT, 1990) and propagator matrix and finite-element (FE) combination (REGAN and HARKRIDER, 1989).

The ideal numerical method for *Lg* propagation simulation should not only be able to deal with complex crustal variations with arbitrary heterogeneities and rough free-surface geometry, but also have a fast speed of computation required for *Lg* modeling with higher frequencies (for a basic *Lg* sensitive window 0–8 Hz) and long regional propagation distances. Practical methods should make optimal compromises in medium complexity, computation accuracy, and computation time. WU *et al.* (1996, 1997) have recently introduced a half-space GSP (Generalized Screen Propagator) for modeling the main characteristics of *Lg* (2-D SH case) in smoothly varying, heterogeneous crustal waveguides. The waveguide is sliced into thin-slabs perpendicular to the *Lg* propagation direction (i.e., horizontal direction). The method is based on the one-way wave equation and small-angle approximation. The one-way propagator GSP neglects backscattered waves, but handles all the forward scattering effects, e.g., focusing/defocusing, diffraction, and interference. The successful application of the one-way approximation to *Lg* is based on the fact that the major *Lg* energy in the crustal waveguide environment is carried by forward propagating waves bouncing up and down between the free surface and the Moho. Beyond the critical angle, these waves become dominant by small wavenumbers and are trapped in the crustal waveguide in the form of guided modes. Therefore the neglect of backscattered waves and small-angle approximation in simulation will not change the main features of regional waves in most cases. The main advantage of GSP is the saving of computing time and memory while preserving main dynamic properties of *Lg*, which makes it a viable method for 3-D *Lg* simulation. However, in the current form, the method cannot handle rough topography in crustal waveguides. Although topographic transforms have been introduced to handle mild to moderate topography (WU *et al.*, 1999), severe topography remains a problem of the GSP method.

Topographic influences on the *Lg* propagation behavior along the path have drawn considerable interest recently, motivated by many observations of correlation between surface features and regional phase behavior. ZHANG and LAY (1994) found linear correlations of the *Sn*/*Lg* and *P*/*Lg* ratios with mean altitude, *rms* roughness, *rms* slope, and skewness of topography along the paths from underground nuclear explosions at Semipalatinsk, Kazakhstan to receivers in Eurasia. Recent studies have shown that these along-path topographic statistics are correlated with regional phase amplitude ratios in the Western United States (ZHANG *et al.*, 1996), western China (FAN and LAY, 1998a,b; HARTSE *et al.*, 1998), and Alibek, Turkmenistan (RODGERS *et al.*, 1997). The above results are based on empirical methods using limited data sets. However, little study has been done for numerical modeling of topographic effects on *Lg*. As illustrated with simple model tests (JIH, 1996), a moderate free-surface topography has relatively weak influence on *Lg* propagation because the surface topography variations are generally small relative to the crustal thickness. This conclusion may be true for relatively low frequency contents and mild topography. However, rough topographic influences and especially the combined effects of surface-topography, sedimentary structure, and small-scale heterogeneities still need extensive study.

Much has been published on modeling free-surface topography. The FD method is widely accepted to simulate wave propagation for complex, realistic media. The implementation of the boundary conditions, however, may cause problems and becomes relatively complicated when applied to the topography of complex geometric shapes (JIH, 1988; HESTHOLM and RUUD, 1994; ROBERTSSON, 1996). In addition, the method is susceptible to significant progressive dispersion errors when simulating wave propagation over distances of hundreds of wavelengths (FORNBERG, 1987; SEI, 1993). The accuracy and stability of solutions might therefore be questioned for high frequencies and long distances. The FE method can model the complicated geometric shapes using irregular grids with elements of different size, however the implementation of the algorithm is not convenient in the presence of a rugged free surface, and the computation is very time-consuming. In contrast, the boundary integral equation method or BE method can be formulated in terms of an integral along boundaries. The traction-free condition for rugged free surface is easy and natural to treat in an accurate and stable manner for the BE method.

The object of this paper is to develop a hybrid method of BE and GSP for the 2-D SH problem to study the effects of both rough topography and heterogeneous media on *Lg* propagation. The BE method is implemented in the frequency domain and has a kernel function compatible with the GSP method. The hybrid method will combine the advantages of the two methods. In the hybrid scheme, the time-consuming BE method can be used to handle all reverberations inside the rugged free-surface structure while the large volume of moderately heterogeneous media can be modeled by the GSP method. As mentioned previously, the *Lg* propagation simulation by the GSP method has errors in the near-source region due to the small-angle approximation. This problem can be circumvented through the hybrid method where the BE method can be used to calculate wave propagation from the source to the critical distance to produce an initial wave field for the GSP method. For an event at far regional distance up to several thousand kilometers, the relatively short sections with severe surface topography in the crustal waveguide can be modeled by the BE method to high frequencies. Subsequently the output will be used as the input to the GSP method for modeling the sections with mild topography. This procedure is repeated to complete the entire crustal waveguide computation.

If the entire waveguide is full of rough topography, the BE method can be used section by section to model the combined effects of rough topography and large-scale structural variation. This modeling of the combined effects becomes practically significant because topographic parameters can be more precisely determined for any real path than other parameters. The large-scale structures of crustal thickness, sedimentary layer thickness, and velocity distribution are also easier to obtain for most areas than small-scale heterogeneities. This method with section by section for long regional waveguides will greatly save computing time and memory in comparison with taking the entire waveguide as one section. It becomes a computationally viable technique for practical *Lg* modeling.

After briefly describing the principles of the BE and GSP methods we develop a boundary connection technique which is a critical step of coupling the BE method with the GSP method. Its validity is tested by numerical experiments. Finally we apply the hybrid method to *Lg* wave propagation in theoretical models and two crustal waveguide models from the Tibet region, one with *Lg* blockage and the other without blockage.

Boundary Element Methods

In many cases wave propagation involving heterogeneous media can be formulated in terms of boundary integral equations. The strict boundary methods include the direct and indirect boundary integral equation techniques. The direct method has been widely used due to the explicit meaning of the unknowns in the formulation, while the indirect method formulates problems in terms of force or force moment boundary densities. These two methods can be implemented either in the space-time domain or in the space-frequency domain. Although the use of the strict boundary methods lags far behind the use of domain methods in engineering, it has been extensively used to study the effects of topography or sedimentary basin structures on ground motions at the surface during the last several decades (e.g.,

WONG and JENNINGS, 1975; SANCHEZ-SESMA and ESQUIVEL, 1979; LEE and LANGSTON, 1983; BENITES and AKI, 1989; SANCHEZ-SESMA and CAMPILLO, 1991), and has gained popularity among theoretical seismologists. For many applications an alternative approach has been developed by the combination of discrete wavenumber expansions for Green's functions with boundary integral formulations (e.g., BOUCHON, 1985; CAMPILLO and BOUCHON, 1985; GAFFET and BOUCHON, 1989; KAWASE and AKI, 1989; KIM and PAPAGEORGIOU, 1993; BOUCHON *et al.*, 1996). In this case the singularity problem of the Green's function is avoided by using wavenumber integration. This technique has been shown to be accurate for any finite frequency and can be extended for multilayered media with irregular interfaces. Because of the computational intensity, boundary integral methods are often limited to the case of low frequencies for *Lg* simulation at regional distances. Applications used to be restricted to simple geometrical structures. Few papers applied these methods to real surface topography modelings.

In this study we select the direct frequency-domain BE method integrated with the screen approach to model the effects of both local irregular topography and complex crustal structure on regional wave propagation. A semi-circular canyon topography problem shown in Figure 1 is used for the formulation of the BE method with an attempt to test the accuracy of the approximate method in comparison with the analytical solution. Consider 2-D SH-wave propagation in the homogeneous half-space Ω with a free surface Γ. The anti-plane displacement $u(\mathbf{r})$ at location $\mathbf{r}$ satisfies the steady-state boundary integral equation

$$C(\mathbf{r})u(\mathbf{r}) + \int_\Gamma u(\mathbf{r}')\frac{\partial}{\partial n}G(\mathbf{r},\mathbf{r}')\,d\mathbf{r}' = \int_\Gamma G(\mathbf{r},\mathbf{r}')\frac{\partial}{\partial n}u(\mathbf{r}')\,d\mathbf{r}' + u^0(\mathbf{r}) \ , \qquad (1)$$

where the coefficient $C(\mathbf{r})$ is equal either to 0.5 for the location $\mathbf{r}$ on Γ or to 1.0 for the location $\mathbf{r}$ inside Ω, $G(\mathbf{r}, \mathbf{r}')$ is the Green's function for the homogeneous region, $\partial/\partial n$ denotes differentiation with respect to the outward normal of the boundary Γ, and $u^0(\mathbf{r})$ is the free-field displacement, i.e., the solution with the irregular boundary Γ absent. $u^0(\mathbf{r})$ can be assumed as an incident harmonic plane wave. For the presence of a point source located at $\mathbf{r}_0$, it can be expressed by $u^0(\mathbf{r}) = S(\omega)G(\mathbf{r}, \mathbf{r}_0)$, where $S(\omega)$ is

Figure 1
Configuration of the semi-circular canyon topography problem considered for comparison of results with the analytical solutions (TRIFUNAC, 1973).

the source spectrum. Applied to topography with the traction-free condition, equation (1) is simplified to

$$C(\mathbf{r})u(\mathbf{r}) + \int_{\Gamma} u(\mathbf{r}') \frac{\partial}{\partial n} G(\mathbf{r}, \mathbf{r}')\, d\mathbf{r}' = u^0(\mathbf{r}) \ . \tag{2}$$

In the strict frequency-space domain BE method, the free surface Γ is discretized into NE linear elements denoted by $\Gamma_e (e = 1, 2, \ldots, NE)$ and NP nodes. By using linear interpolation shape functions $N(\xi)$ in an element Γ_e between the nodes I_1 and I_2 (Fig. 1), we obtain

$$u(\xi) = \sum_{l=I_1}^{I_2} u(\mathbf{r}_l) N_l(\xi) \ , \tag{3}$$

where ξ and l denote the local coordinate and local node index of an element. Then equation (2) for $i = 1$ to NP is discretized into

$$C(\mathbf{r}_i)u(\mathbf{r}_i) + \sum_{e=1}^{NE} \int_{\Gamma_e} \frac{\partial}{\partial n} G(\mathbf{r}_i, \mathbf{r}'(\xi)) \sum_{l=I_1}^{I_2} u(\mathbf{r}_l) N_l(\xi)\, d\mathbf{r}'(\xi) = u^0(\mathbf{r}_i) \ . \tag{4}$$

Adopting the following Kronecker delta function notation relating the local node code l of an element Γ_e to the global node code j

$$\delta_{lj} = \begin{cases} 0, & l \neq j \\ 1, & l = j \end{cases} \ , \tag{5}$$

we have

$$u(\mathbf{r}_l) = \sum_{j=1}^{NP} \delta_{lj} u(\mathbf{r}_j) \ . \tag{6}$$

Let $C_i = C(\mathbf{r}_i)$, $u_i = u(\mathbf{r}_i)$, $u_i^0 = u^0(\mathbf{r}_i)$ and substitute equation (6) into equation (4) to give

$$C_i u_i + \sum_{j=1}^{NP} \sum_{e=1}^{NE} \sum_{l=I_1}^{I_2} \left[\int_{\Gamma_e} \frac{\partial}{\partial n} G(\mathbf{r}_i, \mathbf{r}'(\xi)) N_l(\xi)\, d\mathbf{r}'(\xi) \right] \delta_{lj} u_j = u_i^0 \ . \tag{7}$$

Letting

$$G'_{ij} = \sum_{e=1}^{NE} \sum_{l=I_1}^{I_2} \left[\int_{\Gamma_e} \frac{\partial}{\partial n} G(\mathbf{r}_i, \mathbf{r}'(\xi)) N_l(\xi)\, d\mathbf{r}'(\xi) \right] \delta_{lj} \ , \tag{8}$$

and substituting in equation (7), we get the following simultaneous equations for $i = 1$ to NP

$$C_i u_i + \sum_{j=1}^{NP} G'_{ij} u_j = u_i^0 \ . \tag{9}$$

The coefficient G'_{ij} is obtained by integrating the scalar product of the gradient of Green's function (with respect to the outward normal to the surface) with interpolation shape functions over elements. After solving the linear equation system (9) for u_i at all nodes on Γ, we can compute the seismic response at any receiving point in the medium through back substitutions of equation (9). In general, the coefficient matrix is full-rank. For a piecewise homogeneous media with irregular interfaces (FU and MU, 1994; FU, 1996), the discretization of equation (1) can be done in each subdomain, after which all equations are assembled into a global matrix equation by using the interface conditions of continuity for displacements and stresses across all interfaces. This global matrix is sparse or narrow-banded, depending on the structure of the models. For a crustal waveguide discussed in this paper, the continuity conditions of displacements and stresses are satisfied on the Moho interface

$$\begin{cases} u_1 = u_2 \\ \mu_1 \frac{\partial u_1}{\partial n} = \mu_2 \frac{\partial u_2}{\partial n} \end{cases} , \tag{10}$$

where n is the vector normal to the Moho interface, μ is the shear modulus, u_1 and u_2 are the displacements, respectively, above and below the Moho interface. To improve computation speed a variable element dimension technique is adopted in the program implementation (FU and MU, 1994). The element dimension for each frequency is computed according to the medium velocity and the frequency, and then the model is automatically discretized. This improves the efficiency of the BE method.

To test the accuracy of the computation program, we compare our numerical solution with the exact solution for scattering by a semi-circular canyon of radius a. The analytical solution in the frequency domain has been studied by TRIFUNAC (1973) for the semi-circular canyon topography problem. The two sharp edges at $x = \pm a$ in the model provide a crucial target to test various numerical methods. Figure 2 indicates an excellent agreement between our results (solid lines) and the analytical solutions (dotted lines) for various dimensionless frequencies η and for an incident angle of 0°. The dimensionless frequency is defined as $\eta = 2a/\lambda = a\omega/\pi V$, where ω is circular frequency and V is the velocity of shear-wave propagation. The comparisons for 30° incidence and $\eta = 1.0$ are given in Figure 3 for different values of the element length L used to discretize the arc of the semi-circular canyon. Note that for $\eta = 1.0$, $L = 200$ m (corresponding to 10 elements for a wavelength) yields results of acceptable accuracy. This concludes that L should be small enough to approximate the arc of the semi-circular canyon. The above comparisons confirm the validity of our formulations and computation program. We now demonstrate the applicability of our program by a synthetic example with rough random free surface. Figure 4 demonstrates the comparison of energy attenuations for models with and without a rough random free surface. Energy computation is done first by the calculation of synthetic seismograms at all discrete points, followed by a summation of all frequency components for each synthetic seismogram. The randomness of the random

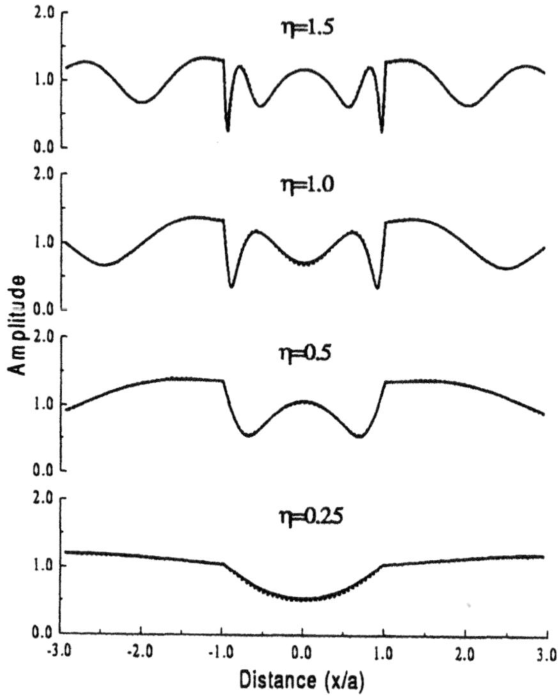

Figure 2
Comparison of results to exact solution for a semi-circular canyon topography with vertical incident SH wave for various dimensionless frequencies η. The solid lines denote our results and the dotted lines denote the exact solutions (TRIFUNAC, 1973).

topography has an exponential correlation function. We see that rough topography can increase energy scatterings and complexity of regional wave propagation.

Boundary integral methods have been used to study the *Lg* blockage problem with limited success. The computation at high frequencies becomes extremely time-consuming because matrices with large sizes must be inverted. In fact, the propagation properties deduced from simulations at relatively low frequencies on the order of 1 Hz show very different characteristics from those with higher frequencies (WU *et al.*, 1997). *Lg* blockage is assumed to be caused by coastlines, mountains and sudden change of crustal thickness. However, numerical simulations of blockage by large-scale crustal structures using boundary methods have not succeeded in matching the observations (CAMPILLO *et al.*, 1993; GIBSON and CAMPILLO, 1994). Geological structures such as grabens and mountain ranges often lead to anomalous attenuation, even extinction of the *Lg* phases. Nevertheless, based on the synthetic seismograms using these methods, the *Lg* waves propagated with a much smaller degree of amplitude reduction than was observed. One of the reasons for these discrepancies between the observed and modeling results is that the models used were oversimplified. Rough topography and small-scale irregularities should be

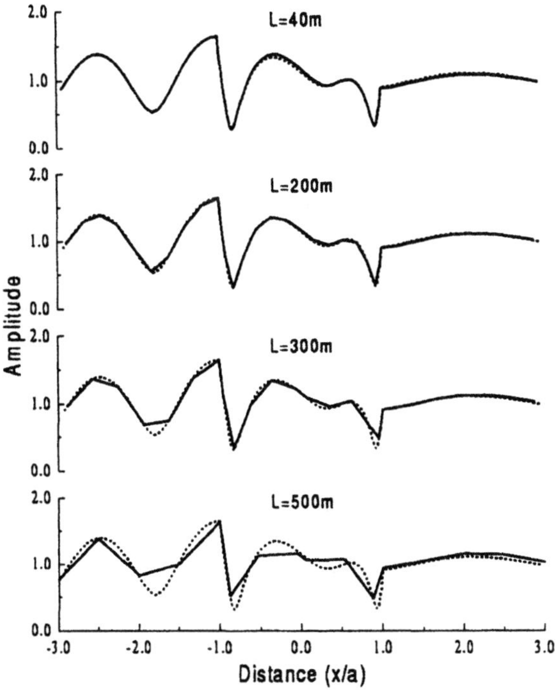

Figure 3
The frequency responses of a semi-circular canyon topography to 30° incident SH wave and $\eta = 1.0$ for various element lengths L. The solid lines denote our results and the dotted lines denote the exact solutions.

included. These problems can be attacked with the use of the hybrid scheme by which the relatively short sections of the local strong heterogeneities, including irregular topography in the crustal waveguide, can be modeled by the BE method to high frequencies, and the exterior field in the relatively weak small-scale heterogeneous media of large volume can be calculated by the generalized screen method.

Generalized Screen Propagator Method

As a modeling method functioning between full-waveform methods and high-frequency asymptotic methods, the phase screen method neglects backscatterings so that the full-wave equation can be simplified as the one-way wave equation. This method has been extensively used to calculate one-way wave propagation (e.g., HARDIN and TAPPERT, 1973; FLATTE and TAPPERT, 1975; MARTIN and FLATTE, 1988; THOMSON and CHAPMAN, 1989; FISK and MCCARTOR,1991). The method is extended into the complex screen method for calculating one-way elastic wave propagation (WU, 1994; WU and XIE, 1994; WILD and HUDSON, 1998). A detailed discussion of various approximations for forward and backward scatterings of different wave types

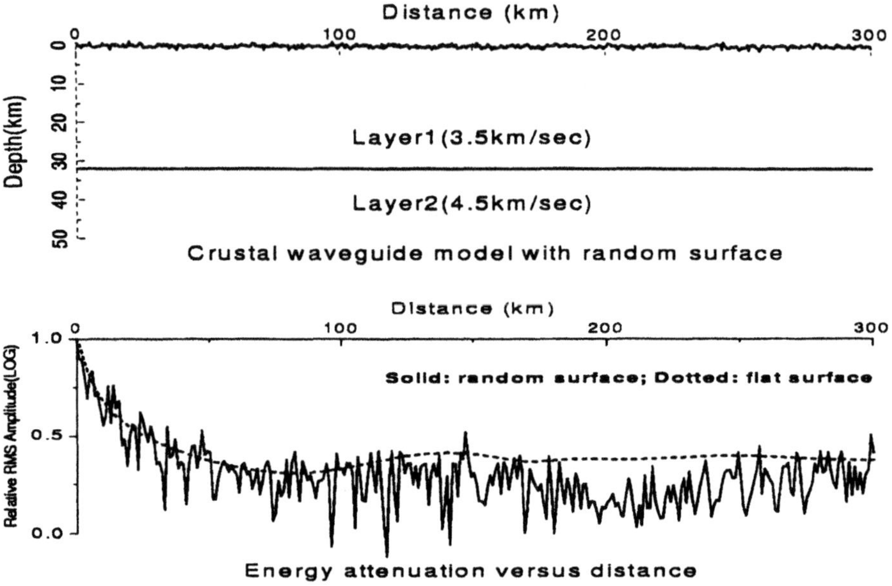

Figure 4
Energy distribution for a random-surface crust model. The random surface has an average thickness of 2 km and a RMS depth perturbation of 0.2 km with correlation length of 2 km. Shown in the top panel is the model with receivers $1 \sim 301$ at 1-km spacing along the free surface, and in the lower panel are the energy attenuation curves. The source depth is 2 km. The source function is a Gaussian derivative with dominant frequency of 0.5 Hz and computation frequency range $0.0 \sim 1.5$ Hz.

is given in WU (1996). The main advantages of the screen method are the fast speed of computation and the large saving in computer memory, especially in calculating the long-range wave propagation.

From the scalar Helmholtz equation of a 2-D SH problem, the seismic response $u(\mathbf{r})$ at location $\mathbf{r}$ in a heterogeneous region Ω can be expressed by the Lipmann-Schwinger equation:

$$u(\mathbf{r}) = u^0(\mathbf{r}) + k^2 \int_\Omega G(\mathbf{r}, \mathbf{r}')F(\mathbf{r}')u(\mathbf{r}')\, d^2\mathbf{r}' \ , \tag{11}$$

where k is the background wavenumber, $u^0(\mathbf{r})$ is the primary wave field, $F(\mathbf{r})$ is the perturbation function of the velocity and density of the medium, and $G(\mathbf{r}, \mathbf{r}')$ is the half-space scalar Green's function in the background medium which can be obtained by the mirror image method (MORSE and FESHBACH, 1953). For the one-way approximation, the half-space model is sliced into thin slabs perpendicular to the propagation direction (i.e., along the x-coordinate direction). Weak scattering condition holds in each thin slab. The thickness of the slab should be made thin enough to ensure the validity of the local Born approximation. Under this condition, the wave field propagating in a thin slab between x' and x_1 can be expressed as

$$u(x_1,z_1) = u^0(x_1,z_1) + k^2 \int_{x'}^{x_1} dx \int_0^{\infty} dz\{G(x_1,z_1;x,z)F(x,z)u^0(x,z)\} \ . \qquad (12)$$

Taking the Fourier transform with respect to z_1 and the small-angle approximation (i.e., $k_z \ll \gamma = \sqrt{k^2 - k_z^2} \approx k$) results in

$$\begin{aligned} u(x_1,k_z) &= e^{i\gamma(x_1-x')} \int_0^{\infty} dz\{2\cos(k_z z)[1 + ikS_s(z)]u^0(x',z)\} \\ &\approx e^{i\gamma(x_1-x')} C\lfloor e^{ikS_s(z)}u^0(x',z)\rfloor \ , \end{aligned} \qquad (13)$$

where $e^{ikS_s(z)}$ is the phase delay operator, $C[f(z)]$ is the cosine transform defined by

$$C[f(z)] = \int_0^{\infty} dz 2\cos(k_z z)f(z) \ , \qquad (14)$$

and $S_s(z)$ is the average *SH*-wave slowness perturbation over the thin slab at depth z. Note that the phase shift between screens is in the wavenumber domain while the phase-screen interaction is in the space domain. Therefore, the dual-domain technique can be used to implement equation (13).

From the above derivation we can see that the GSP approximation consists of the forward scattering, the local Born, and the small-angle approximations. Accuracy analysis for the GSP method can be found in WU and DE HOOP (1996), HUANG and FEHLER (1998), and extensive numerical tests have been given (LIU and WU, 1994; WU *et al.*, 1997) in comparison with various other numerical methods. Fortunately, the GSP approximation works for *Lg* wave propagation quite well since the major part of *Lg* energy consists of small-angle wavenumbers with respect to the horizontal direction. For example, a standard crustal waveguide with $V_{\text{crust}} = 3.5$ km/s and $V_{\text{mantle}} = 4.5$ km/s corresponds to a relative velocity perturbation of 29%. The critical angle of crustal wave for this waveguide is 51° which corresponds to an incident angle of 39° for the GSP method. This leads to a relative phase error of 6% based on the accuracy analysis of the GSP method (HUANG and FEHLER, 1998). This phase error will be translated into a small error in refraction angle, which will not affect the main characteristics of *Lg*. The main disadvantages of the GSP method for *Lg* simulation are (1) the method cannot well model the pre-critical crust wave propagation, especially for nearly vertical incidence with respect to the horizontal direction. This limitation is for a short distance propagation before reaching the critical distance, around 80 km in the case of the standard crust model. (2) The GSP method cannot handle the rough topography in crustal waveguides. These two disadvantages can be circumvented partly by the hybrid approach.

Figure 5 shows the comparison of Moho reflections calculated by BEM, GSP, and a FD algorithm (for FD simulation see XIE and LAY, 1994), respectively. From the lower panel we see that beyond 70 km the three curves agree quite well. Each curve is scaled with respect to a maximum value determined for its simulation. The relative energy variation curves in the top right panel also agree well. The energy curves are

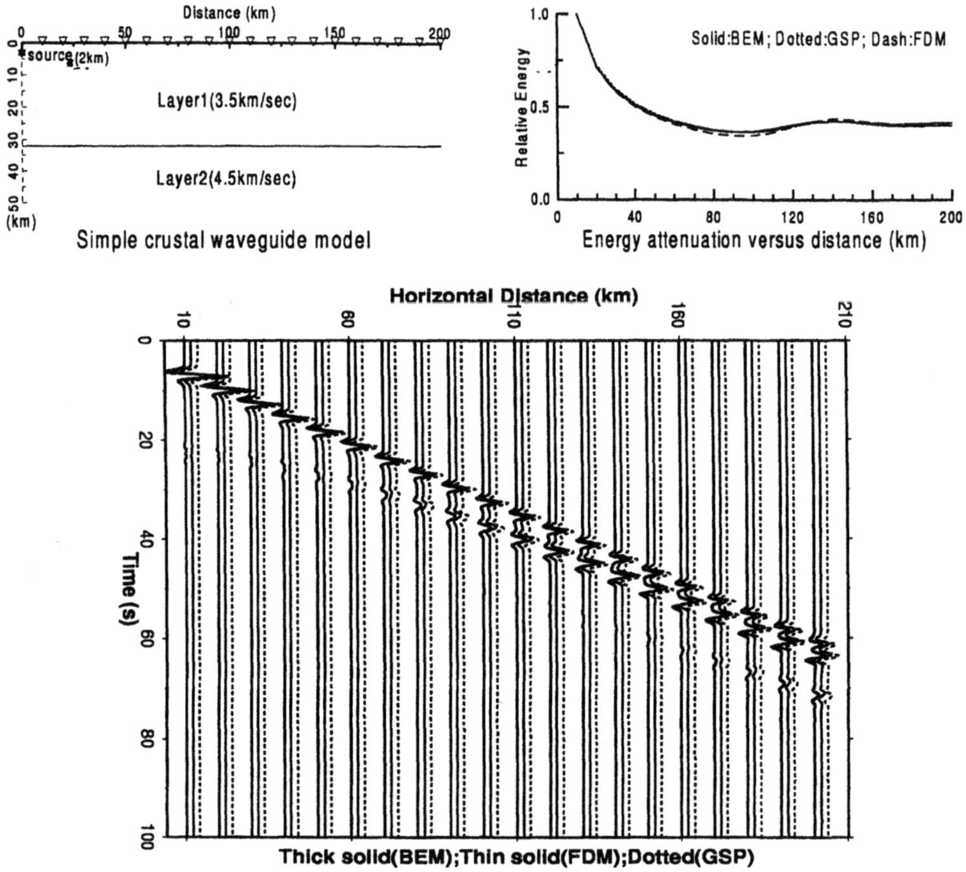

Figure 5
Comparisons among three different numerical methods. The top left panel is a simple crustal waveguide model with receivers $1 \sim 20$ at 10-km spacing along the flat free surface, indicated by small triangles. The top right panel shows the relative energy variation versus distance calculated by BEM, GSP, and FDM, respectively. The bottom panel shows the synthetic seismograms calculated by these three methods, respectively. The source function is a Gaussian derivative with dominant frequency of 0.5 Hz and computation frequency range $0.0 \sim 1.6$ Hz.

scaled in each simulation with respect to the first receiver that is set at 10 km from the source in order to avoid the reflections from nearly vertical incidence. The above comparison proves that the GSP approach can preserve a relative variation in both waveform and energy versus distance for the standard crustal waveguide.

Boundary Connection Technique for The Hybrid Method

The hybrid method will combine the GSP method and the BE method so that the expensive BE calculations can be limited to relatively short sections with severe surface topography. Boundary connection technique to couple the fields calculated by the screen method with those of the BE method will play a crucial role in the hybrid method.

The connection configuration is illustrated in Figure 6a. The model consists of an irregular free surface and an interface as a single layer crustal model. An artificial interface $\Gamma_{AB'}$ is introduced as a wave-field-connection boundary between the screen and BE methods. The entire model is divided into four subdomains, Ω_1, Ω_2 in the crust, and Ω_1', Ω_2' in the mantle. The left boundaries of Ω_1 and Ω_1', and the right boundaries of Ω_2 and Ω_2' are assumed to extend to infinity. The wave fields in Ω_1 and Ω_1' are easy to calculate by the screen method. Its output field $u_0(\mathbf{r})$ on the connection interface $\Gamma_{AB'}$ will be used as the boundary condition when the BE method is used to calculate wave propagation in Ω_2 and Ω_2'. The output field is received along the next connection interface $\Gamma_{CD'}$, and will be used as the input to the screen method in the next propagation.

The total field $u(\mathbf{r})$ on the boundaries of Ω_2 and Ω_2' are composed of

$$u(\mathbf{r}) = \begin{cases} u_0(\mathbf{r}) + u_s(\mathbf{r}), & \mathbf{r} \in \Gamma_{AB'} \\ u_s(\mathbf{r}), & \mathbf{r} \notin \Gamma_{AB'} \end{cases} . \tag{15}$$

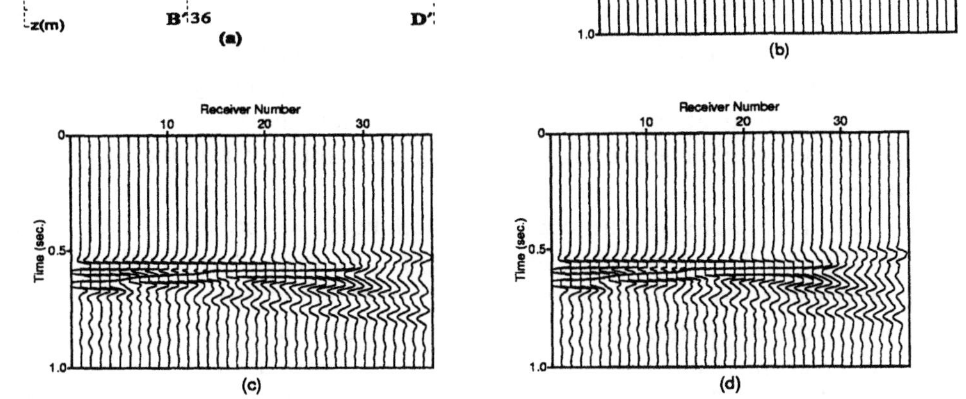

Figure 6

Test of wave field connection. (a) Model geometry with receivers located along $\Gamma_{AB'}$ and $\Gamma_{CD'}$, respectively. (b) The seismograms computed from the source to the connection boundary $\Gamma_{AB'}$. (c) The seismograms computed from the source directly to $\Gamma_{CD'}$. (d) The seismograms on $\Gamma_{CD'}$ computed using the wave fields on $\Gamma_{AB'}$ as the incident field. Comparison of wave fields (c) and (d) shows the validity of the connection technique.

The boundary integral equation for fields on the boundaries of Ω_2 can be obtained from

$$C(\mathbf{r})u(\mathbf{r}) + \int_{\Gamma} u(\mathbf{r}')\frac{\partial G(\mathbf{r},\mathbf{r}')}{\partial n}\, d\mathbf{r}' = \int_{\Gamma} G(\mathbf{r},\mathbf{r}')\frac{\partial u(\mathbf{r}')}{\partial n}\, d\mathbf{r}' \ . \tag{16}$$

The surface Γ surrounding subdomain Ω_2 consists of the connection interface Γ_{AB}, the upper free surface Γ_1, the lower interface Γ_2, and the right boundary Γ_{CD}. Γ_{CD}, in computation, is assumed to be transparent and can be handled using an infinite element absorbing boundary technique (FU and WU, 2000). Therefore the boundary integrals in equation (16) can be decomposed into integrals over segments of the boundary. Considering equation (15) and the free boundary condition on Γ_1, the integral on the right-hand side of equation (16) can be expressed as

$$\int_{\Gamma} G(\mathbf{r},\mathbf{r}')\frac{\partial}{\partial n}u(\mathbf{r}')\, d\mathbf{r}' = \int_{\Gamma_{AB}} G(\mathbf{r},\mathbf{r}')\frac{\partial}{\partial n}[u_0(\mathbf{r}') + u_s(\mathbf{r}')]\, d\mathbf{r}' + \int_{\Gamma_2} G(\mathbf{r},\mathbf{r}')\frac{\partial}{\partial n}u_s(\mathbf{r}')\, d\mathbf{r}' \ . \tag{17}$$

In order to solve $u_s(\mathbf{r})$ and $\partial u_s(\mathbf{r})/\partial n$ on Γ_2, we must build the corresponding boundary integral equation in subdomain Ω_2'.

$$C(\mathbf{r})u_s(\mathbf{r}) + \int_{\Gamma'} u_s(\mathbf{r}')\frac{\partial}{\partial n}G(\mathbf{r},\mathbf{r}')\, d\mathbf{r}' = \int_{\Gamma'} G(\mathbf{r},\mathbf{r}')\frac{\partial}{\partial n}u_s(\mathbf{r}')\, d\mathbf{r}' \ , \tag{18}$$

where $\Gamma' = \Gamma_{BB'} + \Gamma_2' + \Gamma_{DD'}$, and a sufficient long boundary $\Gamma_{BB'}$ is used with its end set as an infinite element. The continuity of displacement and its normal gradient across interface Γ_{BD} is employed when equations (16) and (18) are combined to solve the problem. To solve equation (16) using the boundary condition, the normal gradient of the field on $\Gamma_{AB'}$ must be calculated. It is easy to calculate $\partial u_0(\mathbf{r})/\partial n$ by the screen method. An alternative is to use the Rayleigh integral representation to eliminate the term $\partial u_0(\mathbf{r})/\partial n$. In addition, since there is no discontinuity across $\Gamma_{AB'}$, we can use the transparent boundary condition on $\Gamma_{AB'}$ for $u_s(\mathbf{r})$. By solving the joint boundary integral equations of Ω_2 and Ω_2', we can obtain the wave fields $u_s(\mathbf{r})$ on Γ_1 and Γ_2. The observation field along $\Gamma_{CD'}$ is calculated explicitly from the fields on the boundaries.

To test the validity of the connection technique we present a comparison between the wave field (Fig. 6c) obtained using the BE method to directly calculate wave propagation from the source to the observation surface $\Gamma_{CD'}$ and the one (Fig. 6d) by the connection scheme. In both cases the source wavelet is the same, with the dominant frequency at 20 Hz. First the intermediate wave field (Fig. 6b) on $\Gamma_{AB'}$, calculated from the source, is used as the incident field. Next the BE method is used to calculate wave propagation from $\Gamma_{AB'}$ to $\Gamma_{CD'}$. The dominant arrivals for the incident field at $\Gamma_{AB'}$ consist of three sets of waves. The first arrival is the direct wave propagating from the source to receivers. The second is the reflected wave from the bottom interface, and the third is the reflected wave bounced back from the free surface. From the seismograms at $\Gamma_{CD'}$, more multiply reflected waves between the free surface and the interface can be clearly seen. The agreement between Figures 6c and 6d confirms the validity of the connection technique for the hybrid method.

In Figures 7 and 8 we use two methods (i.e., BEM and BEM-GSP hybrid) to model the 'neck' type and 'belly' type waveguides, respectively. In the BEM-GSP hybrid, BEM is first used to compute wave propagation within the precritical distance from the source over 100 km to produce an initial wave field. Then the initial wave field is propagated over the next 400 km by the GSP method. All the energy curves are scaled in each simulation with respect to the first receiver at 0.0 km. In both figures we see that the curve by BEM-GSP hybrid almost approaches that of BEM. The error in BEM-GSP hybrid may be caused by the irregular Moho which may produce backscatterings and large-angle waves.

Numerical Simulations for A Mixed Waveguide

In this section we apply the hybrid method to regional wave propagation simulations in a synthetic model. Figure 9 displays a laterally varying crust model.

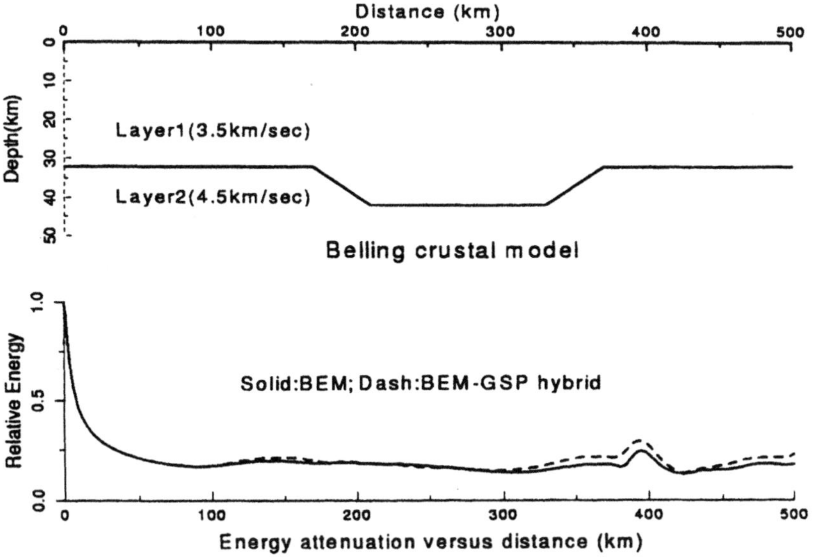

Figure 7

Comparison of energy attenuation curves calculated respectively by BEM and BEM-GSP hybrid for a crust thickening model. Shown in the top panel is a belling crustal model with receivers $1 \sim 501$ at 1-km spacing along the flat free surface, and in the lower panel are the energy attenuation curves. The source depth is 2 km. The source function is a Gaussian derivative with dominant frequency of 1.0 Hz and computation frequency range $0.0 \sim 2.5$Hz.

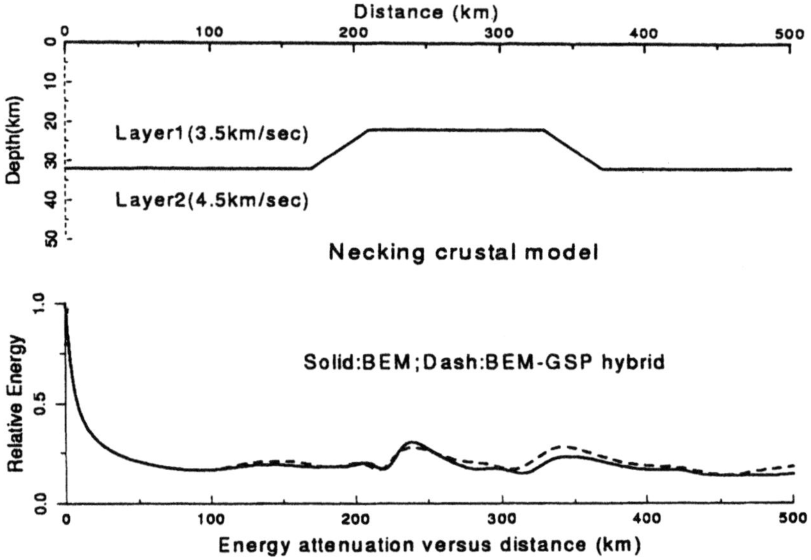

Figure 8
Comparison of energy attenuation curves calculated respectively by BEM and BEM-GSP hybrid for a necking crustal model. Shown in the top panel is a necking crustal model with receivers 1 ~ 501 at 1-km spacing along the flat free surface, and in the lower panel are the energy attenuation curves. The source depth is 2 km. The source function is a Gaussian derivative with dominant frequency of 1.0 Hz and computation frequency range 0.0 ~ 2.5 Hz.

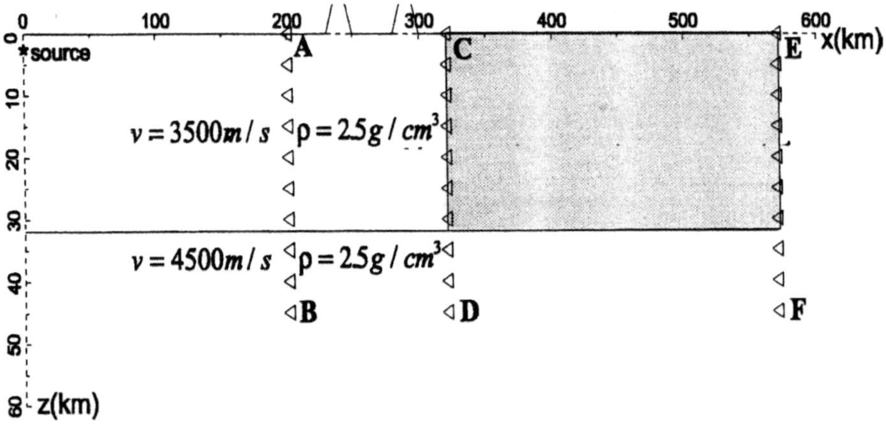

Figure 9
A laterally varying crustal model. The shadowed zone represents a random medium. The receivers are indicated by small triangles. The source depth is 2 km. The source function is a Gaussian derivative with dominant frequency of 1.0 Hz and computation frequency range 0.0 ~ 2.0 Hz.

The first segment is a homogeneous waveguide 200 km in length, 32 km in thickness, with a SH-wave velocity of 3.5 km/s over a half-space (4.5 km/s). The second segment is a complex waveguide consisting of two 6-km high, 20-km long mountain ridges centered at 240 km and 290 km. The last segment represents a 250 km horizontal waveguide of 32-km thickness with small-scale random heterogeneities. The random heterogeneities have an exponential correlation function with a correlation length 3 km in both horizontal and vertical directions, and a velocity perturbation 6%. The point source is located at 2-km depth. The receivers lie along both the free surface and vertical profiles at different distances with a 5-km spacing. Seismograms are computed in the frequency range 0–2 Hz.

First we compute wave propagation from the source over 200 km to produce an initial wave field on the first connection boundary Γ_{AB}. The synthetic seismograms are shown in Figure 10 with the receivers $1 \sim 7$ above the Moho and $8 \sim 10$ under the Moho. Multiple reflections within the crustal waveguide can be seen clearly from the seismograms. Subsequently the wave fields on Γ_{AB} are used as the incident fields, and the BE method is employed to calculate wave propagation for the next 120 km through the segment with the irregular topography. The variable element dimension technique with 6 elements per wavelength is used in the program implementation to improve the computation speed. The synthetic seismograms shown in Figure 11 are recorded along both the free surface and the second connection boundary Γ_{CD} at 320 km. This result is compared to the seismogram in Figure 12, calculated for the

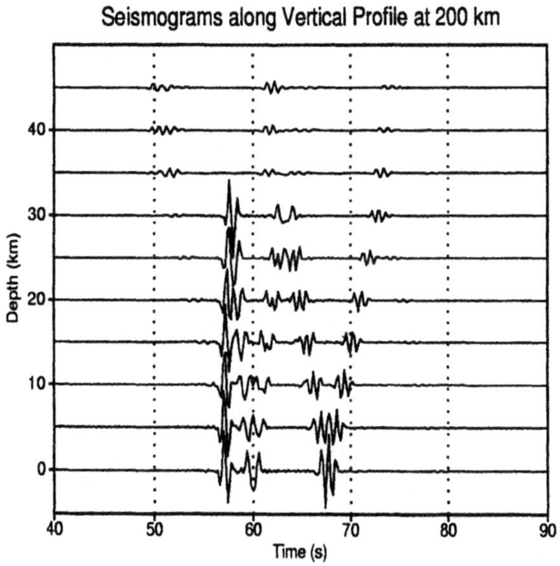

Figure 10
Synthetic seismograms as an initial wave field along the first connection boundary at the distance of 200 km.

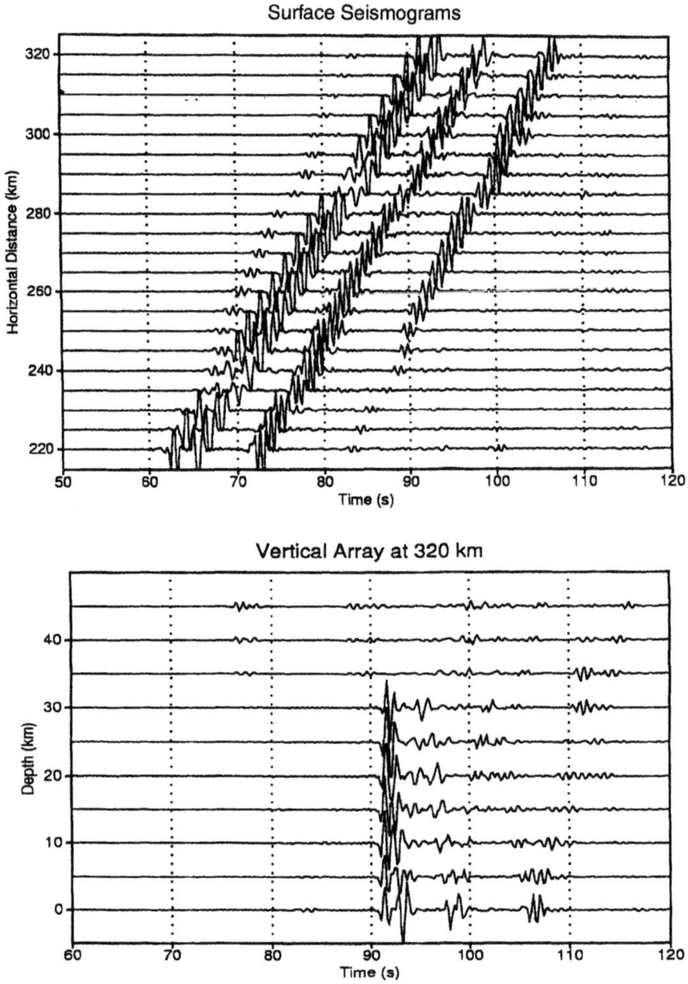

Figure 11
Synthetic seismograms calculated by the BE method using the wave fields in Figure 10 as the incident fields at 200 km. The receivers on the top panel are along the free surface with 5-km spacing, on the bottom, along a vertical profile (Γ_{CD}) at the distance of 320 km.

same segment of the model but with a flat free surface. The synthetic seismograms shown in Figure 11 are more complicated than those in Figure 12. The effects of the irregular topography on the guided waves can be seen from the comparison. The receivers located at or near irregular topographies have anomalous reflected signals due to the focusing/defocusing effects and the contamination by reverberations within these topographic structures (local site effects).

The most characteristic feature of the effects from the irregular topography is the strong scattering by the topographic structures. These scatterings have two

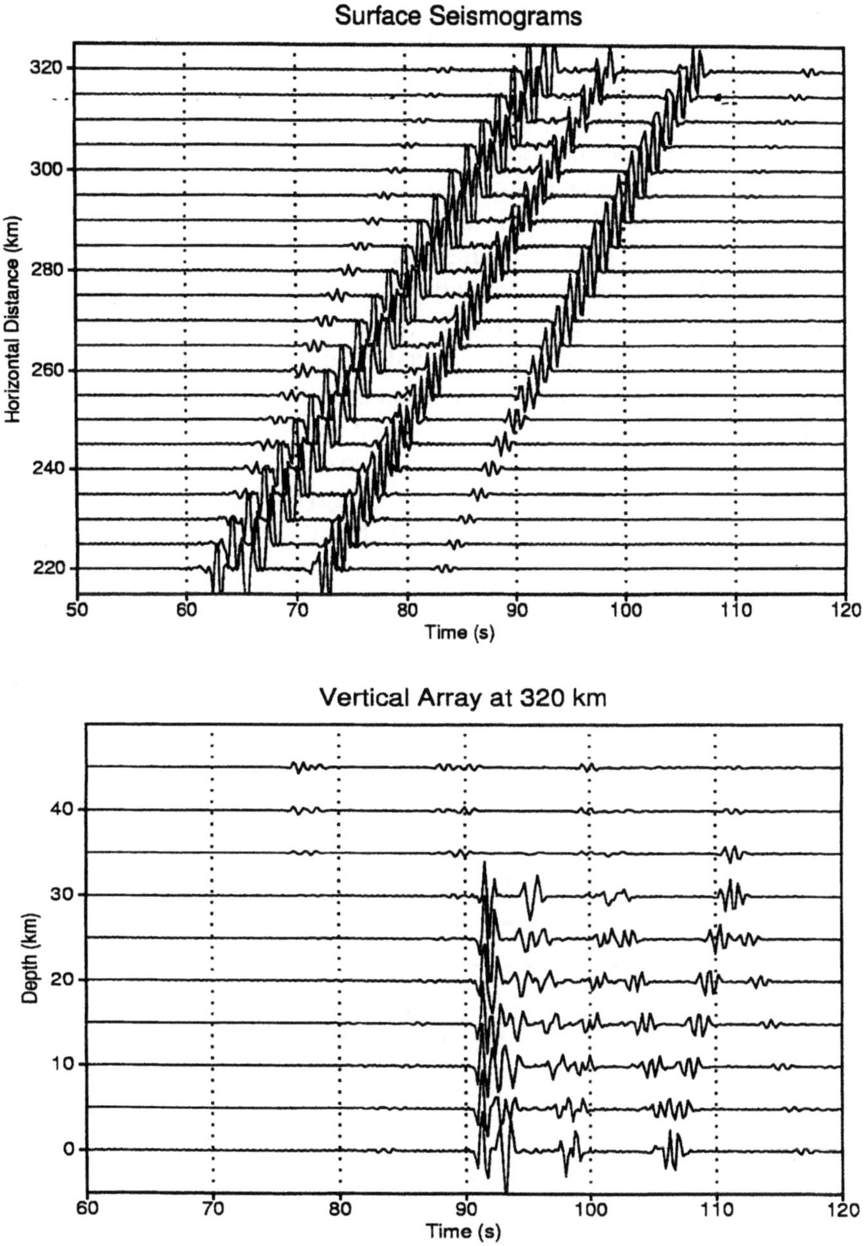

Figure 12
Same as Figure 11 except the free surface is flat.

consequences according to their traveling directions. One part is forescatterings whose active time window is close to that of the wave train. Therefore, they tend to distort the wave train. Another part is backscatterings that add new complexity to the waveguide wave trains in Figure 11. In the surface seismograms there are four groups of strong arrivals after the weak head waves: they are the direct, primary, doubly and triply reflected waves from the Moho. We see very little backscattering for the direct and primary reflected waves, but see clearly the backscattered waves for the doubly and triply reflected waves. Especially for the triply reflected waves, the backscatterings around the two mountain ridges at 240 km and 290 km are rather strong due to the favorable incident angles (steeper angles) in this case. From the vertical profile in Figure 11 (bottom panel) we see more mantle waves compared with Figure 12 (without topography). This is due to the leakage of crustal waves into the mantle by the mountain ridge scattering. In the same time, the multiply reflected crustal waves have been weakened, especially for the quadruply reflected waves (compared to Fig. 12). In summary, the scatterings by local irregular topographies lead to anomalous near-receiver effects and tend to remove energy from the guided waves, which causes decay of amplitude and waveform distortion. These scattered waves partly leak to the mantle and partly merge into the crustal guided waves. It can be expected that rough surface topography with scale length close to the dominant wavelength will be very efficient in attenuating *Lg* waves.

The synthetic seismograms shown in Figure 13 are obtained by using the wave fields on Γ_{CD} in Figure 11 as the input to the screen method and observing along the boundary Γ_{EF} at 570 km. The importance of small-scale random heterogeneities to

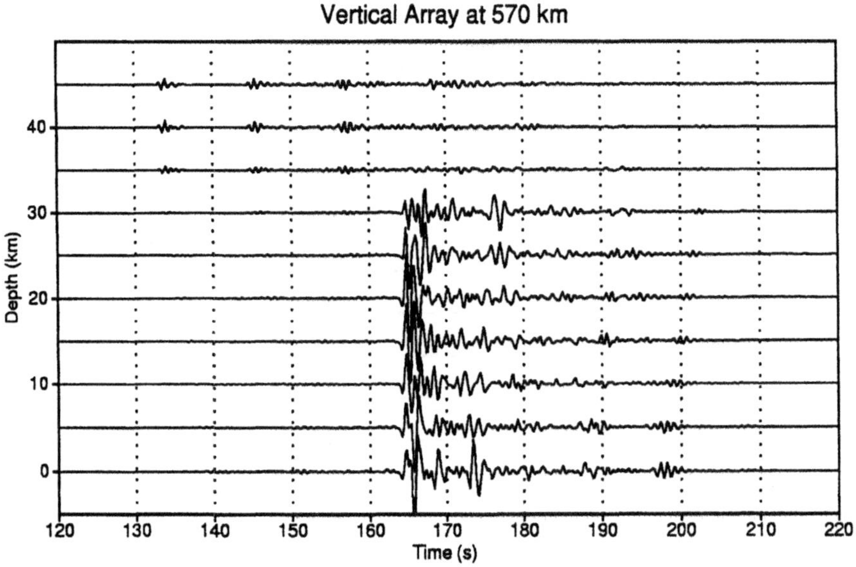

Figure 13
Synthetic seismograms along a vertical profile at the distance of 570 km calculated by the hybrid method. The wave fields in Figure 11 are used as the incident field and the propagation in the random medium is calculated by the screen method.

seismic wave propagation is well known. The role of random heterogeneities in *Lg* wave propagation has been studied by WU *et al.* (1997) using the GSP method, which indicates that random heterogeneities drastically increase the leakage of waves to the mantle and the complexity of the waveforms.

Numerical Simulations with Crustal Waveguide Models of Real Structures

The final models we use are two representative waveguides from Western China. This region has the greatest relief in the world, with frequent earthquakes and underground nuclear test records. Thus it is a region important for quantifying rugged topographic influences on regional seismic discriminants for CTBT monitoring. The topographic relief data of the ray paths linking the earthquakes and WMQ are from the GLOBE (Global Land One-Km Base Elevation) project with a spatial resolution of 30 arc-second (one-km). The database was developed by the National Imagery and Mapping Agency (NIMA) and the National Geophysical Data Center (NGDC) of NOAA. The precisely known topography data make it more reliable for the study of the topographic effects on *Lg* wave propagation. Another attribute of the regional crustal structure used in our modelings is the Moho depth. Moho depth data derive from the database complied by Cornell University. These data were digitized from maps compiled by KUNIN and SHEYKH-ZADE (1983) and KUNIN (1987) based on deep seismic sounding (DSS) profiles and gridded on a 10-km grid (FIELDING *et al.*, 1992). The Moho depth data are relatively coarse, but are fairly useful for low-frequency wave propagation.

In this paper the data of topography and Moho depth along each path from event epicenter to station WMQ are compiled from FAN and LAY (1998). One of the sample cross-section profiles we use is the 29 June 1995 event for ray path from the epicenter (51.923°N, 103.075°E) to station WMQ with magnitude $m_b = 5.6$, and focal depth 15 km. Another is the 2 June, 1990 event for ray path from the epicenter (32.45°N, 92.81°E) to station WMQ with magnitude $m_b = 5.5$ and focal depth 17 km. These two gross crustal models are shown in Figures 14 and 15, respectively. The source is a Ricker wavelet with dominant frequency of 1.0 Hz. The computation is done on a Sun Ultra 5 by BEM using the section-by-section approach in the frequency range $0.0 \sim 2.0$ Hz. Each section is about 500 km long. The division of sections is based on the topographic features. The principle is to minimize the possible multiple scattering between sections so that the transmission between sections can be treated by "one-way" approximation. The receivers are set at 1 km spacing along the surface topography. The total energy for the synthetic seismogram at each receiver is calculated through a summation for all frequency components. The energy curve is scaled with respect to the first receiver at 0.0 km. In each section a variable element dimension technique is adopted in the program implementation to improve the computation speed. The element dimension for each frequency is

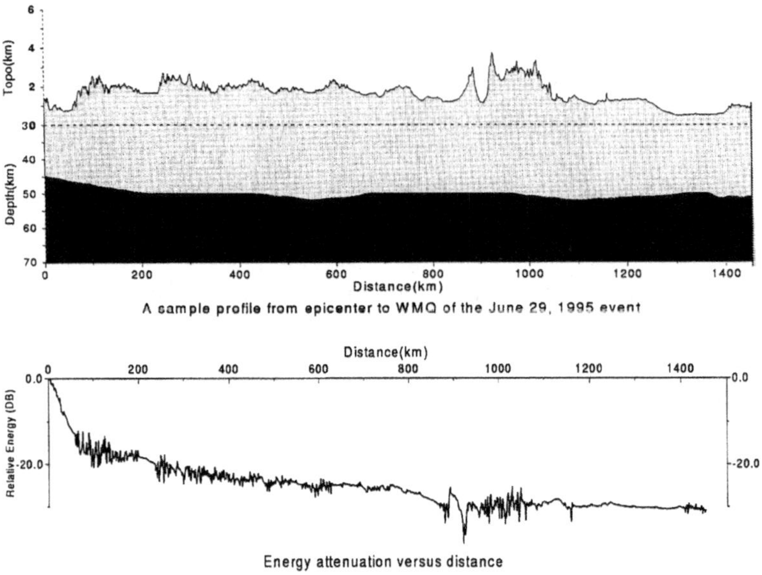

Figure 14
Energy attenuation versus distance for the path without *Lg* wave blockage in the Tibet region. The model is from the event of June 29, 1995, with the location of 51.923 N and 103.075 E, the magnitude of 5.6 and the source depth of 15 km.

computed according to the medium velocity and the frequency, after which the model is automatically discretized. This improves the efficiency of the BE method.

These two waveguides are chosen because one represents the presence of *Lg* blockage (see Fig. 15) and the other, the absence of *Lg* blockage (see Fig. 14) in this area. From Figure 14 we see that the Moho interface is relatively flat. The crustal thickness is large (over 40 km) and rather uniform through the entire waveguide. The surface roughness is locally dramatic (over 2 km), but overall moderate along the path. The energy curve produces strong fluctuations correlated with the topography. This is caused by strong near-receiver topographic scatterings. Studying the correlations of regional wave amplitude ratios with the along-path topographic statistics for different wavelengths may well define the scattering effects. We also see that the overall shape of the energy curve appears to be relatively stable. Especially after 1000 km, the guided energy is trapped in the crust with weak attenuation.

The model with *Lg* blockage in Figure 15 displays completely different waveguide characteristics. The dramatically lateral variation of topography and the up-rise of Moho rapidly change the crustal structure into a necking model after 800 km along the path. The anomalous necking in the crust may lead to *Lg* blockage. Although our computations are restricted to low frequencies (less than 2.0 Hz), the energy curve shows a rapid attenuation in the necking part compared to the case in Figure 14.

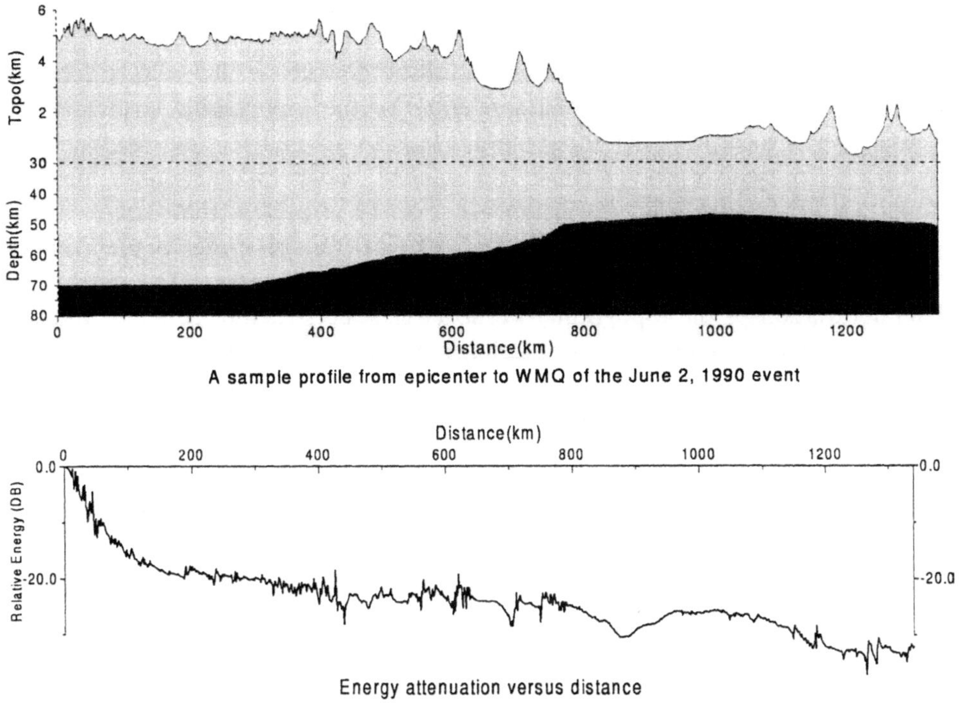

Figure 15
Energy attenuation versus distance for the path with *Lg* wave blockage in the Tibet region. The model is from the event of June 2, 1990, with the location of 32.45 N and 92.81 E, the magnitude of 5.5 and the source depth of 17 km.

Conclusions

We have developed a new hybrid method for the 2-D SH problem to model the combined effects of irregular topography, large-scale crustal variation, and the small-scale heterogeneities on regional wave propagation. In the hybrid scheme, the relatively short sections with severe surface topography in the crustal waveguide can be modeled by the time-consuming BE (boundary element) method to high frequencies, and the exterior field in the relatively weak heterogeneous media of large volume can be calculated by the GSP (Generalized Screen Propagator) method. We present a boundary connection technique which is critical in coupling the BE method with the GSP method. Its validity is tested by numerical experiments. If the entire waveguide consists of rough topography, the BE method can be used section by section via the boundary connection technique to model the combined effects of rough topography and large-scale structural variation on *Lg* wave propagation at large regional distances.

Numerical comparisons with independent methods indicated that our hybrid method is relatively accurate for *Lg* simulation. Numerical simulations with a laterally varying mixed waveguide of rough topography and small-scale random heterogeneities demonstrate the feasibility of the methodology. Through numerical simulations using the known data of topography and crustal structures in the Tibet region, we found that the most characteristic feature of the effects from the irregular topography is the strong scattering by the topographic structures. The scatterings by local irregular topographies lead to anomalous near-receiver effects and tend to remove energy from the guided waves, which causes decay of amplitude and waveform distortion. It can be expected that rough surface topography and random heterogeneities with scale length close to the dominant wavelength will be very efficient in attenuating regional waves. The dramatically lateral variation of the topography-Moho large-scale structure could be the cause of *Lg* anomalous attenuation and blockage, which may explain the observations in this region. More quantitative assessment of the topographic effects must be conducted in the future.

Acknowledgments

We are grateful to X.Y. Wu and S. Jin for their support in screen-modeling code, to X.B. Xie for his finite-difference example, and to G. Fan and T. Lay for providing the data of topography and Moho depth in the Tibet region. We thank anonymous reviewers for valuable comments. This work was supported by the Department of Energy through contract F19628-95-K-0016, administered by the Phillips Laboratory of the Air Force and by the Defense Special Weapons Agency through contract DSWA01-97-1-0004. The facility support from the W. M. Keck Foundation is also appreciated. Contribution number 341 of the Institute of Tectonics, University of California, Santa Cruz.

References

Baumgardt, D. R. (1990), *Investigation of Teleseismic Lg Blockage and Scattering Using Regional Arrays*, Bull. Seismol. Soc. Am. *80*, 2261–2281.

Benites, R., and Aki, K. (1989), *Boundary Integral-Gaussian Beam Method for Seismic Wave Scattering: SH in Two-dimensional Media*, J. Acoust. Soc. Am. *86*, 375–386.

Boore, D. (1970), *Love Waves in Nonuniform Wave Guides: Finite-difference Calculations*, J. Geophys. Res. *75*, 1512–1527.

Bostock, M. G., and Kennett, B. L. N. (1990), *The Effect of Three-dimensional Structure on Lg Propagation Patterns*, Geophys. J. Int. *101*, 355–365.

Bouchon, M. (1982), *The Complete Synthesis of Seismic Crustal Phases at Regional Distances*, J. Geophys. Res. *82*, 1735–1741.

Bouchon, M. (1985), *A Simple, Complete Numerical Solution to the Problem of Diffraction of SH Waves by an Irregular Interface*, J. Acoust. Soc. Am. *77*, 1–5.

Bouchon, M., and Coutant, O. (1994), *Calculation of Synthetic Seismograms in a Laterally Varying Medium by the Boundary Element-discrete Wavenumber Method*, Bull. Seismol. Soc. Am. *84*, 1869–1881.

Bouchon, M., Schultz, C. A., and Toksoz, M. N. (1996), *Effect of Three-dimensional Topography on Seismic Motion*, J. Geophys. Res. *101*, 5835–5846.

CAMPILLO, M., and BOUCHON, M. (1985), *Synthetic SH-seismograms in a Laterally Varying Medium by the Discrete Wavenumber Method*, Geophys. J. Roy. Astr. Soc. *83*, 307–317.

CAMPILLO, M. (1987), *Lg Wave Propagation in a Laterally Varying Crust and the Distribution of the Apparent Quality Factor in Central France*, J. Geophys. Res. *92*, 12604–12614.

CAMPILLO, M., FEIGNIER, B., BOUCHON, M., and BETHOUX, N. (1993), *Attenuation of Crustal Waves across the Alpine Range*, J. Geophys. Res. *98*, 1987–1996.

CAO, S., and MUIRHEAD, K. J. (1993), *Finite Difference Modeling of Lg Blockage*, Geophys. J. Int. *115*, 85–96.

CHEN, X. F. (1995), *Seismogram Synthesis for Multi-layered Media with Irregular Interfaces by the Global Generalized Reflection/Transmission Matrices Method – Part II. Applications of 2-D SH Case*, Bull. Seismol. Soc. Am. *85*, 1094–1106.

FAN, G., and LAY, T. (1998a), *Statistical Analysis of Irregular Waveguide Influence on Regional Seismic Discriminants in China*, Bull. Seismol. Soc. Am. *88*, 74–88.

FAN, G., and LAY, T. (1998b), *Statistical Analysis of Irregular Waveguide Influence on Regional Seismic Discriminants in China: Additional Results for Pn/Sn, Pn/Lg, and Pg/Sn*, Bull. Seismol. Soc. Am. *88*, 1504–1510.

FIELDING, E. J., ISACKS, B. L., and BARAZANGI, M. (1992), *A Geological and Geophysical Information System for Eurasia*, Tech. Report 2, F19601-91-K-DB08, Philips Lab., Hanscom Air Force Base, Massachusetts.

FISK, M. D., and MCCARTOR, G. D. (1991), *The Phase Screen Method for Vector Elastic Waves*, J. Geophys. Res. *96*, 5985–6010.

FLATTE, S. M., and TAPPERT, F. D. (1975), *Calculation of the Effect of Internal waves on Oceanic Sound Transmission*, J. Acoust. Soc. Am. *58*, 1151–1159.

FORNBERG, B. (1987), *The Pseudospectral Method: Comparisons with Finite Differences for the Elastic Wave Equation*, Geophysics *52*, 483–501.

FRANKEL, A., and CLAYTON, R. W. (1986), *Finite-difference Simulations of Seismic Scattering: Implications for Propagation of Short-period Seismic Waves in the Crust and Models of Crustal Heterogeneity*, J. Geophys. Res. *91*, 6465–6489.

FU, L. Y., and MU, Y. G. (1994), *Boundary Element Method for Elastic Wave Forward Modeling*, Acta Geophys. Sinica *37*, 521–529.

FU, L. Y. (1996), *3-D Boundary Element Seismic Modeling in Complex Geology*, 66th Ann. Internat. Mtg., Soc. Expl. Geophys., Expanded Abstracts, 1239–1242.

FU, L. Y., and WU, R. S. (2000), *Infinite Boundary Element Absorbing Boundary for Wave Propagation Simulations*, Geophysics *65*, 596–602.

FURUMURA, T., and KENNETT, B. L. N. (1997), *On the Nature of Regional Seismic Phases-II. On the Influence of Structural Barriers*, Geophys. J. Int. *129*, 221–234.

GAFFET, S., and BOUCHON, M. (1989), *Effects of Two-dimensional Topography Using the Discrete Wavenumber-boundary Integral Equation Method in P-SV Cases*, J. Acoust. Soc. Am. *85*, 2277–2283.

GIBSON, Jr, R. L., and CAMPILLO, M. (1994), *Numerical Simulation of High- and Low-frequency Lg-wave Propagation*, Geophys. J. Int. *118*, 47–56.

HARDIN, R. H., and TAPPERT, F. D. (1973), *Applications of the Split-step Fourier Method to the Numerical Solution of Nonlinear and Variable Coefficient Wave Equations*, SIAM Rev. *15*, 423.

HARTSE, H., FLORES, R., and JOHNSON, P. (1998), *Correcting Regional Seismic Discriminants for Path Effects in Western China*, Bull. Seismol. Soc. Am. *88*, 596–608.

HESTHOLM, S. O., and RUUD, B. O. (1994), *2D Finite Difference Elastic Wave Modeling Including Surface Topography*, Geoph. Prosp. *42*, 371–390.

HUANG, L. J., and FEHLER, M. C. (1998), *Accuracy Analysis of the Split-step Fourier Propagator: Implications for Seismic Modeling and Migration*, Bull. Seismol. Soc. Am. *88*, 18–29.

JIH, R. S., MCLAUGHLIN, K. L., and DER, Z. A. (1988), *Free-boundary Conditions of Arbitrary Topography in a Two-dimensional Explicit Finite Difference Scheme*, Geophysics *53*, 1045–1055.

JIH, R. S. (1996), *Waveguide Effects of Large-scale Structural Variation, Anelastic Attenuation, and Random Heterogeneity on SV Lg Propagation: a Finite-difference Modeling Study*. In Proc. of the 18th Annual Seismic Res. Smp. on Monitoring a Comprehensive Test-Ban Treaty, 182–194.

KAWASE, H., and AKI, K. (1989), *A Study on the Response of a Soft Basin for Incident S, P and Rayleigh Waves, with Special Reference to the Long Duration Observed in Mexico City*, Bull. Seismol. Soc. Am. *79*, 1361–1382.

KENNETT, B. L. N. (1984), *Guided Wave Propagation in Laterally Varying Media-I: Theoretical Development*, Geophys. J. Roy. Astr. Soc. *79*, 235–255.

KENNETT, B. L. N., GREGERSEN, S., MYKKELTVEIT, S., and NEWMARK, R. (1985), *Mapping of Crustal Heterogeneity in the North Sea Basin via the Propagation of Lg Waves*, Geophys. J. Roy. Astr. Soc. *83*, 299–306.

KENNETT, B. L. N. (1986), *Lg Waves and Structural Boundaries*, Bull. Seismol. Soc. Am. *76*, 1133–1141.

KENNETT, B. L. N. (1998), *Guided Waves in Three-dimensional Structures*, Geophys. J. Int. *133*, 159–174.

KIM, J., and PAPAGEORGIOU, A. S. (1993), *Discrete Wavenumber Boundary-element Method for 3-D Scattering Problems*, J. Eng. Mech. *119*, 603–624.

KUNIN, N. Y., and SHEYKH-ZADE, E. R. (1983), *New Data on Lateral Inhomogeneities in the Upper Mantle under Western Eurasia*, Dokl. Akad. Nauk. USSR *273*, 1087–1091.

KUNIN, N. Y. (1987), *Distribution of Sedimentary Basins of Eurasia and the Volume of the Earth's Sedimentosphere*, Int. Geol. Rev. *22*, 1257–1264.

LEE, J. J., and LANGSTON, C. A. (1983), *Wave Propagation in a Three-dimensional Circular Basin*, Bull. Seismol. Soc. Am. *61*, 1637–1653.

LIU, Y. B., and WU, R. S. (1994), *A Comparison between Phase Screen, Finite Difference, and Eigenfunction Expansion Calculations for Scalar Waves in Inhomogeneous Media*, Bull. Seismol. Soc. Am. *84*, 1154–1168.

MARTIN, J. M., and FLATTE, S. M. (1988), *Intensity Images and Statistics from Numerical Simulation of Wave Propagation in 2-D Random Media*, Appl. Opt. *17*, 2111–2126.

MAUPIN, V., and KENNETT, B. L. N. (1987), *On the Use of Truncated Model Expansion in Laterally Varying Media*, Geophys. J. Roy. Astr. Soc. *91*, 837–851.

MAUPIN, V. (1989), *Numerical Modeling of Lg Wave Propagation across the North Sea Central Graben*, Geophys. J. Int. *99*, 273–283.

MORSE, P. M., and FESHBACH, H., *Methods of Theoretical Physics*, Chapter 7 (McGraw-Hill Book Comp., New York. (1953).

PRESS, F., and EWING, M. (1952), *Two Slow Surface Waves across North America*, Bull. Seismol. Soc. Am. *42*, 219–228.

REGAN, J., and HARKRIDER, D. G. (1989), *Numerical Modeling of SH Lg Waves in and Near Continental Margins*, Geophys. J. Int. *98*, 107–130.

ROBERTSSON, J. O. A. (1996), *A Numerical Free-surface Condition for Elastic/Viscoelastic Finite-difference Modeling in the Presence of Topography*, Geophysics *61*, 1921–1934.

RODGERS, A., LAY, T., FAN, G., and WALTER, W. (1997), *Calibration of Distance and Path Effects on Regional P/S Discriminants at Station ABKT (Alibek, Turkmenistan): Statistical Analysis of Crustal Waveguide Effects*, UCRL-JC-129165.

SANCHEZ-SESMA, F. J., and ESQUIVEL, J. (1979), *Ground Motion on Alluvial Valleys under Incident Plane SH Waves*, Bull. Seismol. Soc. Am. *69*, 1107–1120.

SANCHEZ-SESMA, F. J., and CAMPILLO, M. (1991), *Diffraction of P, SV and Rayleigh Waves by Topographic Features: A Boundary Integral Formulation*, Bull. Seismol. Soc. Am. *81*, 2234–2253.

SCHATZMAN, J. C. (1996), *A pseudo-spectral scheme for viscoelastic seismic modeling*. In Proc. of the 18th Annual Seismic Res. Symp. on Monitoring a Comprehensive Test-Ban Treaty, 261–270.

SEI, A. (1993), *Computational Cost of Finite-difference Elastic Waves Modeling*, 63th Ann. Internat. Mtg., Soc. Expl. Geophys., Expanded Abstracts, 1065–1068.

THOMSON, D. J., and CHAPMAN, N. R. (1989), *A Wide-angle Split-step Algorithm for the Parabolic Equation*, J. Acoust. Soc. Am. *74*, 1848–1854.

TRIFUNAC, M. D. (1973), *Scattering of Plane SH waves by a Semi-cylindrical Canyon*, Earthquake Eng. And Struct. Dyn. *1*, 267–281.

WETMILLER, R. J. (1974), *Crustal Structure of Baffin Bay from Earthquake Generated Lg Phases*, Can. J. Earth Sci. *11*, 123–130.

WILD, A. J., and HUDSON J. A. (1998), *A Geometrical Approach to the Elastic Complex Screen*, J. Geophys. Res. *103*, 707–725.

WONG, H. L., and JENNINGS, P. C. (1975), *Effect of Canyon Topographies on Strong Ground Motion*, Bull. Seismol. Soc. Am. *65*, 1239–1257.

WU, R. S. (1994), *Wide-angle Elastic Wave One-way Propagation in Heterogeneous Media and an Elastic Wave Complex-screen Method*, J. Geophys. Res. *99*, 751–766.

WU, R. S. (1996), *Synthetic Seismogram in Heterogeneous Media by One-return Approximation*, Pure appl. geophys. *148*, 155–173.

WU, R. S., and XIE, X. B. (1994), *Multi-screen backpropagator for fast 3D elastic prestack migration.* In *Mathematical Methods in Geophysical Imaging II* (ed. Hassanzadeh, S.), SPIE Proceedings Series *2301*, 181–193.

WU, R. S., and DE HOOP, M. V. (1996), *Accuracy analysis and numerical tests of screen propagators for wave extrapolation.* In *Mathematical Methods in Geophysical Imaging IV* (ed. Hassanzadeh, S.), SPIE Proceedings Series *2822*, 196–209.

WU, R. S., JIN, S., and XIE, X. B. (1996), *Synthetic seismograms in heterogeneous crustal waveguides using screen propagators.* In Proc. of the 18th Annual Seismic Res. Symp. on Monitoring a Comprehensive Test-Ban Treaty, 290–300.

WU, R. S., JIN, S., XIE, X. B., and LAY, T. (1997), *Verification and applications of GSP (generalized screen propagators) method for regional wave propagation.* In Proc. of the 19th Annual Seismic Res. Symp. on Monitoring a Comprehensive Test-Ban Treaty, 552–561.

WU, R. S., XIE, X. B., and WU, X. Y. (1999), *Lg wave simulations in heterogeneous crusts with irregular topography using half-space screen propagators.* In Proc. of the 21th Annual Seismic Res. Symp. on Monitoring a Comprehensive Test-Ban Treaty, 683–693.

XIE, X. B., and LAY, T. (1994), *The Excitation of Explosion Lg, a Finite-difference Investigation*, Bull. Seismol. Soc. Am. *84*, 324–342.

ZHANG, T., and LAY, T. (1994), *Analysis of Short-period Regional Phase Path Effects Associated with Topography in Eurasia*, Bull. Seismol. Soc. Am. *84*, 119–132.

ZHANG, T., LAY, T., SCHWARTZ, S., and WALTER, W.R. (1996), *Variation of Regional Seismic Discriminants with Surface Topographic Roughness in the Western United States*, Bull. Seismol. Soc. Am. *86*, 714–725.

(Received September 9, 1999, revised March 8, 2000, accepted March 29, 2000)

Pure appl. geophys. 158 (2001) 1279–1300
0033–4553/01/071279–22 $ 1.50 + 0.20/0

Pure and Applied Geophysics

Calibration of Regional *S/P* Amplitude-ratio Discriminants

RICHARD D. JENKINS,[1] and THOMAS J. SERENO, JR[1]

Abstract — *S/P* amplitude ratios have proven to be a valuable discriminant in support of monitoring a Comprehensive Nuclear Test Ban Treaty. Regional *S* and *P* phases attenuate at different rates and the attenuation can vary geographically. Therefore, calibration is needed to apply the *S/P* discriminant in new regions. Calibration includes application of frequency-dependent source and distance corrections for regional *Pn*, *Pg*, *Sn*, and *Lg* phases.

JENKINS *et al.* (1998) developed *Pn*, *Pg*, *Sn*, and *Lg* amplitude models for nine geographic regions and two global composite models, stable and tectonic. They determined frequency-dependent source and attenuation corrections from a large data set obtained from the Prototype International Data Center (PIDC). We use their corrections to evaluate calibrated *S/P* discriminants.

Our discrimination data set includes > 1000 amplitude ratios from earthquakes, industrial explosions, chemical explosions, and nuclear explosions from Lop Nor, India and Pakistan. We find that the calibrated *S/P* ratio is largest for earthquakes and smallest for the nuclear explosions, as expected. However, the discriminant is not universally valid. In particular, the *S/P* ratio for the Pakistan nuclear explosion fell within the normal range for the earthquakes. This event was recorded by only a few stations at far-regional distances and appears to have an anomalously high *Sn* amplitude. The industrial explosions overlap with the earthquake population, however the buried chemical explosions generally register lower *S/P* ratio than earthquakes.

Key words: Discrimination, frequency-dependent attenuation, *S/P* amplitude ratios, source-level correction, regional phases, comprehensive Nuclear Test Ban Treaty.

Introduction

Objectives

Results from previous empirical studies indicate that variants of high-frequency *S/P* amplitude ratios are some of the most promising discriminants at regional distances. However, many of these studies are based on co-located earthquakes and explosions in limited geographic regions, and their results must be recalibrated for each new source region. We show distance and source corrections for stations in the primary seismic network of the PIDC and evaluate *S/P* discriminants in regions of interest. We have also generalized the corrections by combining models according to tectonic environment to form stable and tectonic models that appear to be valid for uncalibrated regions.

[1] Science Application International Corporation (SAIC), San Diego, CA 92121, USA. E-mail: jenkins@gso.saic.com

Background

Many empirical studies have indicated that *S/P* amplitude ratios effectively discriminate nuclear explosions from earthquakes (WILLIS *et al.*, 1963; BLANDFORD, 1981; POMEROY *et al.*, 1982; BENNETT and MURPHY, 1986; TAYLOR *et al.*, 1989; BENNETT *et al.*, 1989; 1992) and earthquakes from industrial explosions (BAUMGARDT and YOUNG, 1990; DYSART and PULLI, 1990; WUESTER, 1993; KIM *et al.*, 1997). However, *S* and *P* amplitudes attenuate at different rates in different regions. Most of these studies were able to avoid the attendant bias by restricting the data set to events in the same region. Consequently, *S/P* discriminants developed for one region may not work in another; and this compromises the value of *S/P* discriminants for regions where empirical data are sparse. Unfortunately, this is a problem nearly everywhere since nuclear explosion testing has been conducted in few areas; and many of the best modern stations are (or will be) at sites with little or no historical data from past underground nuclear explosions.

Data Sets

This study used regional waveform data collected at the PIDC for a global network of more than 130 stations during the Group of Scientific Experts Third Technical Test called GSETT-3 (for an overview of GSETT-3, see KERR, 1993). The waveform data and analyst-reviewed event bulletins produced at the PIDC were provided to us by the Center for Monitoring Research (CMR) located in Arlington, Virginia.

The PIDC acquires waveform data from each station and performs automatic processing to detect and characterize signals and form preliminary event hypotheses. These event hypotheses are reviewed by an analyst. Events that are confirmed by an analyst and their associated arrivals form what is called the Reviewed Event Bulletin (REB). Events in the REB must satisfy conformation criteria specified by the PIDC. These criteria can include, for example, the minimum number of stations that must record an event.

We organize our regional attenuation and discrimination data sets according to the stations recording these events. We group the events and stations into what we refer to as regional subnetworks.

Station Selection

The PIDC acquired data from 46 primary and 87 auxiliary stations during the time period examined for this study. As shown in Figure 1, the primary stations are distributed globally. The data from the primary network are sent continuously to the PIDC where they are processed and analyzed. Auxiliary stations expand the coverage of the primary network, but these stations are used by the PIDC only when data are requested to improve the location solutions for events defined by the primary

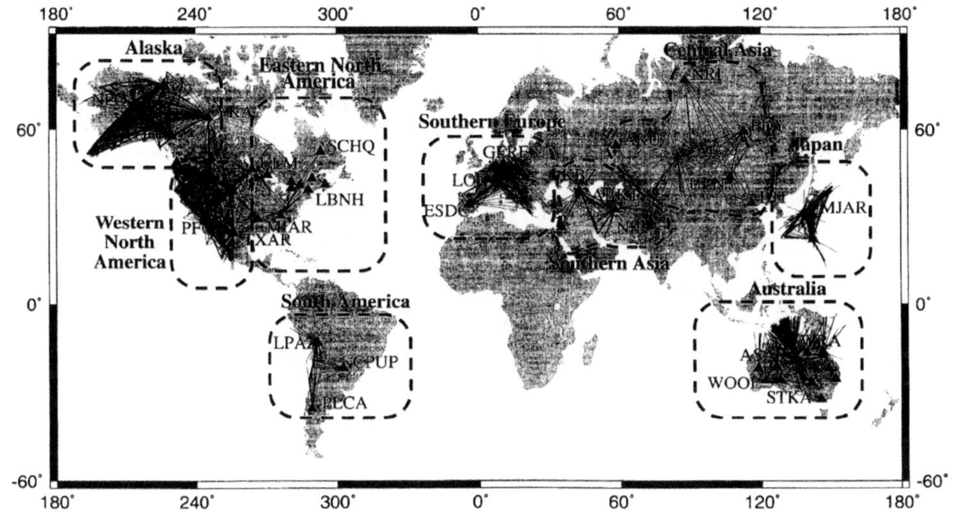

Figure 1
Map showing subnetworks and *Pn* ray paths for events used in this study.

network. We grouped PIDC primary and auxiliary stations into subnetworks based on their proximity to PIDC primary stations, tectonic environment, and the availability of analyst-verified regional phases in the REB.

Stations are grouped into subnetworks if they were within 40° of each other (20° was an upper limit for a regional event). Some PIDC stations may not be included because they do not have a sufficient number of regional phases in the REB. Using these criteria, we defined the nine subnetworks shown in Figure 1.

Event Selection for Attenuation

The data set for our attenuation study is based on events in the REB for the 12-month period ending September 1, 1996. We only use arrivals that were verified by a PIDC analyst. It is difficult for our automated software to screen arrivals otherwise. The review process provided by the analyst provides the necessary level of quality control that is difficult to perform automatically.

We limited our data sets to *Pn*, *Pg*, *Sn*, and *Lg* phases from events ≤33 km depth. We further constrained these phases by distance. *Pn* and *Sn* were limited to 15° or less. *Lg* was limited to 20° and *Pg* was limited to 6°. This reduced the number of events to approximately 3280 out of a total of about 24,000 events available in the REB for this time period. This number was further reduced due to signal-to-noise ratio (*snr*) constraints applied by our inversion software. Table 1 lists each subnetwork name, its tectonic environment name, its component stations, and the number of phases and events actually used in estimating the regional wave amplitude

Table 1

Subnetworks of PIDC primary and auxiliary stations

Subnetwork	Tectonic Name	Stations	*Pn*	*Pg*	*Sn*	*Lg*	Number of Events
Alaska/ NW Canada	Tectonic	DAWY, DLBC, INK, NPO, WHY, YKA	505	80	210	246	211
Western North America	Tectonic	ALQ, BBB, DUG, EDM, ELK, MNV, NEW, PDAR, PFO, PGC, PMB, PNT, RSSD, TUC, TXAR, WALA	819	125	158	386	285
Southern Europe	Tectonic	APL, AQU, BFO, BRG, BUG, CLL, CLZ, ESDC, FUR, GERES, GRFO, HGN, LOR, MOX, PAB, PSZ, TNS, VRAC	1316	666	444	1092	502
Southern Asia	Tectonic	ABKT, BGIO, KBZ, KVAR, NIL	91	7	26	17	75
Japan	Tectonic	MJAR, OGS, TSK	819	41	308	70	564
South America	Tectonic	CPUP, LPAZ, PLCA	62	1	15	4	48
Eastern North America	Stable	ALQ, EDM, EEO, GAC, LBNH, LMN, LMQ, MIAR, PDAR, RSSD, SADO, SCHQ, TBO, TKL, TUL, TXAR, ULM, WALA, YKA	262	36	66	201	100
Central Asia	Stable	ARU, BJT, NRI, PDY, ULN, ZAL	53	11	23	49	38
Australia	Stable	ARMA, ASAR, CTA, FORT, MEEK, QIS, RMQ, STKA, TOO, WARB, WOOL, WRA	323	33	203	157	107
Total	All	All	4250	1000	1453	2222	1930

models. Each subnetwork includes between 3 and 19 stations, and is declared to be in a stable or tectonic region. The events listed in Table 1 range in magnitude from m_L 1.8 to m_L 6.0 as reported in the REB. Most events are presumed to be earthquakes, and some are presumed to be industrial explosions. Neither chemical nor nuclear explosion events are included in this data set.

The coverage provided by the *Pn* arrivals summarized in Table 1 is plotted as ray paths on the map in Figure 1 for each subnetwork. The coverage for Alaska, Australia, eastern and western North America, and southern Europe is extremely good for all regional phases. The South American data set is sparse. However, were able to obtain a stable estimate of the *Pn* attenuation in this region. The coverage for southern Asia is good for *Pn*, marginal for *Sn*, and poor for the crustal phases *Pg* and

Lg. Stations ABKT, KBZ, and KVAR provide most of the data in this region. The coverage for central Asia is good for *Pn* and *Lg*, marginal for *Sn*, and poor for *Pg*. The coverage for Japan is good for *Pn* and *Sn*, but *Pg* and *Lg* are limited to short distances.

Event Selection for Discrimination

Our discrimination data set includes earthquakes, presumed industrial explosions, chemical explosions and nuclear explosions. These events are used to evaluate the effectiveness of calibrated *S*/*P* discriminants. We used events in the REB to build our discrimination data set. The same phase-dependent distance limitations and depth and azimuthal constraints, applied during formation of the attenuation data set, were applied here.

The events are not identified in the REB, and for many events there are no other sources of ground-truth. Therefore, we rely on characteristics (e.g., spatial and temporal distribution and magnitude) of natural and industrial seismicity and previous work of others to identify events in our discrimination data set.

Earthquakes. We presumed all events in seismically-active regions that were not characterized as probable industrial explosions or known nuclear explosions are earthquakes. These events are presumed to be earthquakes because they occur in seismically-active regions and their magnitudes and origin times are consistent with the expected distributions. That is, the number of events as a function of magnitude is consistent with earthquake recurrence models; and the origin times are random with respect to day-of-the-week and local time-of-day.

Industrial Explosions. We presumed events in the REB to be industrial (or mining) explosions on the basis of location, origin time, and magnitude. Events from active mines are expected to be tightly clustered in their location (e.g., numerous events from a small epicentral region). The origin time of mining events is expected to be clustered by day-of-week and time-of-day, consistent with a normal work schedule. Finally, magnitudes are expected to span a limited range, unlike the typical recurrence models for earthquakes.

For location clustering we counted the number of regional events in the REB recorded between September 1, 1995 and September 1, 1996 in 1° latitude by 1° longitude grid cells. If the number of events in a given grid cell exceeded a user-defined threshold (which varied by subnetwork, but typically was set to 10 events per grid cell) and the grid cell was located in an area with low natural seismicity, then this group was identified as a potential source of industrial explosions.

The origin times of the events were used to determine the day-of-week and local time-of-day. Histograms of these parameters were visually inspected for non-random distributions indicative of a cultural source. For example, if the events typically occurred on weekdays between mid-morning and mid-afternoon (local time) in a region of historically low seismicity, then that grid cell was classified as a potential area for industrial explosions.

Based on this procedure, 60 events in North America, 14 in Australia, 317 in southern Europe, and 3 in Asia are presumed to be industrial explosions. These events have a median magnitude of 3.25 m_L. These events will be used to evaluate the S/P discriminant in eastern and western North America, Australia, southern Europe, and central Asia subnetworks.

Chemical Explosions. Four chemical explosions were detonated at the former nuclear test site in Kazakhstan (see Table 2). These tests (Balapan and Degelen) were designed to provide ground-truth information regarding discrimination of underground explosions at various depths as recorded by regional seismic stations. The three Balapan explosions were detonated at burial depths of 50, 300, and 550 m. The Degelen explosion was 140 m underground.

Figure 2 shows the location of the 1997–1998 events and recording stations. Most of these data were obtained from the Center for Monitoring Research. Data from the KNET network in Kyrgyzstan and western Kazakhstan were provided by UCSD/SIO. The eastern Kazakhstan stations are at IRIS/JSP sites and the data were collected by W.Y. Kim of Lamont-Doherty Earth Observatory.

Nuclear Explosions. Four Lop Nor events were confirmed to be nuclear explosions by previous authors (HARTSE *et al.*, 1997; YANG *et al.*, 1997). The REB event solutions and recording stations within a regional distance from these nuclear events are shown in Table 2. At some stations the PIDC analyst only found one

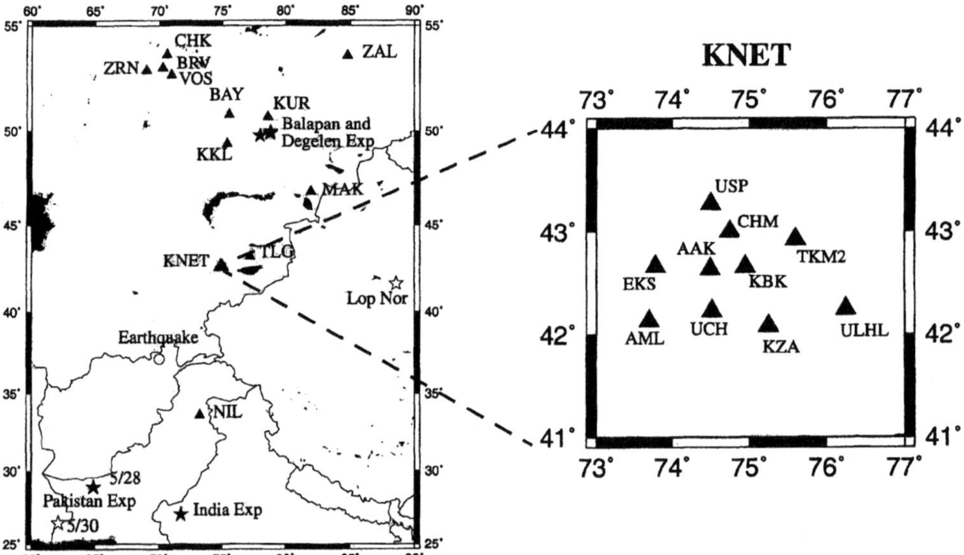

Figure 2
The locations of the 1997–1998 explosions are plotted (filled stars) along with the locations of the stations which recorded them (triangles). Also shown are the 30 May nuclear test in Pakistan (open star), the nearly contemporaneous earthquake in northern Afghanistan (circle), and the Chinese nuclear test site at Lop Nor (open star). The individual stations in the KNET network are shown to the right.

Table 2

Nuclear and chemical explosion event solutions

m_b	Location	Origin Time	Recording stations	Distance (km)
5.73	41.63 N, 88.87 E (Lop Nor)	1995-05-15 04:06:00.0	ULN	1593
5.54	41.60 N, 88.86 E (Lop Nor)	1995-08-17 00:59:59.0	ULN	1596
5.69	41.65 N, 88.76 E (Lop Nor)	1996-06-08 02:55:59.0	NIL	1628
4.71	41.69 N, 88.35 E (Lop Nor)	1996-07-29 01:48:59.0	ZAL, NIL	1353
5.00	27.08 N, 71.72 E (India)	1998-05-11 10:13:45.3	NIL, KNET[a]	743–1800
4.87	28.83 N, 64.95 E (Pakistan)	1998-05-28 10:16:17.0	KNET	1864–1943
–	49.98 N, 78.82 E (Balapan)	1997-08-03 08:07:20.0	ZAL, Balapan[b]	423–699
–	49.88 N, 78.81 E (Balapan)	1997-08-31 07:08:39.3	ZAL, Balapan	255–763
–	49.88 N, 78.76 E (Balapan)	1997-09-28 07:30:15.1	ZAL, Balapan	246–753
3.8	49.76 N, 77.98 E (Degelen)	1998-08-22 05:00:18.9	ZAL, KNET	666–899

[a] The Kyrgyzstan Seismic Network (KNET) consists of 10 broadband stations.
[b] An 8 station network located within Kazakhstan composed of both broadband and short-period stations.

regional arrival acceptable for association for these events. For example, only a *Pn* arrival was associated at the Nilore, Pakistan (NIL) station for the July 29, 1996 Lop Nor event. The location of these events is tightly clustered at the Lop Nor nuclear test site represented by the open star symbol in Figure 2.

We also include the recent nuclear explosions from India and Pakistan (28 May, 1998). One event from each country was satisfactorily recorded for regional discrimination analysis. We do not include the 30 May, 1998 nuclear explosion in Pakistan because the signals at KNET (the only available regional data) are obscured by signals from aftershocks of an m_b 5.5 earthquake which occurred 32 minutes earlier and 1100 km to the northeast (700 km southwest of KNET).

Data Processing

Amplitude Measurements

We estimated regional-wave amplitudes in the time-domain. We experimented with several amplitude measures, but the results summarized here are for our preferred measure. That is, the root-mean-square (*rms*) log amplitude computed on the filtered vertical-component channel for three-component stations or an incoherent beam (filter, rectify, stack) for arrays. The window length was 10 seconds for *Pn*, *Pg* and *Sn* for distances $> 4°$ and 5 seconds for shorter distances. A group velocity window of 3.0–3.6 km/s was used for *Lg*. The filtering was done in eight overlapping one-octave bands extending from 0.5–1.0 Hz to 8–16 Hz. All amplitude measures were corrected for the instrument response for the center of the one-octave wide frequency band.

A sample waveform from the Tuckaleechee Caverns, Tennessee station (TKL) is shown in Figure 3. The m_L 3.8 event is from a local mining region approximately 3

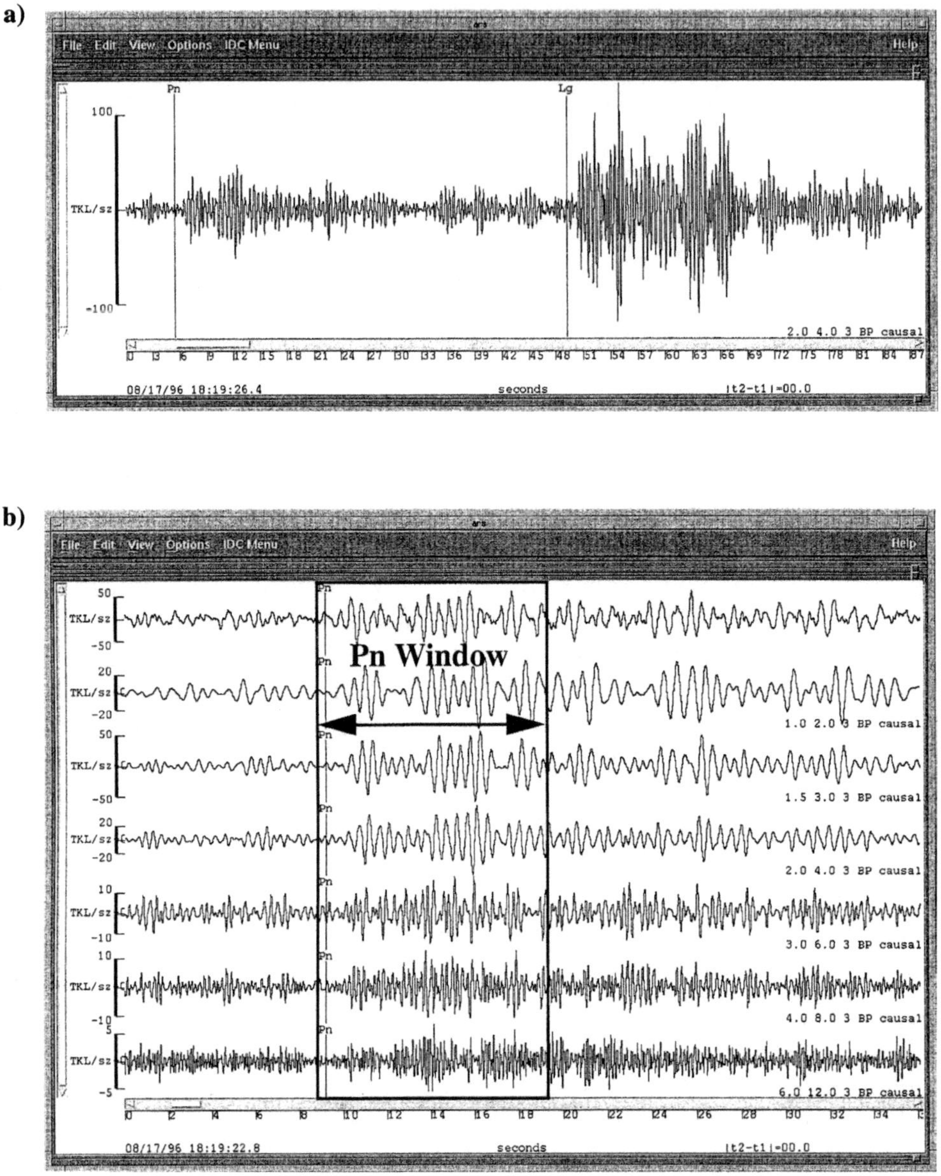

Figure 3

a) Regional waveform of a West Virginia event recorded at station Tuckaleechee Caverns, Tennessee (TKL) at about 327 km epicentral distance. The waveform is bandpass-filtered between 2 and 4 Hz. b) Typical data processing window of 10.0 seconds was used to measure *Pn* amplitude in six pass bands ranging from 1.0–2.0 Hz up to 6.0–12.0 Hz. The waveform in the top panel is unfiltered. *Pg* and *Sn* were also measured in a fixed window. For *Lg* we used a group velocity window of 3.0–3.6 km/s.

degrees away. The top panel shows *Pn* and *Lg* arrivals on the regional waveform bandpass filtered between 2 and 4 Hz. The bottom panel shows the typical data processing window used for *Pn* arrivals. Both unfiltered and filtered channels are shown.

Amplitude Parameterization

We used the method of estimating source and attenuation models developed by SERENO (1990). Briefly the frequency-dependent (either time-domain or spectral) amplitude of the *k*-th wave recorded at the *i*-th station from the *j*-th source is parameterized as:

$$\log A_{ijk}(f) = \log A^0_{jk}(f) + B_k(\Delta_{ij}, \Delta_0, f) + \delta_{ik} \tag{1}$$

where A^0_{jk} (f) is the amplitude at a reference distance Δ_0, $B_k(\Delta_{ij},\Delta_0,f)$ is the attenuation from the reference distance to the epicentral distance Δ_{ij}, and δ_{ik} is a station correction. The reference distance was chosen at a distance where geometrical spreading changes from spherical to a more complicated function based upon the regional phase analyzed.

The amplitude at the reference distance is a combination of the shape of the source spectrum, long-period level, wave-dependent excitation factor, wave-dependent pulse shape, and anelastic attenuation occurring between the event and the reference distance.

The total attenuation from the reference distance to the station is parameterized in terms of a power-law distance dependence with a frequency-dependent exponent:

$$B_k(\Delta_{ij}, \Delta_0, f) = n_k(f) \cdot \log(\Delta_0/\Delta_{ij}) \tag{2}$$

$$n_k(f) = a_k f + b_k \tag{3}$$

The total attenuation includes geometrical spreading, scattering and anelasticity. It is difficult to separate these terms since the geometrical spreading of regional phases is a complicated function of the crustal and upper mantle velocity structure (e.g., SERENO and GIVEN, 1990). Fortunately, it is not important to separate these terms for application to the *S*/*P* discriminant.

Calibration Models

Regional-wave amplitudes were inverted for source- and frequency-dependent attenuation parameters. The system of equations governing the relationship between the data (log amplitudes) and model parameters is formulated by subtracting theoretical data computed from an assumed starting model from the observed data, and solving iteratively for the model perturbations that minimize the data residuals in the least-squares sense.

We estimated source, attenuation and site parameters for each subnetwork from the log *rms* amplitudes. We constrained each regional phase to specific frequency bands. *Pn* and *Pg* amplitudes were used if the center frequency of the passband was between 1 and 15 Hz. *Sn* amplitudes were restricted to the 0.5 to 10 Hz band and *Lg* amplitudes were restricted to the 0.5 to 6 Hz band. Distance constraints based on typical distance ranges for regional phases were also applied.

Phase-dependent *snr* thresholds of 3.5 dB for *Pn* and *Sn*, and 2 dB for *Pg* and *Lg* were applied for all subnetworks except western North America and Japan. For these subnetworks we applied a higher *snr* threshold since there were more data available. The *snr* is the ratio of the signal amplitude and the pre-signal noise amplitude. For example, for *Lg* the *snr* is the *Lg* signal amplitude divided by the pre-*Lg* noise. The pre-*Lg* noise is typically much larger than the ambient (pre-*Pn*) noise because it includes the coda of earlier phases.

We also found it necessary to remove some outliers prior to the inversion based on visual inspection. We examined the waveforms for arrivals with anomalous amplitudes and found that most of the prominent outliers consisted of *Lg* arrivals from events recorded at distances greater than 1000 km that were contaminated by interfering signals. All amplitude measurements related to these arrivals were removed from the data set. All other automated amplitude measurements were retained.

Attenuation

The primary product of this study is the estimate of the frequency-dependent attenuation of *Pn*, *Pg*, *Sn* and *Lg* for each subnetwork and two combined models. These estimates will be used to normalize the discriminants based on *S*/*P* ratios to a common reference distance.

We chose to combine data sets from subnetworks residing in similar environments to form two models we call stable and tectonic. The attenuation results obtained for these two models are typical of that obtained for individual regions (see Jenkins *et al.*, 1998). Examples of the attenuation results for the stable and tectonic models are shown in Figures 4 and 5. Here we show *Pn* attenuation for six frequency bands. The points are the observed log amplitudes minus log A^0 and site response (as determined by the inversion) as a function of distance. The solid curve represents the power-law attenuation model for each frequency. The dashed curves indicate one standard deviation. The rate of attenuation (i.e., $\Delta^{-n(f)}$) and the standard deviation are annotated in each plot. In general, the attenuation increases as expected with

▶

Figure 4
Pn attenuation curves for a stable model. Six frequency passbands are shown: 1–2 Hz, 1.5–3 Hz, 2–4 Hz, 3–6 Hz, 4–8 Hz, and 6–12 Hz.

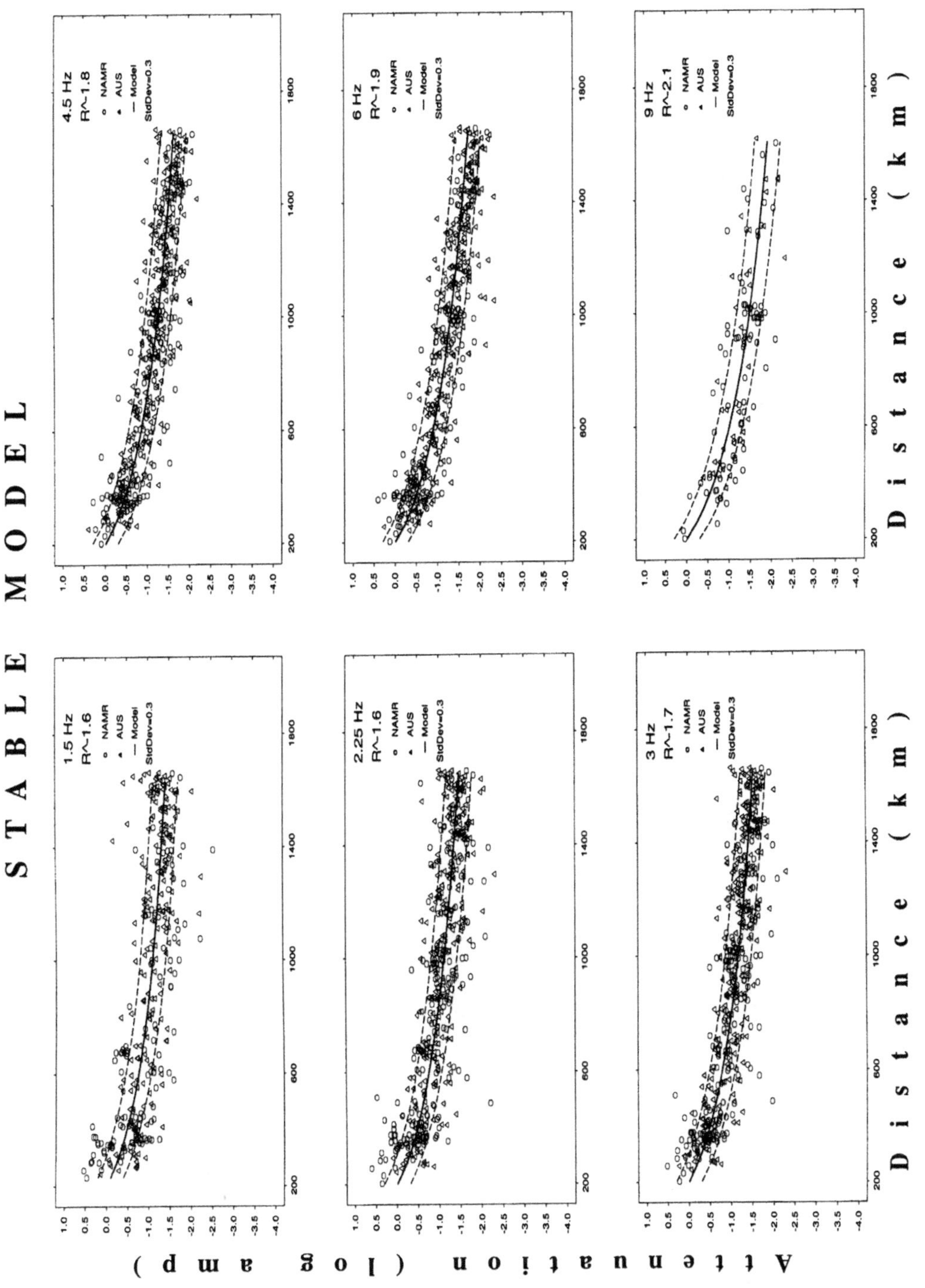
STABLE MODEL
4.5 Hz
R^-1.8
6 Hz
R^-1.9
9 Hz
R^-2.1
1.5 Hz
R^-1.6
2.25 Hz
R^-1.6
3 Hz
R^-1.7
NAMR
AUS
Model
StdDev=0.3
Distance (km)
Attenuation (log amp)

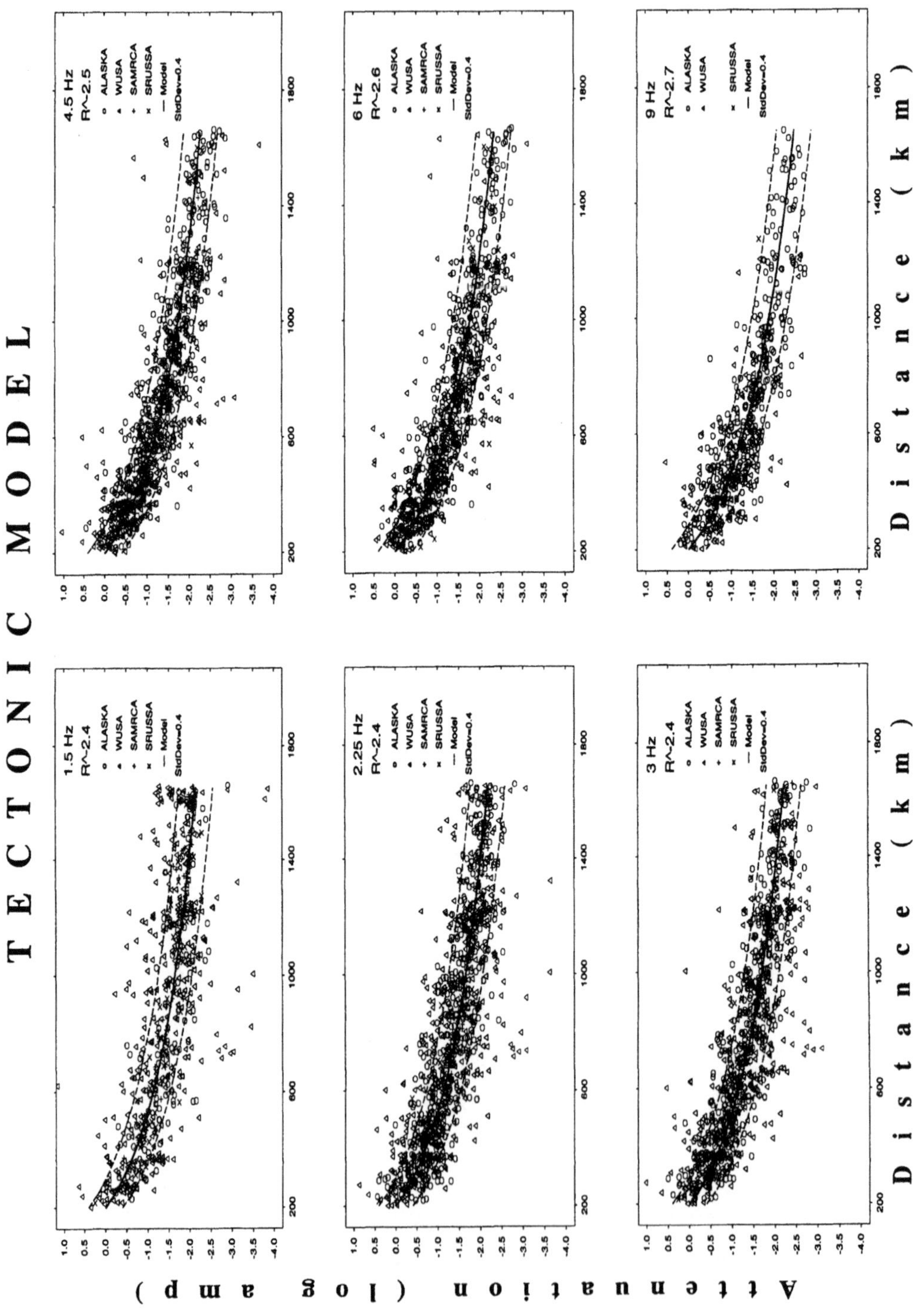
TECTONIC MODEL
1.5 Hz
R^-2.4
2.25 Hz
R^-2.4
3 Hz
R^-2.4
4.5 Hz
R^-2.5
6 Hz
R^-2.6
9 Hz
R^-2.7
ALASKA
WUSA
SAMRCA
SRUSSA
— Model
StdDev=0.4
Distance (km)
Attenuation (log amp)

Table 3

Stable attenuation rates for 2–4 Hz amplitudes

Subnetwork	Tectonic name	Attenuation coefficient (n) at 3 Hz, Δ^{-n}			
		Pn	*Pg*	*Sn*	*Lg*
Eastern N. America	Stable	1.9	3.0	2.4	2.6
Central Asia	Stable	2.1	–	2.6	3.7
Australia	Stable	1.3	1.5	2.1	2.4
Scandinavia[a]	Stable	1.6	2.7	2.0	2.2
Stable	Stable	1.7	2.8	2.2	2.7

[a] The results for Scandinavia are from SERENO (1990).

increasing frequency for each phase and the crustal phases *Pg* and *Lg* attenuate more rapidly (particularly at high frequencies) than the mantle phases *Pn* and *Sn*. Standard deviations of 0.28 and 0.38 were obtained for the stable and tectonic models. We suspect the higher standard deviation of the tectonic model is due to the high variability existing in tectonic regions. Most of the region-specific models fit within 1 standard deviation of the corresponding combined model. As expected, a higher degree of attenuation occurs in the tectonic model as compared to the stable model. This is the case across all frequencies. Both of these combined models are expected to provide adequate path and source corrections in uncalibrated regions.

The rate of attenuation for 2–4 Hz amplitudes at individual subnetworks is shown in Table 3 for stable environments and Table 4 for tectonic environments. A dash in these tables indicates there were too few arrivals (<15) to estimate the attenuation with confidence. The one exception is the *Lg* attenuation rate for Japan which was not included for other reasons as described below. The results for Scandinavia are from SERENO (1990) who used the same method to estimate regional-wave attenuation.

Our estimates of the attenuation rates for *Pn*, *Sn* and *Lg* for Alaska, western North America, and South America are relatively high, as would be expected for tectonic regions. The South America subnetwork had substantially fewer arrivals, most of which were classified as *Pn*. The *Pg* attenuation for all regions is difficult to determine because it is recorded over a very limited distance range. The attenuation rates in southern Europe are surprisingly low for a tectonic region. However, much of the data was from events recorded along stable paths at distances less than 600 km, suggesting this subnetwork should be classified as a "*mixed*" environment. The *Pn* and *Sn* attenuation in southern Asia is also low for a tectonic region. Although we could not determine the *Lg* attenuation rate in southern Asia, we expect

◀

Figure 5
Pn attenuation curves for a tectonic model. Six frequency passbands are shown: 1–2 Hz, 1.5–3 Hz, 2–4 Hz, 3–6 Hz, 4–8 Hz, and 6–12 Hz.

Table 4

Tectonic attenuation rates for 2–4 Hz amplitudes

Subnetwork	Tectonic name	Attenuation coefficient (n) at 3 Hz Δ^{-n}			
		Pn	*Pg*	*Sn*	*Lg*
Alaska/NW Canada	Tectonic	2.2	2.3	3.3	3.5
Western N. America	Tectonic	2.4	3.1	3.6	3.4
Southern Europe	Tectonic	1.9	2.4	2.4	2.2
Southern Asia	Tectonic	1.7	–	2.6	–
Japan	Tectonic	1.8	1.3	1.9	–
South America	Tectonic	2.1	–	–	–
Tectonic	Tectonic	2.4	2.7	3.4	3.3

that it is very high since few arrivals were associated in the REB. This may be related to the Moho topography.

Japan appears to have low *Pn* and *Sn* attenuation rates. However, this is a complicated tectonic region that may not be properly characterized by our simple model. Also, it is possible that some of the REB events included for this subnetwork are deeper than 40 km. The *Lg* attenuation rate for Japan is not meaningful because *Lg* was only reported in the REB for distances less than about 350 km. This is probably because *Lg* is blocked for many paths which include an oceanic segment.

The attenuation rates for *Pn*, *Sn* and *Lg* in eastern North America and Australia are relatively low, as expected for stable regions. They are also quite similar to the attenuation rates determined by SERENO (1990) for Fennoscandia. Central Asia has a higher attenuation rate (especially for *Lg*), but some of the ray paths appear to cross tectonic regions.

Source

We estimated the long-period level for each event by constraining a scaling parameter to yield the desired corner frequency of approximately 10 Hz for m_L 3.0 events in all subnetworks (SERENO *et al.*, 1988).

The long-period levels estimated from the inversion were evaluated for each subnetwork by comparing them to magnitude. As expected, our long-period level estimates increase with magnitude at a rate between 2/3 and 1 (BUNGUM *et al.*, 1982). The median rate of change of source as a function of m_L for all subnetworks is 0.83. Examples of the source and magnitude relationship for the stable and tectonic models are shown in Figure 6.

Site

A phase-dependent site response was estimated for each station providing regional data. We imposed the constraint that all site responses for one phase must

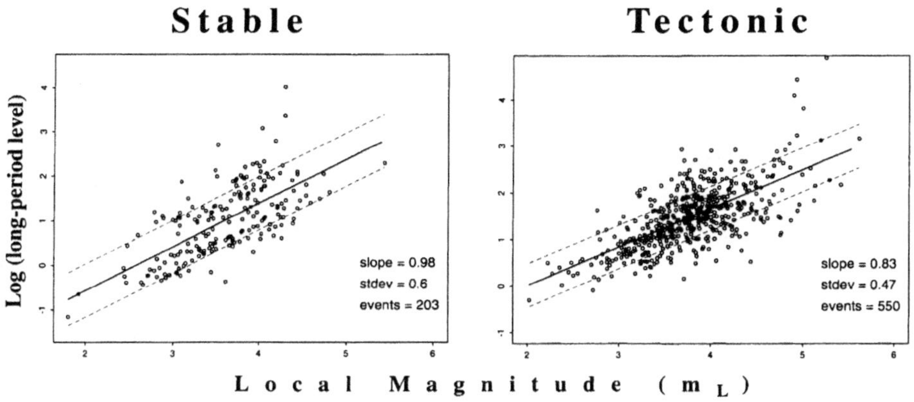

Figure 6
Magnitude and source relationship for the stable and tectonic models. Slope of the best-fit line shown in the stable and tectonic model plot is 0.98 and 0.83.

sum to zero. The average of the median absolute values of the site responses of regional phases in all subnetworks combined is 0.24. The estimated values for *Pn* and *Pg* were slightly lower than those found for *Sn* and *Lg*. We verified the need for site corrections by comparing the data variances of two inversions: one with site corrections fixed at zero for all stations, and the other with site corrections as free parameters. The data variance increased with the site corrections fixed, however the attenuation models were not much different.

Discrimination

The regional discriminants used in this study are based on the regional phases *Pn*, *Pg*, *Sn*, and *Lg* as defined in the REB. That is, we only considered phases that were confirmed by a PIDC analyst to be associated with an event at a regional distance. We normalized the *S*/*P* amplitude ratio by applying both distance and source corrections. Cross-spectral ratios such as *Lg*(0.5–1 Hz)/*Pn*(2–4 Hz) required normalization for the source corner frequency (e.g., TAYLOR and HARTSE, 1998). We examined normalized *S*/*P* amplitude ratios using a variety of frequency bands, subject to the constraint that the *P*-wave frequency is greater than or equal to the *S*-wave frequency. Three types of amplitude ratios were evaluated: *Lg*/*Pn*, *Sn*/*Pn*, and Largest *S*/Earliest *P*. *Pg* is not included in the first two discriminants because there are too few *Pg* phases in our data set. The frequency bands for *Lg* are 0.5–1, 1–2, 2–4, 3–6, and 4–8 Hz. The same bands were used for *Sn* plus an additional one at 6–12 Hz. The frequency bands for *Pn* are 1–2, 2–4, 3–6, 4–8, 6–12, and 8–16 Hz. We consider all phase ratios (e.g., 1–2 Hz *Lg*/*Pn*) and cross-spectral ratios (e.g., 1–2 Hz *Lg*/3–6 Hz *Pn*). Results are compared for earthquakes, presumed industrial explosions, chemical explosions, and a few nuclear explosions.

Calibrated S/P Ratios

Our model estimates provide both distance and source correction capability across all valid frequencies. The attenuation models for each regional phase can be used to compute distance corrections for any variant of the *S/P* discriminant. Other studies derive distance corrections directly from the amplitude ratios (HARTSE *et al.*, 1997). That approach can be effective and simple for studying a particular discriminant in one region. However, distance corrections must be developed for each variant of the discriminant, and the results are difficult to transport to uncalibrated regions.

The source correction is necessary for evaluation of cross-spectral ratios since the *S*-wave amplitude is measured at a different frequency than the *P*-wave amplitude. It corrects for the source corner frequency and is defined as:

$$\text{Source Correction} = \log s(f_S) - \log s(f_P) \tag{4}$$

where $s(f)$ defines the shape of the source function, f_S is the frequency of the *S*-wave amplitude measurement, and f_P is the frequency of the *P*-wave amplitude measurement. The shape of the source function is assumed to decay as f^2 beyond a corner frequency that scales inversely with the cube root of the long-period level.

Both source and distance corrections are necessary to obtain a fully-calibrated *S/P* discriminant. An example is shown in Figure 7 of the correction surface based on our tectonic model between 1–2 Hz (i.e., referred to as 1.5) *Sn* and 6–12 Hz (i.e., referred to as 9) *Pn* amplitudes. We indicate both source and distance corrections for a range of events between m_L 2 and 5 occurring at epicentral distances between 200 and 1600 km.

Discriminant Results

The performance of calibrated *S/P* discriminants is evaluated using our set of earthquakes, presumed industrial explosions, chemical explosions, and nuclear explosions. We require the *snr* of either the *P* wave or *S* wave to be above a threshold. We use a threshold of 1.5 for *Pn* or *Pg* and 1.3 for *Sn* or *Lg*. A lower threshold was used for *S* phases because the noise is the pre-signal noise which includes the coda of earlier arrivals. Our *S/P* ratio is a lower bound if the *S* wave exceeds the *snr* threshold but the *P* wave does not. This is observed for some earthquakes. Similarly, our *S/P* ratio is an upper bound if the *P* wave exceeds the *snr* threshold and the *S* wave does not. This is typical for some nuclear explosions which generally do not have an *Sn* or *Lg* arrival above the *snr* threshold in our data set.

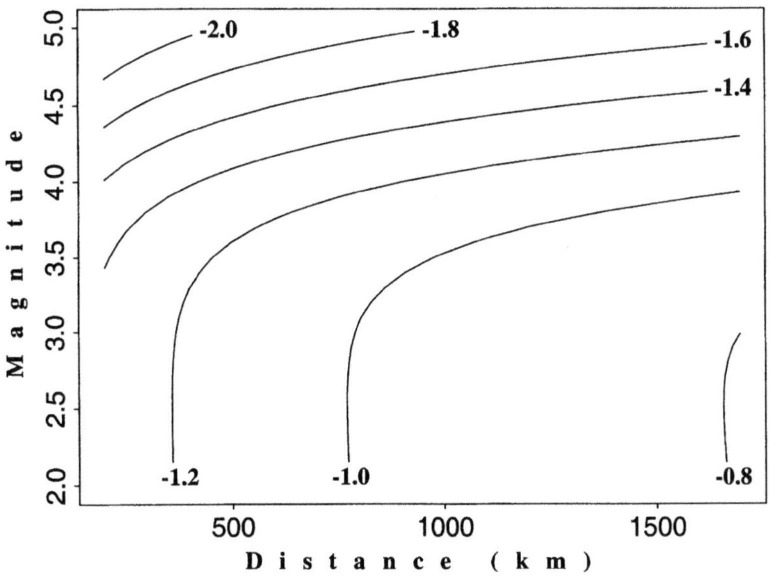

Figure 7
Correction surface for ratio between 1.5 Hz *Sn* and 9 Hz *Pn* amplitudes. The surface is based on the tectonic model generated for this study. Units are Log amplitude.

Central Asia. Results of the normalized *Lg*/*Pn* ratio are summarized in Figure 8 for the central Asia subnetwork. Similar results were obtained for the *Sn*/*Pn* and Largest *S*/First *P* ratios. We use this subnetwork as an example because it includes Lop Nor nuclear explosions, presumed industrial explosions, and earthquakes. We do not include the India, Pakistan, Balapan, or Degelen explosions in these results. We find that the *Lg*/*Pn* ratio tends to be lowest for the nuclear explosions. The mean ratio is only slightly lower for the industrial explosions than for the earthquakes. Unfortunately there is considerable overlap between all three populations in some cases.

Good separation occurs between earthquakes and industrial explosions by ratios with *Lg*(4–8 Hz). The best separation between nuclear explosions and earthquakes is provided by the *Lg*(1–2 Hz)/*Pn* (2–4 Hz) and *Lg*(1–2 Hz)/*Pn*(3–6 Hz) ratios. The ratios that include *Pn*(1–2 Hz) or *Lg*(4–8 Hz) are poor discriminants. Some of the ratios with *Pn* (6–12 Hz) are good discriminants, but they are limited to fewer data because of the *snr* criteria. The actual *Lg*/*Pn* ratios for the nuclear explosions are probably considerably lower than our estimates which are upper bounds, since *Lg* did not satisfy the s*nr* criterion. We used the amplitude in the *Lg* time window which, for the Lop Nor events, is *P* wave coda. We expect the *Lg*/*Pn* discriminant to perform better at closer distances where a meaningful *Lg* amplitude could be measured.

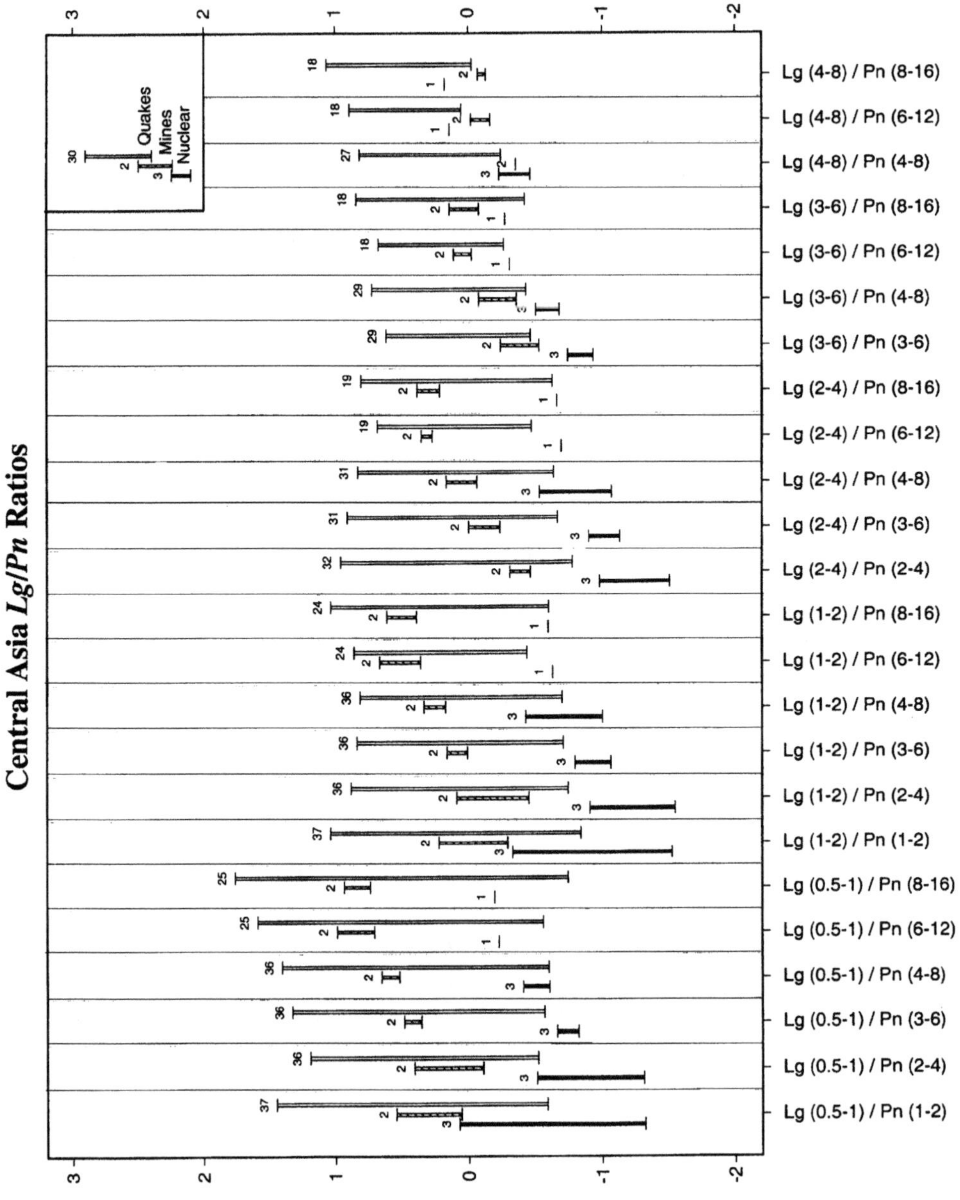

Figure 8

Summary of normalized *Lg*/*Pn* amplitude ratios for earthquake and explosion populations found in central Asia. Bars represent range of *Lg*/*Pn* ratios. The number of events in each population is indicated.

The fact that the *Lg*/*Pn* ratio does not provide unambiguous separation of nuclear explosions and earthquakes implies that there must be earthquakes with low *Lg*/*Pn* ratios. A multivariate approach, as used by some authors (HARTSE *et al.*, 1997), is recommended to obtain the best available discrimination of regional events.

All Regions. The consistency of the earthquake and industrial explosion populations for many regions suggests that the *S*/*P* discriminant is likely to be transportable, as long as attenuation and source models are available for each region.

We applied distance and source corrections to all events in our discrimination data set. We used region-specific models for all events except the India and Pakistan nuclear explosions, for which we used the tectonic model. Network-averaged and path-corrected 2–4 Hz log amplitudes of *Sn* and *Pn* are plotted in Figure 9. *Sn* was used instead of *Lg* to increase the number of observations available. The path corrections were derived from earthquakes so, on average, *Pn* equals *Sn* for earthquakes. The events span a magnitude range of approximately 1.5 to 5.5.

The earthquake and industrial explosion populations have similar variance and considerable overlap. The *S*/*P* ratio does not appear to discriminate between these classes of events.

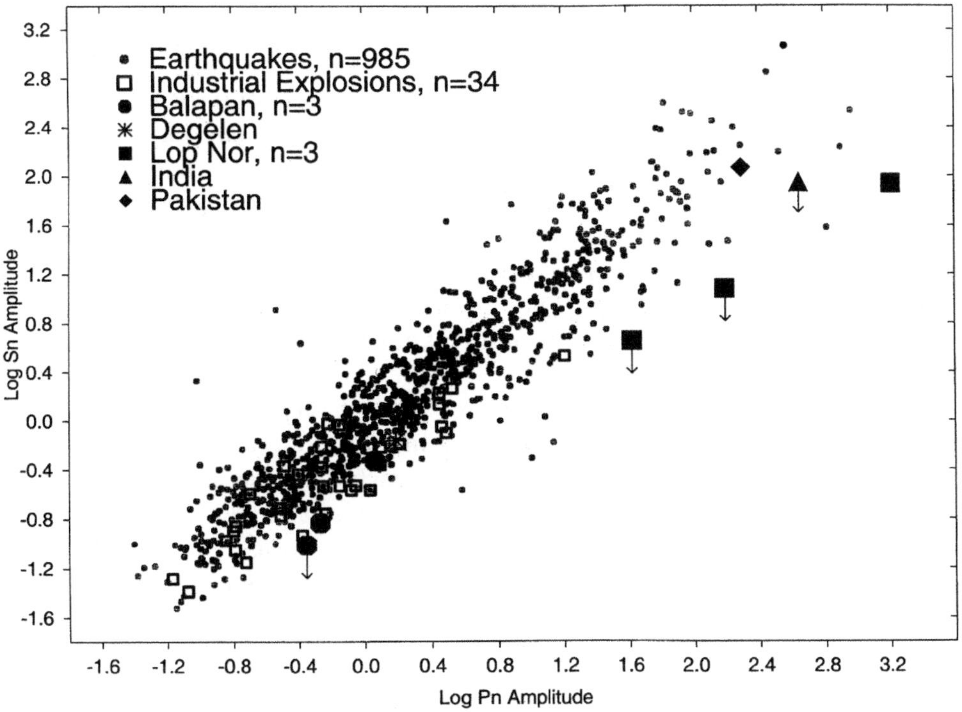

Figure 9

Sn amplitude is plotted versus *Pn* amplitude in the 2–4 Hz band. Both are distance corrected. The data set includes 1475 *Pn*, *Sn* amplitudes for the 985 earthquakes plotted and 52 *Pn*, *Sn* amplitudes for the 34 industrial explosions. For events observed at multiple stations we plot the average log(*Pn*) versus the average log(*Sn*). The downward arrows indicate the *Sn* amplitude is an upper bound measurement.

Four chemical explosions were evaluated; three at Balapan and one at Degelen. The two deeper Balapan explosions (which are grossly overburied compared to typical nuclear tests) can be discriminated from earthquakes with high confidence and from industrial explosions with somewhat lower confidence. However, the shallow Balapan and Degelen events fall near the center of the industrial explosion population and cannot be discriminated from earthquakes with high confidence. These two events generated considerably larger *S* phases.

The Lop Nor nuclear explosions (ranges between 1353 and 1628 km) are well outside the earthquake and industrial explosion populations. The *S*/*P* discriminant also works well for the India explosion (ranges between 743 and 1801 km), but not as well as for the Lop Nor events. Based on our observations at KNET stations (ranges about 1900 km), the Pakistan event cannot be discriminated from earthquakes. Though markedly larger in magnitude, it is actually on the earthquake side of the industrial explosion population. This event generated extraordinarily large *S* waves which caused the *Sn*/*Pn* ratio to be anomalously large.

Conclusions

Regional-wave amplitudes recorded by nine subnetworks of the PIDC in varying geologic and tectonic environments were used to determine source and attenuation models. These models were used to normalize and evaluate variants of the *S*/*P* regional discriminant using a data set of earthquakes, presumed industrial explosions, four chemical explosions, and a few nuclear explosions. We evaluated the performance of *S*/*P* discriminants for each subnetwork and two global models for stable and tectonic environments. Future work could explore the use of two global models, based on subnetworks with consistent attenuation rates rather than similar tectonic environments.

The calibrated *S*/*P* regional discriminant was not very successful at separating industrial explosions from earthquakes. The median *S*/*P* was generally lower for industrial explosions than for earthquakes, although there was substantial overlap. However, it was considerably more successful in separating earthquakes from deeply buried chemical explosions or nuclear explosions. Perhaps the most significant result of this study is the parameterization of the calibrated *S*/*P* ratio for many earthquakes worldwide. The primary challenge for a Comprehensive Test-Ban Treaty is identifying small nuclear explosions. However, statistical properties of such events are not available. Therefore, outlier methods seek to identify these events as anomalous with respect to earthquakes (TAYLOR and HARTSE, 1997; FISK *et al.*, 1993). The distribution of the calibrated *S*/*P* ratio for earthquakes can be used as the basis for these methods. An improvement for CTBT monitoring of small nuclear explosions would be to increase the use of auxiliary stations during regional event discrimination analysis. In many cases this will provide the necessary regional data which may not be available from the primary CTBT network.

References

BAUMGARDT, D., and YOUNG, G. (1990), *Regional Seismic Waveform Discriminants and Case-based Event Identification Using Regional Arrays*, Bull. Seismol. Soc. Am. *80*, 1874–1892.

BENNETT, T. J., and MURPHY, J. (1986), *Analysis of Seismic Discrimination Capabilities Using Regional Data from Western United States Events*, Bull. Seismol. Soc. Am. *76*, 1069–1086.

BENNETT, T. J., BARKER, B., MCLAUGHLIN, K., and MURPHY, J. (1989), *Regional Discrimination of Quarry Blasts, Earthquakes, and Underground Nuclear Explosions*, Final Rep. GL-TR-89-0114, 146 pp., ADA223148.

BENNETT, T. J., CAMPANELLA, A., SCHEIMER, J., and MURPHY, J. (1992), *Demonstration of Regional Discrimination of Eurasian Seismic Events Using Observations at Soviet IRIS and CDSN Stations*, Final Rep. PL-TR-92-2090, 122 pp., ADA253275.

BLANDFORD, R. (1981), *Seismic Discrimination Problems at Regional Distances*. In *Identification of Seismic Sources – Earthquake or Nuclear Explosion* (E. S. Husebye and S. Mykkeltveit, eds.) (D.Reidel Publishing Co., Dordrecht, The Netherlands) pp. 695–740.

BUNGUM, H., VAAGE, S., and HUSEBYE, E. (1982), *The Meløy Earthquake Sequence, Northern Norway; Source Parameters and their Moment Scaling Relations*, Bull. Seismol. Soc. Am. *72*, 197–206.

DYSART, P., and PULLI, J. (1990), *Regional Seismic Event Classification at the NORESS Array: Seismological Measurements and the Use of Trained Neural Networks*, Bull. Seismol. Soc. Am. *80*, 1910–1933.

FISK, M. D., GRAY, H. L., and MCCARTOR, G. D. (1993), *Applications of Generalized Likelihood Ratio Tests to Seismic Event Identification*, Final Rep. PL-TR-93-2221,96 pp., ADA279479.

HARTSE, H. E., TAYLOR, S. R., PHILLIPS, W. S., and RANDALL, G. E. (1997), *A Preliminary Study of Regional Seismic Discrimination in Central Asia with Emphasis on Western China*, Bull. Seismol. Soc. Am. *87*, 551–568.

JENKINS, R. D., SERENO, T. J., and BRUMBAUGH, D. A. (1998), *Regional Attenuation at PIDC Stations and the Transportability of the S/P Discriminant*, Final Rep. AFRL-VS-HA-TR-98-0046, 174 pp.

KERR, A. (ed.) (1993), *Overview GSETT-3*, Report prepared by the GSE Working Group on Planning, 9 pp., October.

KIM, W.-Y., AHARONIAN, V., LERNER-LAM, A. L., and RICHARDS P. G. (1997), *Discrimination of Earthquakes and Explosions in Southern Russia Using Regional High-frequency Three-component Data from the IRIS/JSP Caucasus Network*, Bull. Seismol. Soc. Am. *87*, 569–588.

POMEROY, P., BEST, W., and MCEVILLY, T. (1982), *Test-ban-treaty Verification with Regional Data – A Review*, Bull. Seismol. Soc. Am. *72*, S89–S129.

SERENO, T. J. (1990), *Attenuation of Regional Seismic Phases in Fennoscandia and Estimates of Arrival Time and Azimuth Uncertainty Using Data Recorded by Regional Arrays*, Tech. Rep. SAIC-90/1472 Science Applications International Corporation, 115 pp.

SERENO, T., and GIVEN, J. (1990), *Pn Attenuation for a Spherically Symmetric Earth Model*, Geophys. Res. Lett. *17*, 1141–1144.

SERENO, T. J., BRATT, S. R., and BACHE, T. C. (1988), *Simultaneous Inversion of Regional Wave Spectra for Attenuation and Seismic Moment in Scandinavia*, J. Geophys. Res. *93*, 2019–2035.

TAYLOR, S. R., and HARTSE, H. E. (1998), *A Procedure for Estimation of Source and Propagation Amplitude Corrections for Regional Seismic Discriminants*, J. Geophys. Res. *103*, 2781–2789.

TAYLOR, S. R., and HARTSE, H. E. (1997), *An Evaluation of Generalized Likelihood Ratio Outlier Detection to Identification of Seismic Events in Western China*, Bull. Seismol. Soc. Am. *87*, 824–831.

TAYLOR, S., DENNY, M., VERGINO, E., and GLASER, R. (1989), *Regional Discrimination between NTS Explosions and Western United States Earthquakes at Regional Distances*, Bull. Seismol. Soc. Am. *79*, 1142–1176.

WILLIS, D., DENOYER, J., and WILSON, J. (1963), *Differentiation of Earthquakes and Underground Nuclear Explosions on the Basis of Amplitude Characteristics*, Bull. Seismol. Soc. Am. *53*, 979–987.

YANG, X., NORTH, R., and ROMNEY, C. (1997), *CMR Nuclear Explosion Database*, Tech. Rep. CMR-97/19, Center for Monitoring Research, 81 pp., May.

(Received July 14, 1999, revised January 11, 2000, accepted January 18, 2000)

Pure appl. geophys. 158 (2001) 1301–1338
0033–4553/01/071301–38 $ 1.50 + 0.20/0

Pure and Applied Geophysics

Calibration of the Regional Crustal Waveguide and the Retrieval of Source Parameters Using Waveform Modeling

CHANDAN K. SAIKIA[1], BRADLEY B. WOODS[1] and H. K. THIO[1]

Abstract—Regional crustal waveguide calibration is essential to the retrieval of source parameters and the location of smaller ($M < 4.8$) seismic events. This path calibration of regional seismic phases is strongly dependent on the accuracy of hypocentral locations of calibration (or master) events. This information can be difficult to obtain, especially for smaller events. Generally, explosion or quarry blast generated travel-time data with known locations and origin times are useful for developing the path calibration parameters, but in many regions such data sets are scanty or do not exist. We present a method which is useful for regional path calibration independent of such data, i.e. with earthquakes, which is applicable for events down to $M_w = 4$ and which has successfully been applied in India, central Asia, western Mediterranean, North Africa, Tibet and the former Soviet Union. These studies suggest that reliably determining depth is essential to establishing accurate epicentral location and origin time for events. We find that the error in source depth does not necessarily trade-off only with the origin time for events with poor azimuthal coverage, but with the horizontal location as well, thus resulting in poor epicentral locations. For example, hypocenters for some events in central Asia were found to move from their fixed-depth locations by about 20 km. Such errors in location and depth will propagate into path calibration parameters, particularly with respect to travel times. The modeling of teleseismic depth phases (*pP*, *sP*) yields accurate depths for earthquakes down to magnitude $M_w = 4.7$. This M_w threshold can be lowered to four if regional seismograms are used in conjunction with a calibrated velocity structure model to determine depth, with the relative amplitude of the P_{nl} waves to the surface waves and the interaction of regional *sPmP* and *pPmP* phases being good indicators of event depths. We also found that for deep events a seismic phase which follows an *S*-wave path to the surface and becomes critical, developing a head wave by *S* to *P* conversion is also indicative of depth. The detailed characteristic of this phase is controlled by the crustal waveguide. The key to calibrating regionalized crustal velocity structure is to determine depths for a set of master events by applying the above methods and then by modeling characteristic features that are recorded on the regional waveforms. The regionalization scheme can also incorporate mixed-path crustal waveguide models for cases in which seismic waves traverse two or more distinctly different crustal structures. We also demonstrate that once depths are established, we need only two-stations travel-time data to obtain reliable epicentral locations using a new adaptive grid-search technique which yields locations similar to those determined using travel-time data from local seismic networks with better azimuthal coverage.

Key words: Location, depth determination, waveform modeling.

[1] URS Group, Inc., 566 El Dorado Street, Pasadena, CA 91101, U.S.A.
E-mail: chandan_saikia@urscorp.com

Introduction

Recent studies have demonstrated that regional path calibration is essential for obtaining accurate locations of seismic events at all magnitudes (RUSSIAN FEDERATION/UNITED STATES CALIBRATION WORKING GROUP, 1998; SAIKIA *et al.*, 1996a, b; THIO *et al.*, 1999; ZHU *et al.*, 1997; RYABOY, 1999). Accurate locations are integral to Comprehensive Test-Ban Treaty (CTBT) seismic monitoring efforts, particularly for smaller events, and are also crucial to many types of calibration studies. RYABOY (1999) has demonstrated that applying corrections for regionalized velocity models to the travel times of regional phases from nuclear explosions reduces location errors significantly, putting the relocations on par with announced event locations. Smaller ($M < 5$) events may only be recorded by a few stations, often with poor azimuthal coverage; consequently their hypocenter error ellipses can be extremely large, especially when located by only a few teleseismic observations (BONDÁR, 1998). At even lower magnitude levels the issue of signal-to-noise ratio (SNR) poses a problem for picking first arrivals. SAIKIA *et al.* (1996a) have demonstrated that the arrival times of teleseismic P waves may have timing errors up to several seconds for small events ($m_b \approx 4$), even after the stacking of array data, although such events are well recorded at closer regional distances ($\Delta \leq 10°$). Therefore regionalized velocity models for locating events with closer-in observations become necessary.

In addition, a crustal velocity model which can generate the recorded characteristics of regional phases is also useful for inverting regional waveforms in order to retrieve source parameters of earthquakes (e.g., LANGSTON and HELMBERGER, 1975; SAIKIA and HERRMANN, 1985; ZHAO and HELMBERGER, 1994; DREGER and HELMBERGER, 1993; SAIKIA and HELMBERGER, 1993; ZHU and HELMBERGER, 1996; RODGER and SCHWARTZ, 1998), particularly for smaller events ($M < 5$), which are difficult to analyze teleseismically because of the SNR thresholds. Such source information can directly be used for event identification. For example, seismic moment (M_0) which is a resultant product of the source inversion, is an essential parameter in some discriminants such as m_b:M_0 (PATTON and WALTER, 1993) and M_L:M_0 (WOODS *et al.*, 1993). M_0 is also an important parameter for accurate single-station path calibrations of regional seismic-phase amplitudes used for discrimination (PRIESTLEY and PATTON, 1997) and reliable magnitude measurements.

Furthermore, regional analysis of shallow crustal events can improve source depth estimates as teleseismic P waves can be contaminated by source-end upper-crustal reverberations, making it difficult to identify emergent first arrivals (STEIN and WIENS, 1986; THIO *et al.*, 1999). Well calibrated regional crustal waveguides are also important to modeling or inversion of teleseismic P waves (P, pP and sp) to determine source depth (NELSON *et al.*, 1987; WOODS *et al.*, 1998), as teleseismic P-wave based hypocenters can have significant error in depth, even for larger events ($M_w > 5$). Thus modeling the regional crustal waveguide is not only critical to accurate source analysis of smaller events ($4 \leq M_w \leq 5$), but is also useful for locating events of all sizes, particularly with respect to depth. Both hypocentral location and source parameter information are important ground-truth information for various calibration efforts, as well as for understanding earthquake source phenomenology.

In general, the crustal waveguide of a region can readily be calibrated, with respect to travel time and amplitude of predominant regional seismic phases, provided there exists a set of master events for which ground-truth (GT) locations and origin times are available (BONDÁR, 1998). The best events for travel-time calibration are announced nuclear explosions or quarry blasts for which reliable shot times and locations are available. Unfortunately for most regions, explosions or quarry blasts only cover a few source-specific station paths and often are not available at all for many areas. In lieu of such events, well located earthquakes are required. Earthquakes whose hypocenters are determined using the P and S arrivals from the local network stations with adequate azimuthal coverage are ideal because they are based on the velocity models that are consistent with travel times of local and regional seismic phases. However, not all regions have such comprehensive networks and so other means must be used to obtain the near GT source information. In such cases the modeling of broadband regional waveforms can be very useful in constraining hypocentral locations as well as other source information. Finding accurate depths is essential for ground-truth events because errors in depths can also result in inaccurate estimates of calibration parameters such as the station-specific-source corrections (SSSC).

Regional records, however, are generally more complicated than teleseismic signals due to the greater influence of upper-crustal heterogeneities (DREGER and HELMBERGER, 1993; SAIKIA, 1994) which gives rise to additional reflected and converted body-wave phases, particularly at shorter periods and thus can require considerable modeling. Additionally, lateral variations in the crustal structure give rise to significant travel-time residuals relative to a single uniform crustal waveguide model for a large area such as North America (ROMNEY *et al.*, 1962). Regional surface waves are also sensitive to lateral variations in the crustal waveguide, particularly Moho depth and the thickness of low velocity sedimentary layers in the upper crust, indicating the importance of possible sub-regionalization of such models (SAIKIA and HELMBERGER, 1993; THIO and KANAMORI, 1995; WOODS *et al.*, 1998) for areas in which one crustal model will not suffice.

In this paper we will review methods that use regional seismograms for calibrating regional waveguides in areas where there is a lack of earthquakes with reliable hypocentral locations. These methods were successfully used in calibrating crustal structure and locating earthquakes in the Pamir Hindu-Kush region (ZHAO and HELMBERGER, 1992; SAIKIA *et al.*, 1996a, b; ZHU *et al.*, 1997), western Mediterranean and North Africa (THIO *et al.*, 1999), the former Soviet Union (SAIKIA and HELMBERGER, 1993), southern California (DREGER and HELMBERGER, 1993; ZHU and HELMBERGER, 1996) and central Asia (WOODS *et al.*, 1998). This modeling of potentially complicated regional waveforms to obtain a crustal waveguide structure is facilitated by independent estimates of earthquake source depth, orientation and seismic moment so that these effects can be correctly included in the synthetic seismograms. The accuracy and applicability of this information from various techniques and sources shall also be discussed. We will describe a procedure for modeling crustal structure using regional waveforms in conjunction with source constraints from other seismic methods, and provide examples of results from our more recent studies as well as comparisons with results from other sources.

Methodology

In order to model the crustal waveguide with regional seismograms, it is important to have accurate focal mechanisms and source depths for the master events so that these effects can be correctly accounted for generating synthetic seismograms to match the observed waveforms. The calibration, or master, earthquakes used for this purpose are preferably larger events in the $5 \leq M_w \leq 6$ range. Events of this size should be well recorded teleseismically, thereby providing locations which can meet the GT5, GT10 or GT25 criteria (BONDÁR, 1998). Such large events will also provide necessary seismograms for determining focal depth and source parameters at all ranges. Earthquakes much larger than $M_w = 6$, though well recorded, tend to have more complex source processes for which the seismograms can be complex to model unless the source process is successfully deciphered, whereas events smaller than this have relatively simple source processes which can be approximated with a simple source-time function; hence are useful as master events. The events that have been modeled to data are small in magnitude ($M_w \leq 5.5$) and we have used just triangular source functions in modeling their waveforms.

In general, earthquakes ($M_w \geq 5$) which satisfy the criteria for master events have their source parameters, namely the seismic moment and focal mechanisms, readily available. Surface-wave centroid-moment-tensor (CMT) solutions are often available for $M_w \geq 5.5$ events globally from the Harvard catalog (DZIEWONKSI *et al.*, 1981). Recently ARVIDSSON and EKSTRÖM (1998) have extended the global CMT analysis to moderate-sized ($M_w \geq 4.5$) earthquakes using shorter period (40–150 sec) waveforms. Regional results are also being routinely compiled by a number of groups (RITMESA and LAY, 1993; THIO and KANAMORI, 1995; GEE *et al.*, 1996) and as more broadband seismic networks become operational, the number of regional CMT catalogs should also grow. In addition, literature reviews of regional source studies which make use of regional surface waves as well as regionalized earth models should also provide a wealth of information for potential master events. Teleseismic body-wave CMT inversions (SIPKIN, 1982) also provide well constrained solutions, particularly if they include analysis of S waves; although such catalogs are less extensive. However, there are numerous regional source studies in the literature which use similar teleseismic analysis. Furthermore, teleseismic studies which model, or invert, the phases P, pP, and sP (ZHAO and HELMBERGER, 1991a; STEIN and WIENS, 1986) provide the best estimates of focal depth; other methods are more susceptible to error. This subject is discussed in more detail below.

Teleseismic location and depth estimates are known to contain appreciable errors. ISC locations are determined primarily from first arrival P waves (ADAMS *et al.*, 1982). Secondary phases, in particular pP and sP, are used to constrain depth only when the P-wave inversion does not converge. This depth phase analysis, however, is not done with regionalized models, but with a single generic global velocity model. Therefore, the depth is sometimes not well determined and is often assigned a fixed value for many events. RÖHM *et al.* (1999) found significant delays in ISC travel-time data which they attribute to errors in the picking of arrival times or the timing of data. BILLINGS *et al.* (1994) conclude that the sensitivity of P arrivals to changes in depth are swamped by model errors and, therefore, the inclusion of depth phases such as pP is highly recommended. They also found that the effect of picking phase arrival errors on location is considerably smaller than the mislocation caused by lateral variations when only the direct P arrivals are used. RÖHM *et al.* (1999) found temporal variations in median station delays on the order of 0.5–1.0 seconds which, because of its systematic nature, may introduce biases in locations, particularly since the biases are often shared by stations in a given regional network.

It is well established that ISC and PDE depth estimates can contain major errors, particularly for small and shallow crustal earthquakes. Recent relocation studies have shown significant differences in location and depths relative to the corresponding ISC values. In a relocation of 60 Tibetan earthquakes using short- and long-period teleseismic waveform modeling, ZHAO and HELMBERGER (1991a) found the majority of events' depths moved to the upper 20 km of the crust, whereas the ISC depths had a more random distribution throughout the top 60 km of the crust. Although the hypocentral locations did not move significantly, the origin times decreased by an average of 3 seconds due to the trade off in the focal depth and station corrections in the travel times. Using regional waveforms, ZHU *et al.* (1997) found crustal earthquakes in the Hindu-Kush region on average to be 17 $\pm$ 19 km shallower than their ISC estimates, while mantle earthquakes (h > 100 km) only moved by about 8 $\pm$ 18 km upwards relative to their ISC depths. Teleseismic P-wave depth estimates for shallow earthquakes are generally considered to have larger errors as that is where the largest variations in source structure are, and crustal events tend to generate more scattered phases which can make picking the initial P wave and other depth phases more difficult, especially for smaller events and for records with emergent P-wave first arrivals (STEIN and WIENS, 1986). In a comprehensive relocation of $M_w \geq 5.2$ events, using available secondary phases, ENGDAHL *et al.* (1998) found that the average depth decreased by less than 2 km relative to ISC values, however, the standard deviation was 12 km, suggesting considerable scatter between the two sets of depth estimates. Further, events for which depth shifts were greater than 40 km (approximately 2% of the data) were not included in the analysis, thus the overall average depth shift may well be larger. Hypocenters moved by less than 7.5 ± 5.5 km, on average, which is within the estimated ISC location errors.

Surface-wave centroid-moment-tensor (CMT) solutions are also susceptible to errors in depth estimates which, in turn, can give rise to errors in M_0. A significant portion of global CMT solutions for crustal earthquakes have depths fixed at 10, 15 or 33 km (SAIKIA *et al.*, 1997). On average the CMT hypocenters are mislocated by over 30 ± 20 km relative to the relocations of ENGDAHL *et al.* (1998). Globally, the focal depths are on average only 2 ± 13 km shallower than the relocated depths. However, again the variance is large and regional variations can be more significant. For example, in the Tien Shen region of northwest China, regional surface-wave spectra-inversion for 26 earthquakes relocated the events 2 km shallower on average than the CMT depths (PATTON, 1998). In a study of the same events, WOODS *et al.* (1998) found that depths determined from a combination of *P*-wave depth-phase waveform modeling and regional waveform inversion yielded source depths that were respectively 8 and 5 km shallower, on average, than the CMT depths. The ISC focal depths were generally deeper. Another regional waveform source inversion study found similar depth differences for the available CMT comparisons (GHOSE *et al.*, 1998).

Accurate source depth helps to constrain the epicentral location by minimizing the trade-off between source depth and origin time, so that absolute travel-time calibrations are improved. Source depth has a strong influence on the impulsiveness and amplitude of regional body waves, with deeper events having sharper, more impulsive arrivals and fewer scattered phases. The excitation functions of fundamental-mode surface waves are also depth dependent, with deeper events generating relatively smaller amplitude waves. Thus a priori depth information is important for determining regional waveguide Green's functions. Ground-truth depth, of course, is an important parameter in the source discrimination studies (WALTER *et al.*, 1995; TAYLOR, 1995; PRIESTLEY and PATTON, 1997; VÖGFORD, 1997; GUPTA *et al.*, 1997) as many source phenomena have direct or indirect depth dependence such as source medium material properties (WALTER *et al.*, 1995).

M_0 is an important parameter for calibrating amplitude-distance corrections. Absolute amplitude information is required of any single-station calibration method, including empirical distance corrections, measures of attenuation, waveform modeling, and spectral amplitude fitting. As surface-wave excitation functions are source-depth dependent, there exists a trade-off between focal depth and seismic moment. Furthermore, M_0 is also earth-model dependent. Consequently, discrepancies between teleseismic Harvard CMT moments (DZIEWONSKI *et al.*, 1981), which use a generic PREM crust ($h = 24.4$ km) to determine the surface wave source excitation functions, and regional M_0 estimates, based on more appropriate earth models with thicker crusts, are expected to arise. Such biases in CMT moment estimates have been found in several regions. The surface wave-spectra inversion using a regional crustal model also yielded seismic moments 0.27 log units lower than Harvard CMT solutions (PATTON, 1998). The cause was attributed to the difference in source depth as discussed above and to the difference between the PREM and the regional crustal models for the area, for which the Moho depth ranges between 45 and 60 km. Similarly, regional moments of Hindu-Kush earthquakes were also lower than the CMT values by 0.33 units (ZHU *et al.*, 1997), with the largest differences for the crustal events, as was noted by PATTON (1998). Here the crust is estimated to be 70 km thick.

Focal mechanisms of CMT solutions tend to be well determined, generally agreeing with teleseismic first-arrival and waveform modeling solutions (WOODS *et al.*, 1998) as well as well-constrained regional estimates (ZHU *et al.*, 1997; ZHAO and HELMBERGER, 1991b). In regions of extreme lateral variation in the crust, regional and teleseismic CMT source mechanisms are less compatible as was found to be the case in the western Mediterranean (THIO *et al.*, 1999). But, in general, these long-period teleseismic CMT solutions are adequate for correcting for mechanism effects in the regional waveform modeling. When possible, source mechanism information determined from regional data should be complemented with teleseismic body-wave waveform inversion solutions. For large events ($M_w \geq 5.5$) and, in some cases, slightly lower magnitude ones (down to $M \geq 5$), teleseismic source determinations combining P- and S-wave waveforms can provide well constrained solutions whereas, for example, first-motion mechanisms can differ by as much as 45° in strike and rake for the same earthquake, depending on the data if local station data are unavailable (NELSON *et al.*, 1987).

In order to obtain the reliable source information for master events to calibrate a regional waveguide, it is important to use teleseismic body wave observations to determine depth and source parameters. The relative time separation of pP and sP waves from the P waves helps to constrain focal depth, while their amplitudes along with S waves provide accurate estimates of focal mechanism and M_0. Although the character of these teleseismic waveforms is dependent on the source and receiver crustal structures, it is less sensitive to the details of regional waveguide than that of surface waves and regional body waves. Using the standard technique of generalized ray theory (LANGSTON and HELMBERGER, 1975), synthetic teleseismic seismograms can be generated in the range of 30° to 90° for which the ground motion displacement is approximated by

$$W_{Z,R}(t) = R_{(pZ,pR)}\left[\frac{\mathrm{d}\phi}{\mathrm{d}t} + R_{pP}\frac{\mathrm{d}\phi}{\mathrm{d}t}\cdot H(t-\Delta t_1) + R_{sP}\frac{\mathrm{d}\psi}{\mathrm{d}t}\cdot H(t-\Delta t_2)\right]$$

where $\mathrm{R}_{(pZ,pR)}$ represents the vertical or radial receiver function, R_{pP} and R_{sP} are the respective P- and S-wave free-surface reflection coefficients, Δt_1 and Δt_2 are time lags of pP and sP relative to the mantle P waves, η_α and η_β are vertical slowness for P and S waves, and $H(t-\Delta t)$ is the time-lagged heavy-side step function; ϕ and ψ are displacement potentials and their detailed expressions can be found in LANGSTON and HELMBERGER (1975), as well as the similar expressions for the S-wave ground motions. The synthesized ground motions are convolved with the source time function and attenuation model. The source radiation affect is also added prior to comparing with the instrument-deconvolved observations.

Another approach to generate teleseismic seismograms is utilization of the Thomson-Haskell propagator matrix method (HASKELL, 1962; AKI and RICHARDS, 1980) to propagate the wavefield through the source and receiver regions. We developed a code to compute synthetic Green's functions for the fundamental faults following this approach in which geometrical spreading of the seismic rays and the anelasticity of the earth are accounted for separately, using the formulation presented in BEN-MENAHEM *et al.* (1965) and TENG and BEN-MENAHEM (1962). We propagate the wavefield using the formulation presented in HUDSON (1969) where the Haskell layer 4×4 matrices have been modified to the of format 6×6 compound matrices (SAIKIA, 1994) and used a complex frequency. Unlike in the generalized ray approach, this method needs no specification of rays as it includes the total wavefield. The code calculates the ray parameters for the teleseismic paths using the J-B travel-time table and requires only the receiver and source locations.

The teleseismic source mechanisms of master events are inverted using the Green's functions generated as described above by performing a grid search sweeping through focal-mechanism parameter space of strike, dip, and slip to obtain the best mechanism at each depth. Depth is constrained by fitting travel times of *pP* and *sP* relative to the travel time of the *P* onsets. Green's functions and the observed waveforms are cross-correlated (WALLACE *et al.*, 1981; ZHAO and HELMBERGER, 1994). This is implemented in a code "teleINV" which uses Green's functions generated by the above teleseismic method and finds the best fitting solution by sweeping through the entire focal parameter space. Figure 1 illustrates an application of this teleseismic modeling using the recent southern India earthquake of May 21, 1997 in Jabalpur of Madhya Pradesh. Depth is particularly well constrained in such teleseismic inversions. The error in depths can be minimized and sometimes depths become resolvable to within ± 1 km using accurate source-region crustal structure (WOODS *et al.*, 1998); even with more general velocity models absolute depth error is likely to be at most ± 5 km (NELSON *et al.*, 1987).

Once the master events' source parameters and locations are determined, their regional records are forward modeled by perturbing the crustal velocity structure to

▶

Figure 1
Example of the fit of teleseismic *P* waves used to determine source parameters and depth for an earthquake in India. The observed (thick line) and synthetic (thin line) displacement waveforms have been convolved with a Wood-Anderson long-period instrument. Station name, azimuth and distance, in degrees, are given to the left of the vertical seismograms; the peak amplitude (cm) of each trace is given at the right of each seismogram pair. The synthetic seismograms were computed for the best-fitting focal mechanism solution ($\delta = 65°$, $\lambda = 68°$, $\Phi = 70°$) and event depth of 35 km, shallower than the PDE depth only by 1 km, which were obtained using the grid-search technique. Both vertical and radial *P*-wave seismograms were modeled and only one receiver crust was used for all 27 stations.

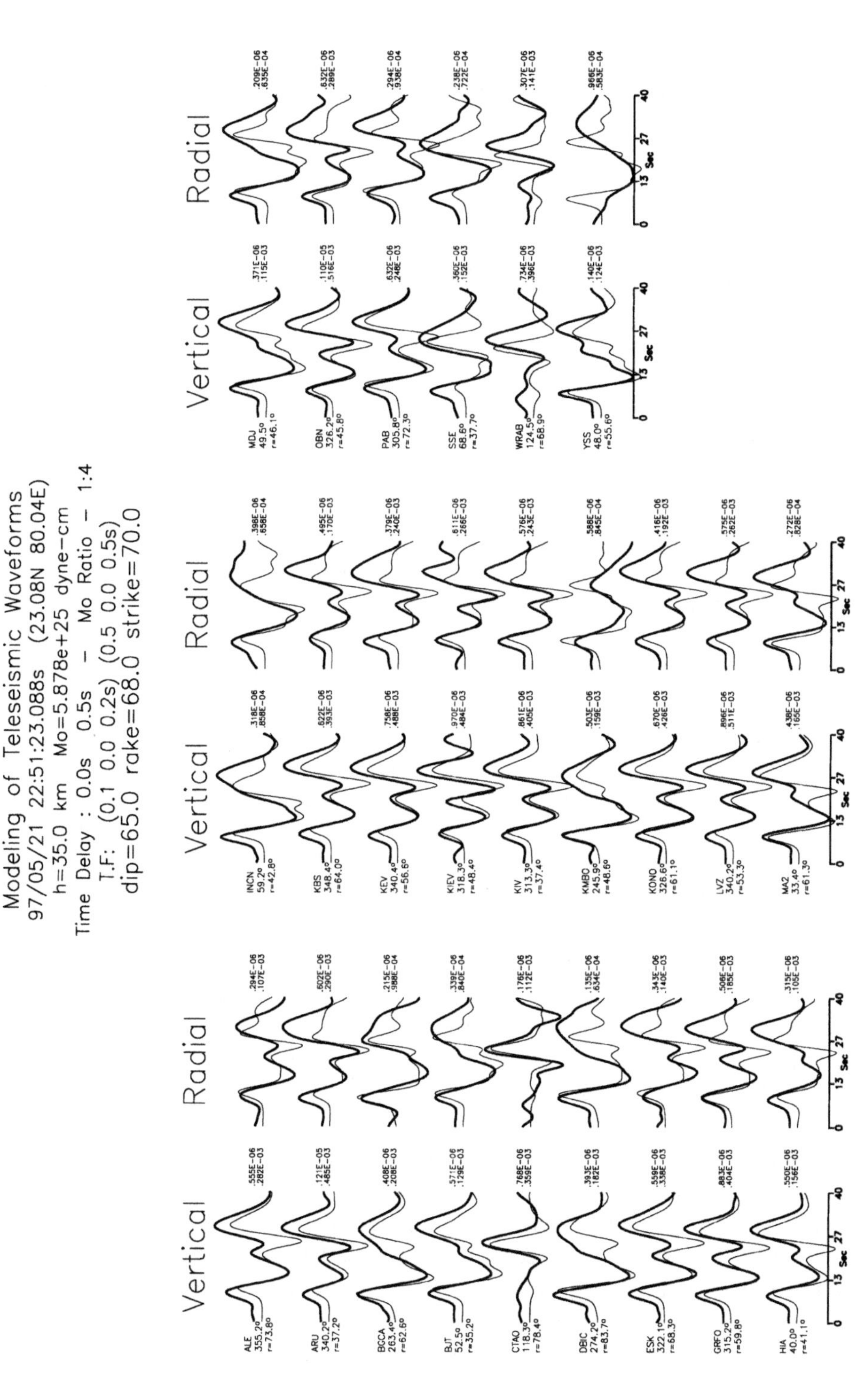
Modeling of Teleseismic Waveforms
97/05/21 22:51:23.088s (23.08N 80.04E)
h=35.0 km Mo=5.878e+25 dyne-cm
Time Delay : 0.0s 0.5s - Mo Ratio - 1:4
T.F: (0.1 0.0 0.2s) (0.5 0.0 0.5s)
dip=65.0 rake=68.0 strike=70.0
Vertical
Radial
Sec

better fit the observed waveforms. In conjunction, other geophysical information such as the P_n velocity, Moho depth determined by reflection studies or other methods, teleseismic receiver functions, and surface-wave dispersion data, is used to constrain the model.

The approach taken to develop the regional waveguide is to begin with a relatively simple 1-D model with a couple of crustal layers over a mantle half space; only incorporating additional layers to include the significant discrepancies between observed and synthetic waveforms. Generally this is done by modifying the overall thickness of the crust and velocities of the crustal layers and the upper-most mantle. The primary upper layer, corresponding to the sedimentary structure, strongly affects the Rayleigh waves and the character of the *PL* (extended *P*-wave) and *S* wavetrains (SONG *et al.*, 1996; SAIKIA, 1994; SAIKIA and BURDICK, 1991). If observations suggest a strong gradient in the velocity structure – that is, the initial *P* wave is sharp and impulsive with the character of turning ray rather than the longer period ramp-like head-wave character – a detailed model with a gradient in the mantle structure must be used (SAIKIA, 1994).

Figure 2 is a modeling example showing the agreement between data and synthetic seismograms obtained for a set of three-component regional seismograms from the southern India earthquake, described previously, recorded 645 km away at HYB Hyderabad (HYB), a Geoscope broadband station in India. We employed as an initial regional model one developed by GAUR and PRIESTLEY (1997), using teleseismic receiver function analysis at HYB. The top seismograms are the deconvolved broadband displacement records and the two waveforms immediately below correspond to synthetic seismograms generated using the respective teleseismic and regional mechanism solutions determined by waveform inversion. Both solutions yield similar mechanisms and seismic moment, suggesting a well-constrained stable solution.

Another factor to be accounted for in modeling the crustal waveguide is attenuation, or Q. The amplitude and frequency content of regional seismograms are strongly influenced by the anelasticity of the propagating medium. As attenuation affects absolute amplitudes, it thereby directly affects seismic moment and magnitude estimates. The varying degree of attenuation between NTS and the Russian test sites is found to influence the frequency content of the regional seismograms, thus impacting various regional discriminants (TAYLOR, 1995). As the frequency-wavenumber integration method used to generate the regional Green's functions can directly incorporate the effects of intrinsic Q (SAIKIA, 1994), it can be directly accounted for by forward modeling the absolute amplitudes of the regional waveforms, provided that accurate source information, mechanism and moment, are known – preferably from teleseismic estimates of *P*- and *S*-wave source inversions. With the M_0 and mechanism from the teleseismic data modeling, a trial-and-error approach can be used to best fit the amplitude of various groups of waves in the regional records by modifying the Q values as needed. It is possible that there will be some trade-off between attenuation and source time function, especially more

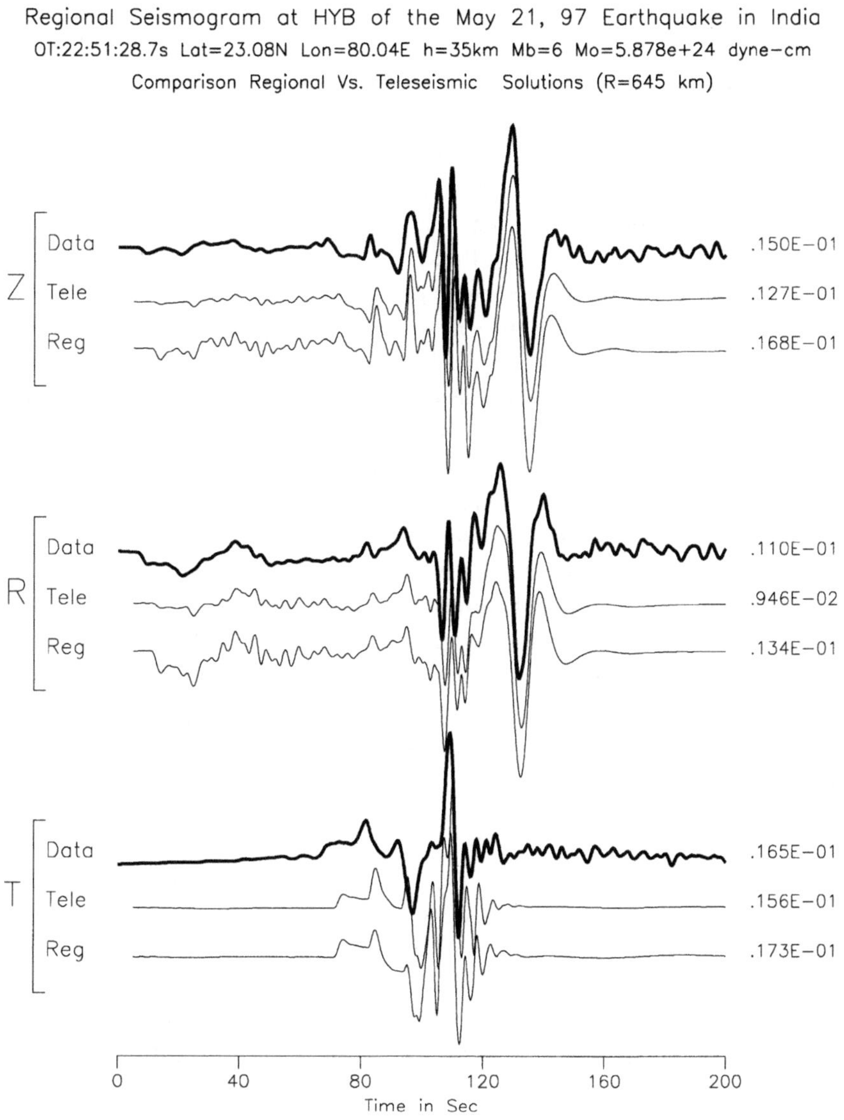

Figure 2
Comparison of observed and synthetic regional seismograms (for the same Indian earthquake shown in Fig. 1); both teleseismic and regional focal mechanism results are shown. The two solutions are similar. A seismic moment of 5.9×10^{24} dyne-cm ($M_w = 5.9$) and depth of 35 km were used. Note the agreement in the amplitude of the P_{nl} waves relative to the surface waves, including the peak amplitudes (cm).

so when the events are large. However, the trade-off can be reduced when both P_{nl} and surface waves are modeled together because the P_{nl} waves attenuate less compared to the short-period surface waves due to the Q effect for the same source function. In an attenuating medium, Q plays an important role in shaping the frequency content of the surface waves relative to the P_{nl} waves, especially for the shallow events. Of course it is helpful to have a priori Q constraints from other seismological methods or to use as an initial Q model appropriated from other previously studied regions with similar tectonic features. Attenuation related to scattering is best accounted for in regional modeling by developing crustal models for particular paths. In some cases this may be done by simply modifying the Q model for a single crustal model, but often will require the modification of the velocity model as well, sometimes employing thin layers of alternating high and low velocity material in that portion of the waveguide giving rise to the scattering (SAIKIA, 1994).

Regional seismic waves have proven to be amongst the most difficult to analyze or model in deterministic fashion in the field of seismology. At high frequencies the effects of scattering in the crust become so intense that only the statistical properties of waveforms are meaningful. One exception to this generality involves long-period P_n and shear-coupled P (PL) mode body waves at regional distances (1°–12°). HELMBERGER (1972, 1973), and HELMBERGER and ENGEN (1980) have shown these P-wave phases, which comprise the first 1–2 min of the regional wavetrain, to be very stable and straightforward to model with synthetic seismograms. In subsequent studies, WALLACE *et al.* (1981) demonstrated that long-period P_{nl} waveforms are stable and deterministic enough to be utilized in automated inversion schemes to determine source mechanism. LEFEVRE and HELMBERGER (1989) and HOLT and WALLACE (1989) demonstrated that the P_{nl} analysis procedures can be applied without substantial modification to observations from virtually any tectonic region. In a later study SAIKIA and BURDICK (1991) extended the application to shorter period P_{nl} waves and demonstrated that short-period P_{nl} waves also have many stable features which can be deterministically modeled to characterize the regional crustal waveguide. This characteristic of P_{nl} waves is due to their primarily sampling the lower crust and the Moho interface, and consequently being relatively insensitive to lateral crustal variations, which occur mainly in the upper crust. Even for mixed paths traversing continental and oceanic crust in the western Mediterranean, THIO *et al.* (1999) found P_{nl} waveforms to remain quite stable. Hence this portion of the wavetrain can be adequately modeled over long distances with a single model, although its relative timing with respect to the rest of the wavetrain may vary.

Surface waves are also susceptible to lateral variations in the crustal waveguide as they are strongly effected by the lateral variations in the upper crust and mountain topography as well as the moho depth. This is especially true for Rayleigh waves (WOODS and HARKRIDER, 1995; SONG *et al.*, 1996; THIO *et al.*, 1999). Therefore, it becomes difficult to model the entire wavetrain waveform with a single 1-D model. An example of such behavior is provided in Figure 3, taken from THIO *et al.* (1999), which compares the vertical and tangential displacement records for an earthquake in Gafsa in Tunisia (June 12, 1992, origin time: 19 h 16 m 45.7 s) recorded by the stations MEB (Medea, Algeria) and VSL (Villasalto, on the Island of Sardina). The path to MEB is a pure continental one whereas the one to VSL consists of both

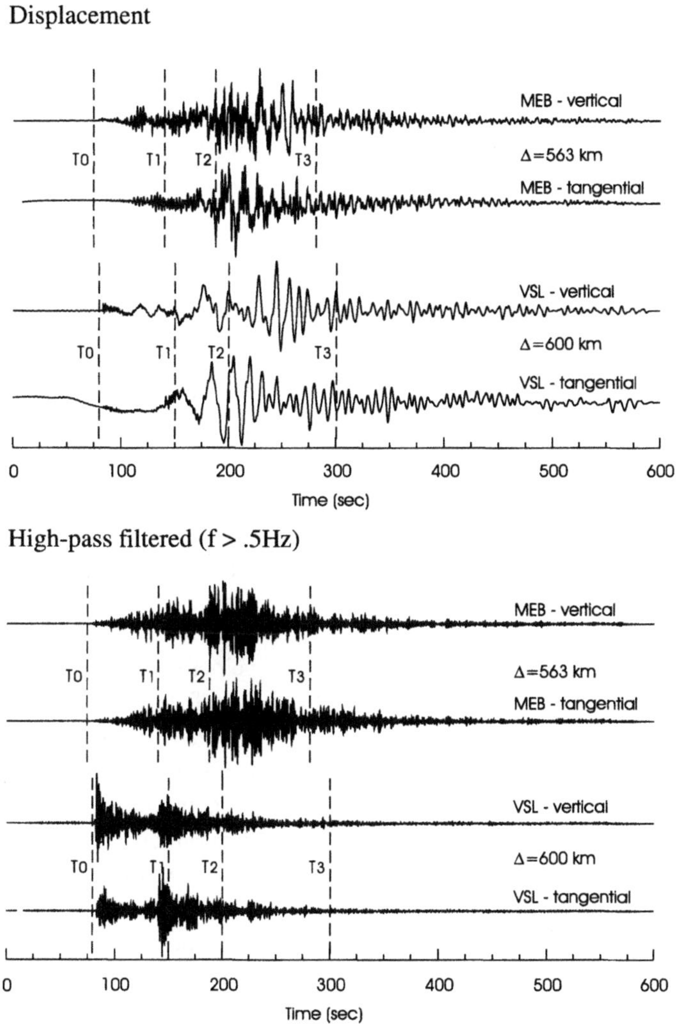

Figure 3
Broadband displacement (upper panel) and high-pass filtered (lower panel) seismograms (vertical and tangential components) for an event in North Africa at Gafsa, Tunisia recorded at MEB and VSL (units are in "cm"). The path to VSL traverses is the Mediterranean and causes strong Lg wave blockage. The path to MEB is continental. (Notations: the lines **T0**, **T1**, **T2** and **T3** correspond to the travel times for group velocities of 7.5, 4.0, 3.0 and 2.0, respectively.

continental and oceanic sections almost equal in length. The two recordings are at comparable distances, yet the waveforms are very different in character and the surface-wave arrival time is considerably sooner for VSL. A striking feature is found in the frequency content, with the record for the oceanic path to VSL being deficient in high frequencies relative to the one for the pure continental path to MEB.

Another example of the variation in regional waveform propagation characteristics is shown in Figure 4 which compares three pairs of vertical component displacement seismograms, recorded at different azimuths from an earthquake in central Asia. For each pair of seismograms the lower trace is the short-period filtered version of the upper broadband waveform. The top pair of waveforms is for a path crossing the Junggar basin to the station WMQ in China, while the other two seismogram pairs are for paths traversing the Tien Shen mountains to the stations AAK in Kyrgzystan and WUS in China. There is no significant difference in short-period seismograms besides the expected amplitude decay with distance, however the surface waves crossing the basin are modulated to more dispersive ringing signals than are the other two surface-wave observations which, in fact, have longer paths. Generally, more dispersion would be expected from the more distant observations at WUS and AAK, but this is not the case. Modeling these surface-wave observations well with one crustal waveguide is not possible. The use of a crustal model, which consists of a relatively fast upper crust to emulate the effects of the mountain structure recorded at WUS and AAK, to also model the basin effects recorded at WMQ would tend to underestimate the source depth and consequently the seismic moment as well for this observation.

For regions with large lateral variations in crustal waveguide, surface waves are difficult to model using a single 1-D crustal velocity model. One approach to overcome this difficulty is to model the P_{nl} and surface waves separately using a hybrid method in which the surface waves are generated using normal-mode synthesis so that mixed-path synthetics can be produced. A simple and computationally efficient way to account for surface waves traversing a laterally varying structure is by the method of mixed-path synthetics which can account for two or more velocity models along a single path (WOODS and HARKRIDER, 1995; THIO *et al.*, 1999). LEVSHIN (1985) showed that this approximation holds for cases in which $\lambda \gg \ell$, where λ is the horizontal wavelength and ℓ is the scale length of the velocity variation. The cut-and-paste source inversion method (CAP, ZHAO and HELMBERGER, 1992) has the advantage of accommodating P_{nl} and surface-wave Green's functions generated from two or more separate crustal models, including mixed paths (THIO *et al.*, 1999).

Another approach to account for the varying crustal structure along the propagation path is to regionalize the phase velocity model to correct for the surface-

▶

Figure 4
(a) Map showing the location of event 96072 and the three recording stations: AAK, WMQ, and WUS, (b) vertical component displacement records for event 96072. Three pairs of traces are shown, the top one corresponds to the recorded broadband displacement and the lower one to the Wood-Anderson short-period instrument play-out (units are in "cm"). Note the strong dispersive ringing character of the WMQ displacement record, which traverses primarily basin structure, compared to the other two observations which traverse the Tien Shen mountains.

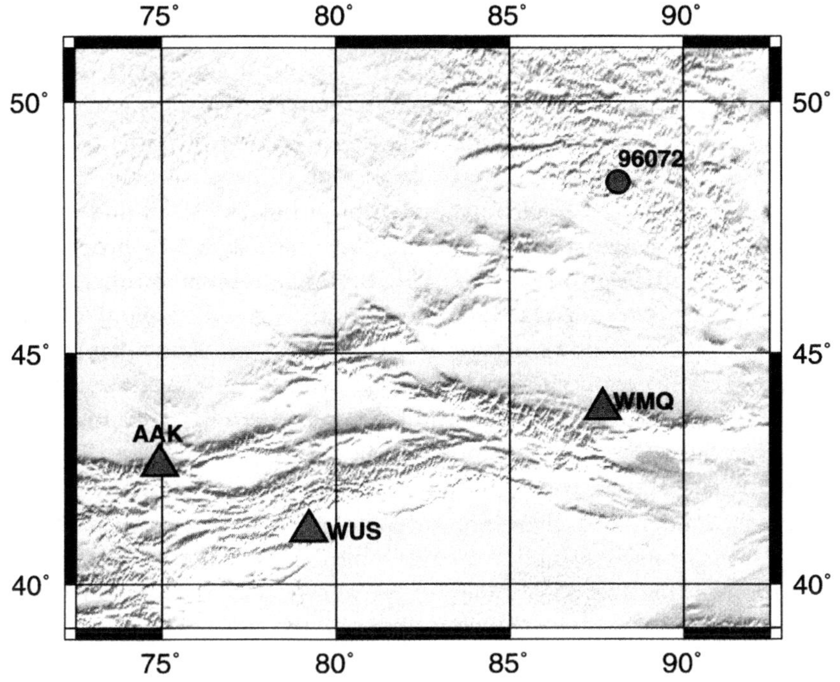
75°
80°
85°
90°
50°
45°
40°
96072
WMQ
AAK
WUS

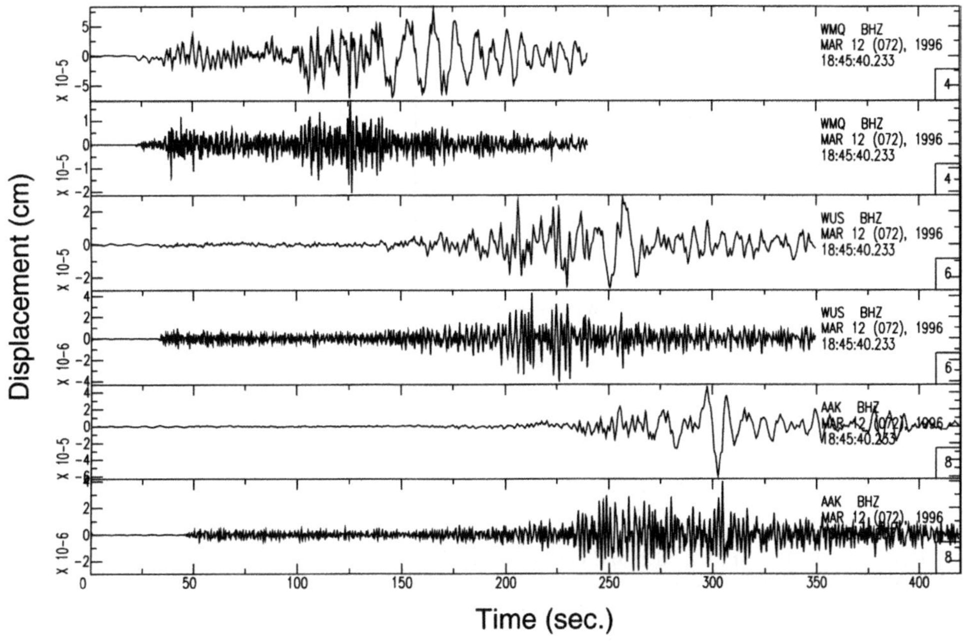
Displacement (cm)
Time (sec.)
WMQ BHZ
MAR 12 (072), 1996
18:45:40.233
WUS BHZ
AAK BHZ
X 10-5
X 10-6
0
50
100
150
200
250
300
350
400

wave path effects. This method is based on the fact that to a first-order perturbation in structure to a homogeneous velocity model is primarily in the phase velocity and is widely used in the surface-wave studies (DZIEWONSKI and STEIM, 1982; NAKANISHI and KANAMORI, 1982). THIO and KANAMORI (1995) applied successfully a regionalized phase velocity model for moment tensor inversion in southern California using short-period (10–60 s) surface-wave data. One drawback to this method is that, unlike the method used by WOODS and HARKRIDER (1995), it does not account for the amplitude changes that can occur when surface waves propagate from one medium to an adjacent one with varying crustal waveguide properties. However, it provides a means to adequately model surface waves in regions of strong lateral variations. Figure 5 shows a suite of synthetic seismograms generated using the approach of THIO *et al.* (1999) in which Rayleigh waves have been generated for a suite of mixed paths, starting with a purely Spanish (continental) path and sequencing through a range of mixed paths with increasing contributions from the Mediterranean (oceanic) path, to a purely Mediterranean path. The panel to the right displays the models. The change in the composition of the mixed-path model clearly has a strong influence on the arrival times as well as on the waveforms. An application of this approach to surface wave modeling is shown in Figure 6, which compares observed and synthetic waveforms at four stations, MEB, VSL, TAM and

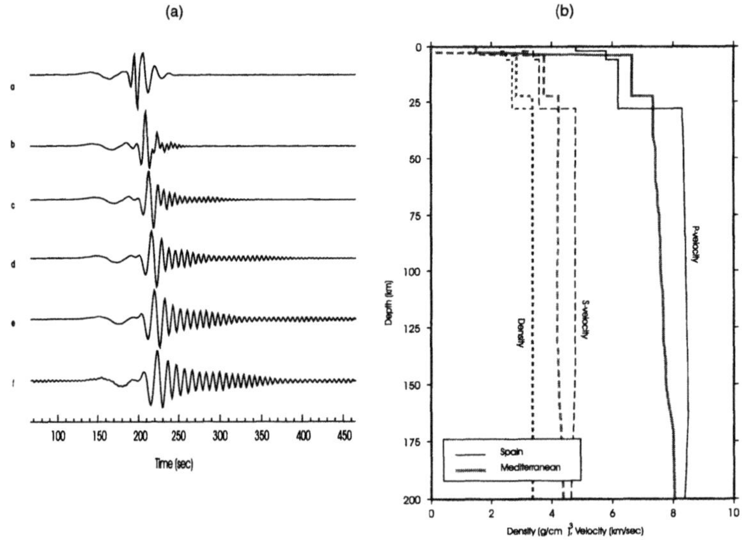

Figure 5
(a) Vertical-component Rayleigh waves simulated using the mixed-path approach for differing ratios of purely continental (Spanish model) and purely oceanic (Mediterranean model) paths: i) 100% Spain path, ii) 80% Spanish/20% Mediterranean, iii) 60%/40%, iv) 40%/60%, v) 20%/80%, and vi) 100% Mediterranean path. (b) Regionalized velocity models used to generate the hybrid surface-wave synthetics shown in Figure (a).

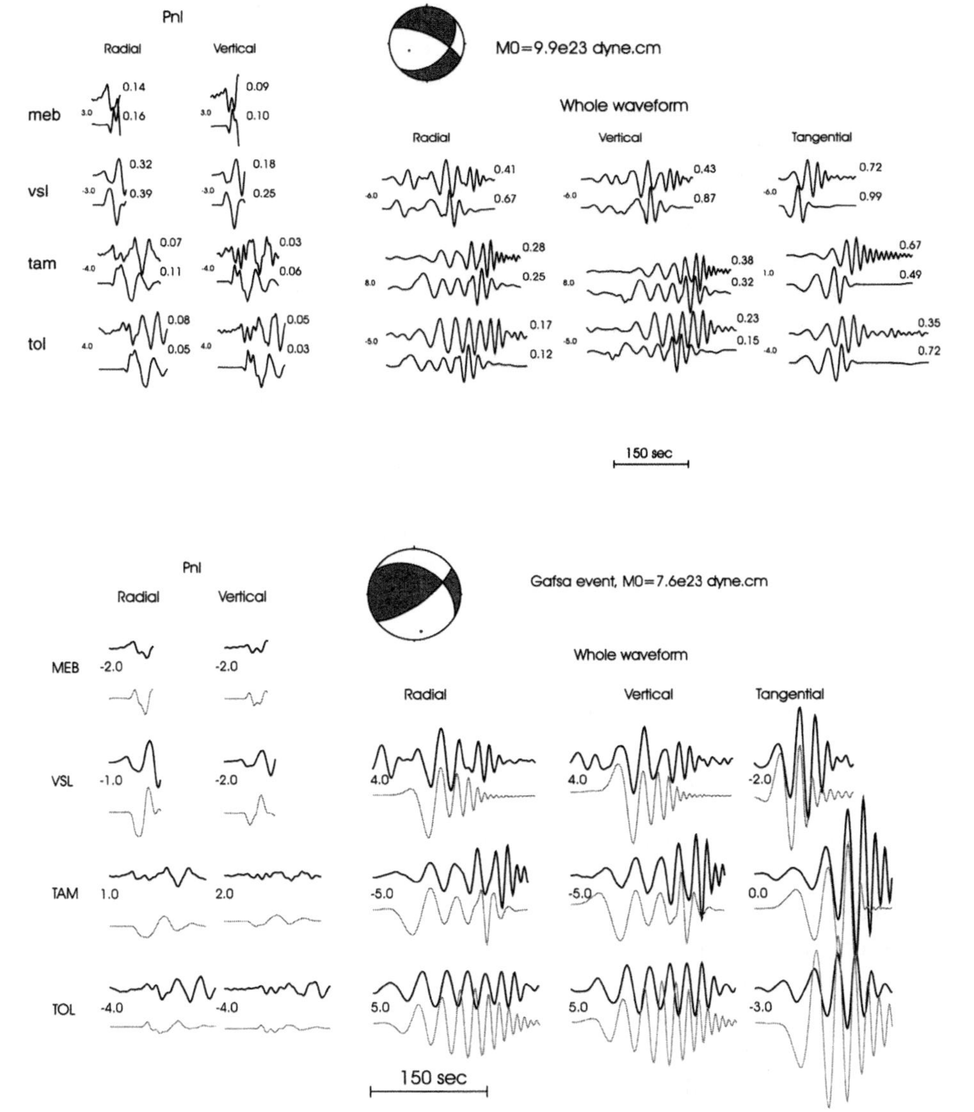

Figure 6

Source inversion results for the Gafsa, Tunisia earthquake (6/12/92). (a) Results using a single 1-D model. The top trace in each pair is the observed waveform; the lower trace is the synthetic. The time shift between the two seismograms is printed to the left of each pair, with a positive number indicating a fast synthetic; peak amplitudes (cm) given to the right of each trace. (b) Source parameter solution for the same event using mixed-path (continental and oceanic crust portions) synthetics for the surface waves; P_{nl}-wave synthetics were computed using a single 37 km thick crust.

TOL, for an earthquake which occurred on June 12, 1992 (19 h 16 m 45.7 s, $M_w = 5.2$) near Gafsa, Tunisia, in North Africa. The top panel displays inversion results using a single 1-D model for all stations, with peak amplitudes given to the right of each trace and the observed waveform above the synthetic in each seismogram pair. These results exhibit inconsistencies in waveform fitting. The mechanism solution is also inconsistent with the Harvard CMT as well as with that of the moment tensor solution obtained from mixed-path surface-wave inversion, results of which are presented in the panel below. It is clear that one 1-D model cannot even fit the surface wave data even with a longer period, whereas the mixed-path synthetics yield satisfactory results. The lower panel displays the results of the mixed-path synthetics for the surface waves (the P_{nl} wavetrain was modeled with a single 1-D model). The waveform fits are significantly better and the source mechanism is consistent with the Harvard CMT solution.

Once the crustal waveguide model is established for a region or province, it can be used to locate events and to estimate their source parameters from waveform inversion. This event analysis of regional seismograms can be routinely applied to smaller events ($4 \leq M_w \leq 5$) for which teleseismic observations are too small to be analyzed.

Source Inversion Methodology Using Regional Seismograms

In this section we present a brief summary of a methodology for inverting regional seismograms. We discuss the effectiveness of the method by comparing results obtained by inverting regional seismograms from two stations versus regional seismograms from two full networks consisting of multiple stations with a better azimuthal coverage. The mathematical details have appeared in several other publications (ZHAO and HELMBERGER, 1994; ZHU and HELMBERGER, 1996) therefore we shall only discuss the salient features of this source inversion method.

There are two issues which make it difficult to use the entire seismogram for source estimation. First, the whole-seismogram inversion is mainly controlled by the surface waves because of their large amplitude relative to that of P waves, except in the case of deep events. However the P waves, especially the long-period P_{nl} waves, are very useful for source estimation as they are stable and insensitive to variations in the velocity structure (HELMBERGER and ENGEN, 1980; SAIKIA and BURDICK, 1991). Further, the amplitude ratio between the P_{nl} and surface waves is a good indicator of source depth, consequently both wavetrains should be properly weighted in the inversion. Secondly, as these two parts of the wavefield propagate through different portions of the regional waveguide, the crustal velocity model must be able to accommodate these effects.

Waveforms at regional distances are generally complex due to propagation effects and the source rupture processes, especially in the case of larger events. Because of the path effects alone it is often difficult to fit the relative timings between the P_{nl}, S_{nl} and surface waves, especially in regions known to contain laterally varying crustal structure; fitting the relative timing between the Rayleigh and Love waves can also be quite difficult. Therefore, to invert regional waveform data, ZHAO and HELMBERGER (1994) developed an approach in which the P_{nl} and surface waves are split into two separate wave groups in order to mitigate timing differences between the two wavetrains in the synthetic and observed seismograms. This approach is popularly known as the cut-and-paste (CAP) method, which allows the fitting of the P_{nl} and surface-wave waveforms together while disregarding their relative timings. A similar approach was also applied by SAIKIA and HERRMANN (1985) for inverting local seismograms in which they separated the P seismograms from the S seismograms to obtain focal mechanisms and event depths. The CAP approach makes the source inversion relatively insensitive to the velocity model and lateral crustal variations as well as proving quite useful when the initial event locations are poorly constrained. However, as the waveform misfit errors are normalized by the data and synthetics, the amplitude information of the data is not fully utilized to constrain the source orientation and event depth.

To fully utilize the amplitude information, ZHU and HELMBERGER (1996) used a distance scaling factor to normalize amplitudes in the inversion, to avoid bias in the estimated source parameters introduced by the nearer observations. Both P_{nl} and surface waves are normalized to a reference distance r_0 allowing the P_{nl} waves to decay as $1/r$ and surface waves as $1/r^{1/2}$. A general expression for the misfit error e between a segment of observed $u(t)$ and synthetic $s(t)$ waveform is given by

$$e = (r/r_0)^p ||u(t) - s(t)|| \ ,$$

where the scaling factor p is 1 for the P_{nl} waves and 0.5 for the surface waves. These scaling factors may be adjusted corresponding to the actual amplitude decay rate of these waves after a certain amount of data has been accumulated for a particular area. For the whole waveform the total misfit error is a weighted sum of the two errors estimated for the P_{nl} and surface waves using the above expression. Surface waves, in general, show larger waveform misfits compared to those of body waves as the surface waves are more affected by lateral heterogeneity in the crustal waveguide, whereas P_{nl} waves are relatively insensitive to structural variations. However since surface waves from an earthquake are generally larger in amplitude, by a factor of 2–4, the inversion can be weighted more heavily for the P_{nl} waves to improve the source estimate. Because of this scaling, the relative P_{nl} and surface wave amplitude information of stations at extended distances is also preserved.

As in the teleseismic inversion, the regional source inversion applies a grid search through the focal mechanism parameter space for a suite of depths, and the synthetic and observed seismograms are cross-correlated for the best alignment (WALLACE *et al.*, 1981; DREGER and HELMBERGER, 1993). For many crustal earthquakes which are recorded by only one or two regional stations and are too small to be recorded at teleseismic distances with good signal to noise ratios, their depths are poorly constrained by current location procedures. Therefore, for these types of events it is necessary to have reliability checks on these locations that are obtained using only a few observations. Previous studies in the Hindu-Kush region of the Pamir Himalaya have shown that it is possible to obtain reliable depth estimates for such earthquakes using only a few regional observations (SAIKIA *et al.*, 1996a, b; ZHU *et al.*, 1997). Figure 7 provides examples of source depth vs. error curves for earthquakes in this region, which illustrates that depths determined from regional waveform inversion of two stations (dashed lines) are similar to those using many stations (solid lines). The left panel presents results for five shallow earthquakes, while the right panel is for five deep earthquakes. The observations used in the source inversion were taken from PAKN (a PASSCAL network consisting of 11 broadband stations which operated in Pakistan near Nilore, NIL towards its northeast in 1992) and KNET (a permanent array in Kyrgyzstan) networks. The two-station inversion used one station from each of the networks. For the shallow crustal earthquakes, solutions are stable over a broad range of depths and have well resolved waveform misfit errors. The two inversions (two station vs. two networks) yielded similar source depths for all events. These focal depths are notably shallower than their depths given in the ISC bulletin. Figure 8 compares observed and synthetic waveforms for two earthquakes, one shallow and one deep, from 10 earthquakes studied in Figure 7. The agreement in these waveforms suggests that the regional depth solutions are reliable.

To establish the reliability of the regionally determined depth estimates, teleseismic *P*-wave seismograms from these events, taken from the short-period array, were examined for their *P*, *pP* and *sP* phases. The teleseismic short-period array seismograms clearly exhibit depth phases, especially when the seismograms are stacked. Figure 9 presents examples of short-period array data recorded at Alice Springs (ASP) and Waurramunga (WRA) in Australia, Chiang Mai (CHG) in Thailand, and Gauribidanur (GBA) in India for a Hindu-Kush earthquake of (date: 11/19/92, origin time: 02 h 38 m 50.1 s) along with a suite of synthetic seismograms for various depths; the best-matching synthetics are for a 10 km source depth, which is consistent with that determined from regional waveform modeling. Similar agreement was found for the other events so analyzed. The fact that the regional source inversion yields a depth compatible to the observations for four teleseismic arrays, establishes the strength of this source inversion method. As even with stacked array data, the teleseismic observation magnitude (M_w) threshold is about 4.2, thus the regional modeling may be quite useful for smaller events.

Depth Resolution Based on Regional and Teleseismic Depth Phases

The depth of an earthquake is a critical source parameter for the path calibration purposes as it helps to constrain the epicenter and obtain reliable origin times.

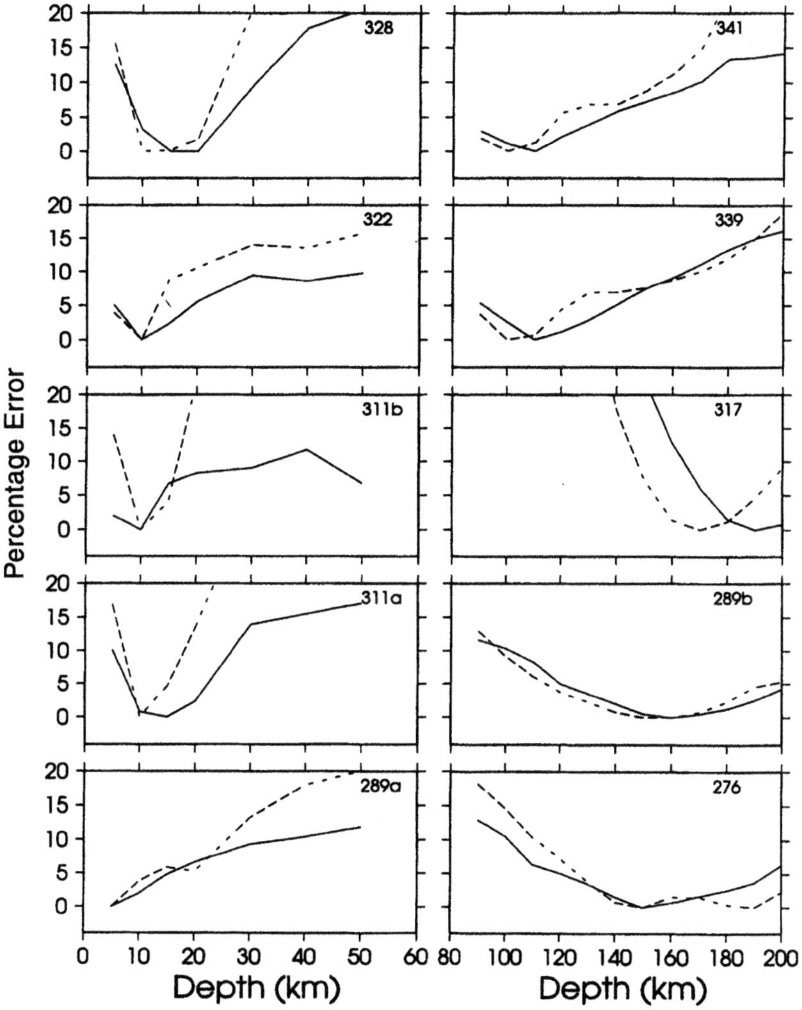

Figure 7
Waveform misfit error as a function of focal depth for 10 Hindu-Kush earthquakes. Solid lines correspond to the misfit error when the solutions were obtained using all seismograms from both PAKN and KNET networks and dashed lines correspond to the solutions when only one station from each network was used (taken from SAIKIA *et al.*, 1996; ZHU *et al.*, 1997). Note that the two minima for each event occur at the same depth.

Criteria adopted for the selection of the ground-truth catalog must be based on the reduced level of uncertainty as discussed by ENGDAHL (1998). However, earthquakes occur in regions where many of those criteria are not met, especially for depth and origin times. If depths are not known and regional crustal models are not adequate, events can be mislocated significantly away from their true locations. In general, the

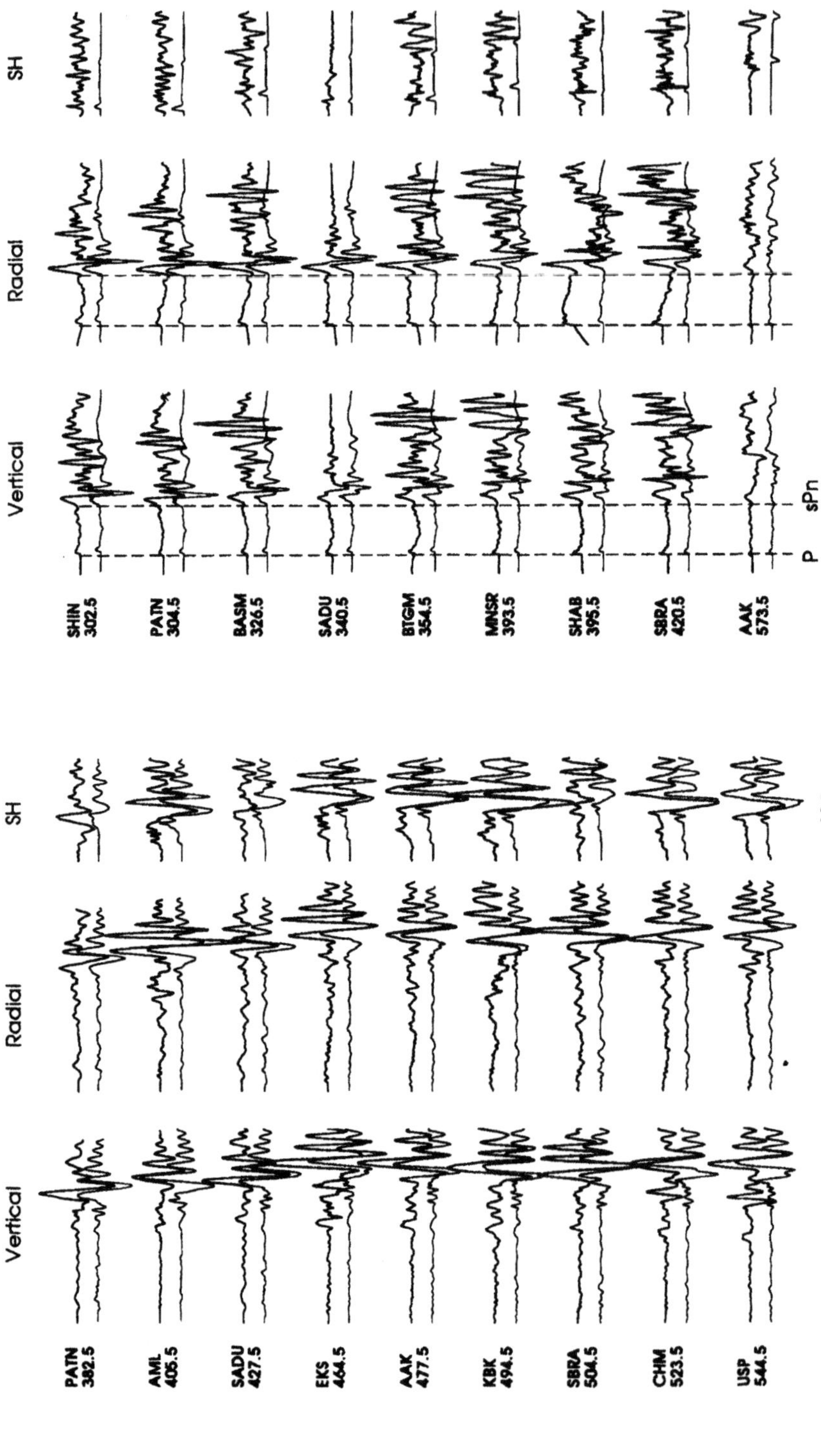

Vertical
Radial
SH
SHIN 302.5
PATN 304.5
BASM 326.5
SADU 340.5
BTGM 354.5
MNSR 393.5
SHAB 395.5
SBRA 420.5
AAK 573.5
P
sPn
sec
0
30
60
PATN 382.5
AML 405.5
SADU 427.5
EKS 464.5
AAK 477.5
KBK 494.5
SBRA 504.5
CHM 523.5
USP 544.5

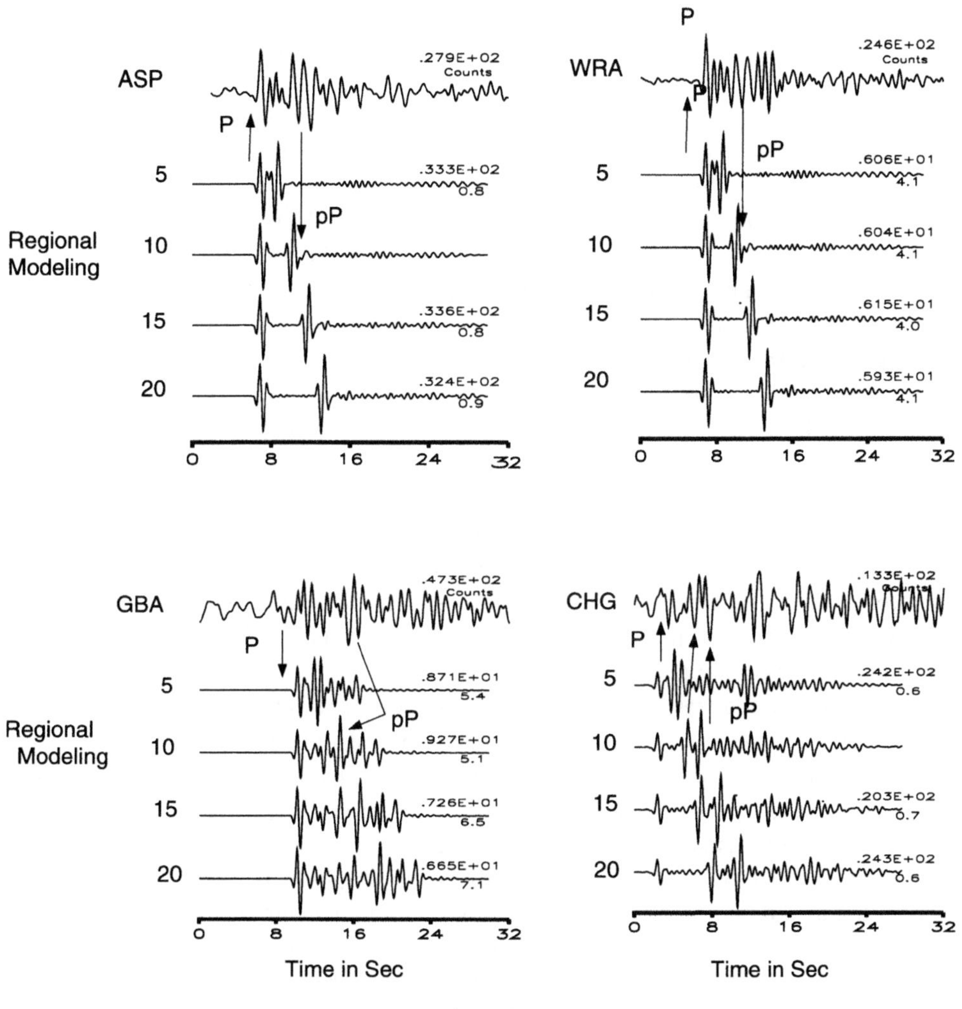

Figure 9

Modeling of the P-wave depth phases for a Hindu-Kush earthquake (Nov. 17, 1992; origin time-02 h 38 m 50.1 s; $M_0 = 1.41 \times 10^{23}$ dyne-cm) observed at four short-period teleseismic arrays: Alice Springs (ASP) and Warramunga (WRA) in Australia, Gauribidanur (GBA) in India, and Chiang Mai (CHG) in Thailand. In each panel the top seismogram is the observation; below this is a suite of synthetic seismograms at depths ranging from 5 to 20 km using its regional focal mechanism solutions. Synthetics computed for a depth of 10 km clearly produce the best fit to the recorded *pP* phase relative to the *P* waves; this was the depth determined from regionally modeling as well.

◄

Figure 8

Regional waveform fits for two Hindu-Kush events (taken from SAIKIA *et al.*, 1997; ZHU *et al.*, 1997), one shallow (left: $\delta = 45°$, $\lambda = -35°$, $\Phi = 200°$, and $h = 15$ km) and one deep (right: $\delta = 35°$, $\lambda = -55°$, $\Phi = 290°$, and $h = 110$ km), shown at several stations from the PAKN and KNET networks. For each station the synthetic seismogram is plotted immediately below the observed seismogram; displacement is in "cm". Synthetic seismograms were computed using the frequency-wavenumber integration algorithm (SAIKIA, 1994).

depths of earthquakes reported in various catalogs are biased (ZHAO and HELMBERGER, 1991a; ZHU *et al.*, 1997; SAIKIA *et al.*, 1996a, b; THIO *et al.*, 1999) because of the errors in travel-time picks of the *P* and *S* phases, inappropriate crustal models used for locating events, and misidentification of the depth phases (*sP* or *pP*).

Figure 10 depicts relocation of sixty moderate-sized Tibetan earthquakes (ZHAO and HELMBERGER, 1991a); the top panel shows a map view of the earthquakes investigated, the middle panel shows the cross-sectional view of the depths of the relocated earthquakes and the bottom panel shows the ISC depths for the same earthquakes. The new depths were estimated by modeling waveforms of teleseismic depth phases. It is interesting to note that all events have moved to shallower depths relative to the ISC depths except for one event. Errors in location result primarily from errors in the estimation of depths, especially for shallow earthquakes and also from the lack of regional models. ZHU and HELMBERGER (1996) observed similar variation changes in source depths in southern California wherein regional waveform inversion yielded deeper focal depths than the local network's. It has been argued that the *P*-wave model used by the local network is too fast and thus forces the earthquakes to shallower depths (DI LUCCIO *et al.*, 1998). Thus regional broadband waveform modeling and source inversion can provide more accurate hypocentral locations than those from extensive local networks, which hitherto were thought to provide the most reliable results.

We present similar results, in Figure 11, for twenty seven Hindu-Kush earthquakes whose depths were estimated by minimizing the misfit between the observed and synthetic regional waveform using regional seismograms from the PAKN and KNET networks, and the station NIL in Pakistan. The regional depths are marked by blue circles which were also confirmed by a separate waveform modeling of the teleseismic depth phases (SAIKIA *et al.*, 1996a). The corresponding ISC Harvard CMT depths are designated by red circles. The top panel shows the relative depths for the events which have ISC depths larger than 90 km. These events were relocated with respect to their epicentral location by first constraining their depths to the regional-waveform inverted values, and then applying an adaptive grid-location scheme to search for the best epicentral location over a 1° by 1° area, centered about the ISC location, by minimizing the travel-time residuals for each event (ZHU *et al.*, 1997; SAIKIA *et al.*, 1996a). We relocate these events by minimizing regional travel-time residuals of the *P* and *S* waves for the fixed source depth using the object function defined by

$$\mathbf{\Phi}(x,y) = \sqrt{\frac{1}{N}\sum_{i=1}^{N}(\delta T_i - \overline{\delta T})^2 + \lambda \Delta r} \ ,$$

where x and y are the horizontal coordinates and δT_i is the travel-time residual of a first P or S arrival at station i. The average residual over all stations, δT_{ave}, is subtracted to remove the uncertainty of origin time; Δr is the offset between the new and the original location and the weighting constant λ is set to 0.01 s km^{-1}. This

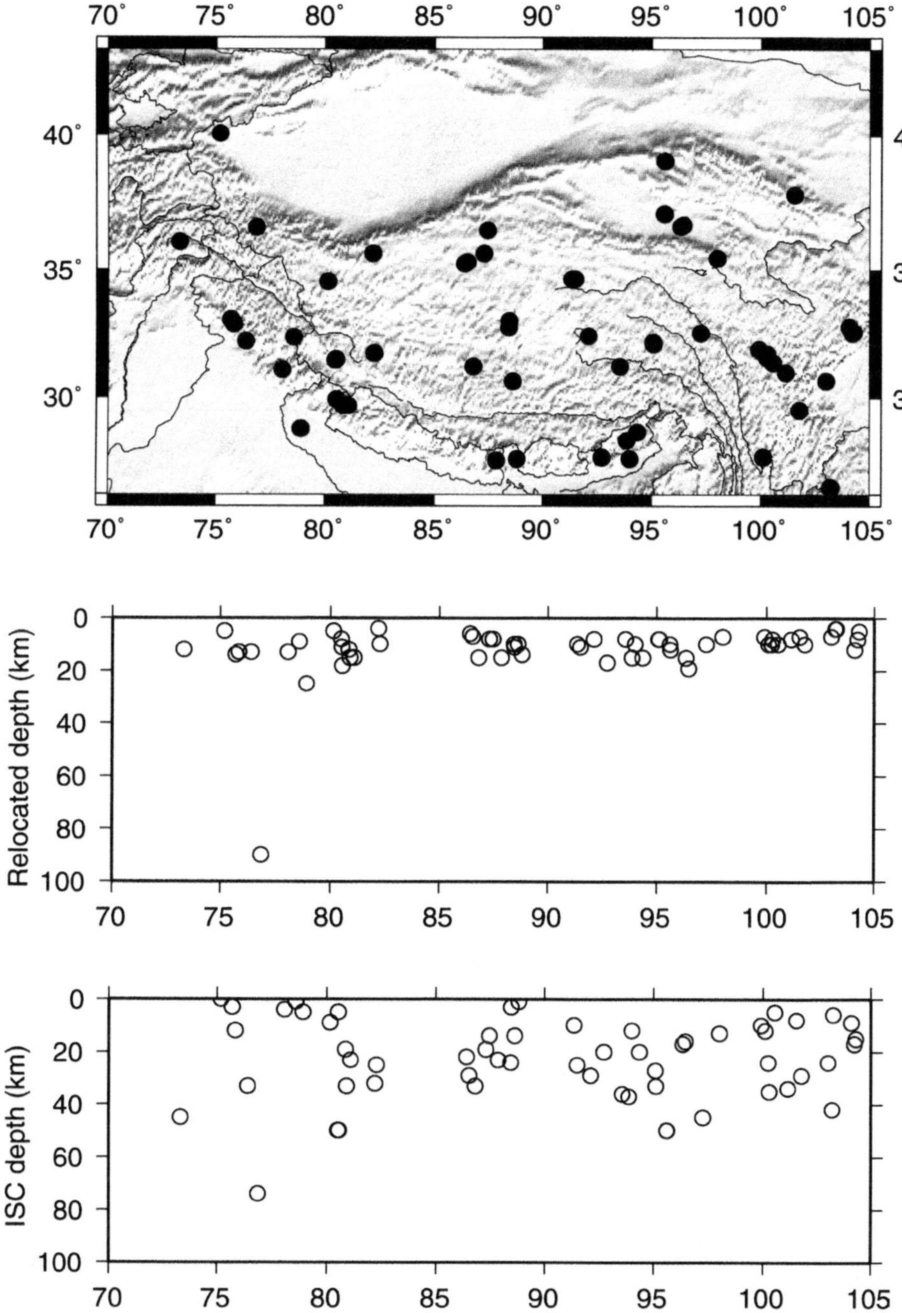

Figure 10
Relocation of 60 earthquakes in Tibet (taken from ZHAO and HELMBERGER, 1991). Top panel shows the epicentral locations of these events. The bottom and middle panels show the ISC and the corresponding relocated depths of these events, respectively. Note that, except for one event, all events moved to shallower depths.

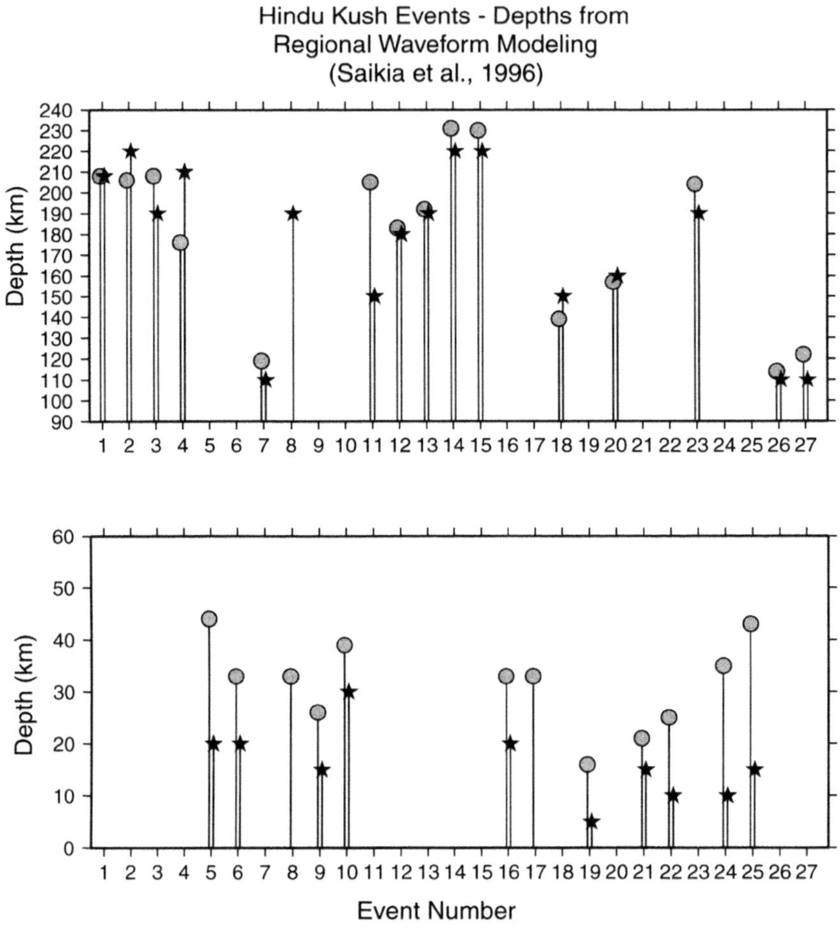

Figure 11
Depths determined by modeling regional waveforms (black stars) recorded at NIL (Nilore) and SBRA (surrogate station for NIL) in Pakistan and AAK in Kyrgyzstan are compared to depths reported in the ISC bulletin (grey circles). All crustal events (lower panel) have moved to shallower depths. These events were relocated using the adaptive grid-search technique (ZHU *et al.*, 1997; SAIKIA *et al.*, 1997) by constraining their depths to the regional solutions; for events which are small ($M_w \leq 4.2$) the new epicenters shifted by more than 30 km relative to the ISC locations.

algorithm was also extended to many smaller ($M_w < 4$) events using regional seismograms recorded by the PAKN and KNET network and applying the same regional crustal model (SAIKIA *et al.*, 1997). We found, in general, that the locations of smaller events moved the most relative to the ISC solutions – up to 30 km in some cases. Table 9 in SAIKIA *et al.* (1996a) summarizes the locations of several Hindu-Kush events determined using *P*- and *S*-wave travel times and waveform modeling of the PAKN and KNET network seismograms with those of the ISC and AFTAC (Air

Force Technical Application Center) locations. The table gives the locations determined using distant *P*-wave travel-times (teleseismic), locations using *P*- and *S*-wave travel times from the PAKN and KNET network stations for both free and fixed depths (local and local fixed-depth locations, respectively). For fixed-depth locations, depths were constrained to the depths estimated by the waveform modeling. For these events, locations determined by applying the adaptive-grid search technique by minimizing the travel-time residuals using travel-time data only from two stations, and constraining the depths to the waveform-modeling depths, were found to be similar to those locations determined using the local *P* and *S* travel times for fixed depths using travel-time data from all stations of the PAKN and KNET networks. In this paper we summarize the results of Table 9 of SAIKIA *et al.* (1996a) in Figure 12 (for details refer to the figure caption).

Mislocation of source depth has also been encountered with events occurring in central Asia (WOODS *et al.*, 1998), and Egypt (THIO *et al.*, 1999). Figure 13a compares depth estimates for 26 central Asian earthquakes, determined by WOODS *et al.* (1998) using regional broadband waveform inversion of P_{nl} and surface waves (purple circles) and teleseismic *P*-wave modeling (light blue), with those of PATTON (1998), who used regional surface wave spectral inversion (green, Patton), and with Harvard CMT (dark blue) and PDE (red) catalog values. Although the depth estimates are quite consistent between all methods for 8 of the events (#'s 6, 8, 9, 11, 18, 19, 23 and 25), there is considerable variation in the other events, with the Harvard CMT or PDE depth generally being the greatest. However, the depths determined by the inversion of regional P_{nl} and surface waves (yellow), and by modeling of the teleseismic depth phases (red) are quite consistent, with relative differences of less than 10 km, suggesting validation of the two sets of results. Some of the spectral depths (green, Patton) are particularly shallow relative to the other results.

Source depth not only can trade off with origin time but can also effect the epicentral locations. An example of this is shown in Figure 13b, which indicates the results of relocating 17 of the above 26 central Asian earthquakes by Aaron Velasco (personal communication, Los Alamos National Laboratory, 1998) using regional and teleseismic travel-time data and a velocity model of MOONEY and LI (1997) for the region. The upper panel shows the fixed (constrained to the average value of WOODS *et al.*, 1998) and free depths for the relocations and the bottom panel plots the shift in epicentral distance, ΔR, relative to the EDR locations for the fixed-depth and unconstrained hypocentral locations. Some events, for example events 7 and 16, moved significantly away from the EDR locations for the unconstrained hypocenter relocations. Assuming the EDR locations are accurate, then the constrained-depth relocations are more accurate. In either case, the location of an earthquake will trade-off with how well its depth is determined or constrained. In general, depth is the least controlled parameter and establishing its ground-truth value is essential for accurate epicentral locations.

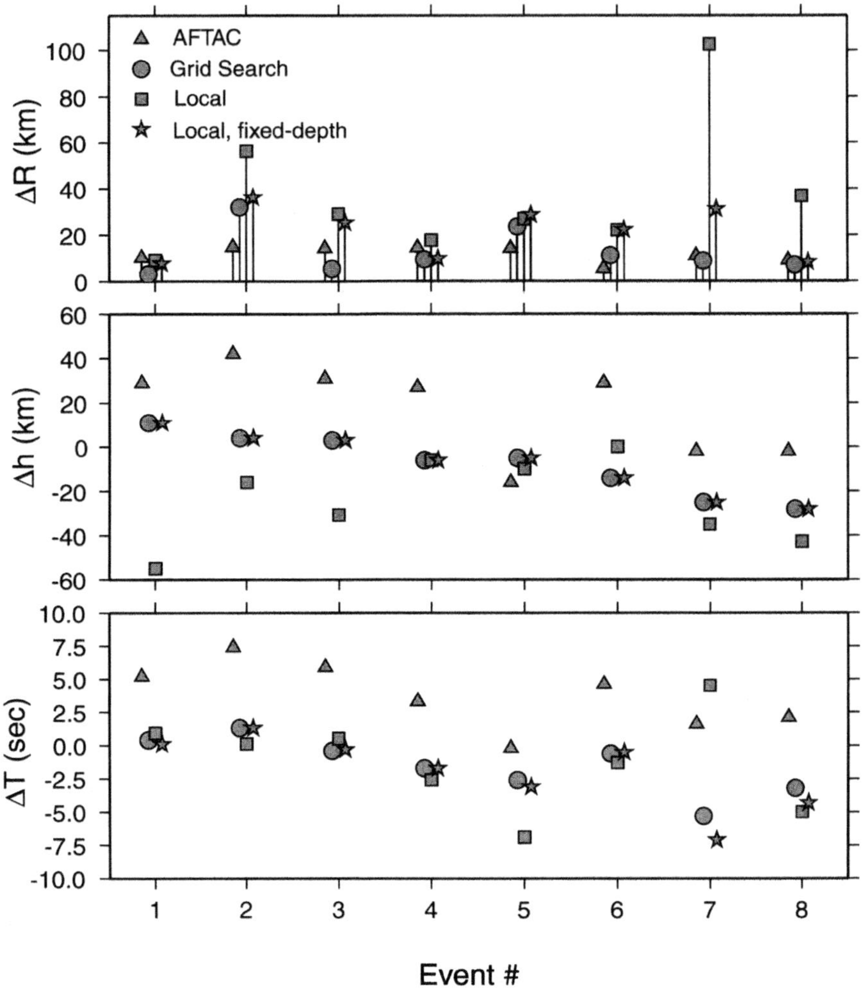

Figure 12
Comparison of various location results for 8 Hindu-Kush earthquakes; all values are relative to the ISC values. Local (squares) and local fixed-depth (stars) locations were obtained by using *P*- and *S*-waves travel-time data from the PAKN and KNET networks. For local fixed-depth locations depths were constrained to the regional waveform modeling depths (SAIKIA *et al.*, 1996). Grid search locations (circles) were obtained using travel-times from the PAKN and KNET networks for a minimum required number of stations applying the adaptive grid-search technique, and triangles correspond to the locations provided by AFTAC. *Top*: relative shift (Δr) in epicentral locations; *Middle*: relative shift in depth (Δh), with negative values indicating a shallower depth than the ISC value; *Bottom*: difference in origin time (ΔT), with a negative value indicating an origin time earlier than the ISC value. Note the similarity between the results for the local fixed-depth and the grid-search locations.

For crustal earthquakes, the ratio of the P_{nl} wave amplitude relative to the surface wave amplitude is diagnostic of source depth, as shallow earthquakes generate larger amplitude surface waves whereas the primary body-wave phases tend to be sharper

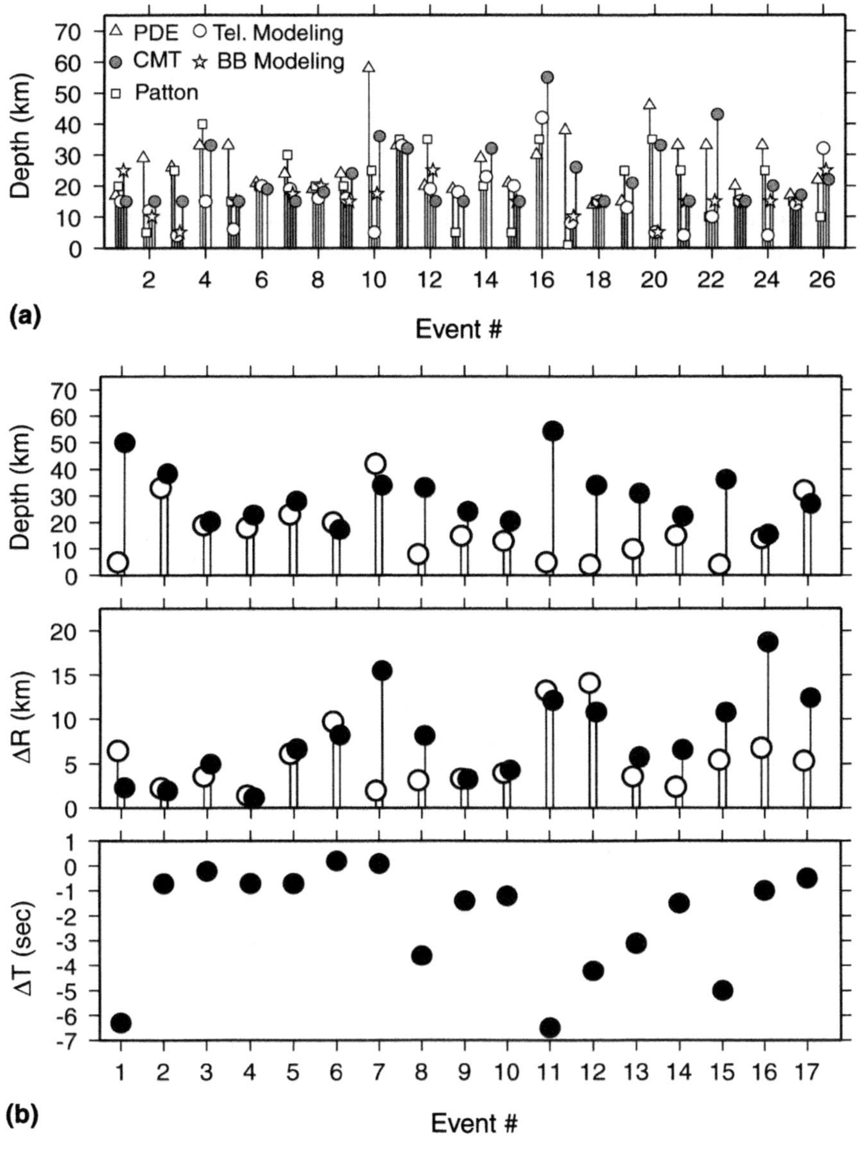

Figure 13
(a) Comparison of source depths, determined by five different methods, for 26 central Asian earthquakes: PDE values determined from teleseismic *P*-wave first arrivals (triangles), Harvard CMT solutions (filled circles), PATTON's (1998) solutions using regional surface wave spectra (squares), and teleseismic *P*-wave waveform modeling (open circles) and BB regional waveform modeling (stars) results from WOODS *et al.* (1998). Generally the PDE depths, which are similar to ISC values, yield the deepest focal depths. (b) Of these events, seventeen were relocated using regional and teleseismic travel times and a regional crustal model. This was done for two cases, one for unconstrained depths (filled circles) and one whereby the depth was fixed (open circles), based on regional and teleseismic modeling. The top panel is a plot of the two sets of depths; the middle panel shows the relative shifts of epicenters from the EDR epicenters after relocations using fixed and free depths; and the bottom panel gives the relative origin times ($T_{\mathrm{fixed}}-T_{\mathrm{free}}$).

and larger in amplitude as source depth increases. For crustal earthquakes, interaction of the *sPmP* phase, which leaves the source as a *S* wave and is converted to a *P* wave at the free surface before being reflected at the Moho boundary with the *PmP* and *Pn* phases, is also a good indicator of depth (SAIKIA and HELMBERGER, 1993). In addition, as the earthquakes move into the mantle, the high-frequency waves trapped by the crustal waveguide begin to disappear and the primary body wave phases become dominant, thus making the seismograms increasingly simple for deeper – particularly mantle – earthquakes (SAIKIA *et al.*, 1996b). Generally, for deep events a strong long-period phase (*sPn*) appears on regional seismograms as shown by the dashed line in Figure 14, which is diagnostic of deep mantle earthquakes. The seismograms shown in this figure were recorded by the PAKN and KNET networks. The phase *sPn*, shown in Figure 14, follows an *S*-wave path to the surface and becomes critical developing a head wave by *S*-to-*P* mode change. It begins slowly at large on the PAKN network and has a sharper onset on the KNET at similar distances see top panels). The detailed characteristic of this phase is controlled by the crustal waveguide. This phase is less impulsive at PAKN because there exists a sharp crust-to-mantle transition with uniform velocities. Beneath KNET there exists a more gradual crust-to-mantle transition zone along with a mantle gradient which causes *sPn* to be sharp. This long-period feature is quite stable and, given its broadband nature jointly with that of *sPmP*, is more important than the direct *S* waves for source depth estimation. Recently a number of studies have investigated this phase in order to constrain event depth (SINGH *et al.*, 1995; LANGSTON, 1996; ZANDT *et al.*, 1996). However, it can also suffer from *S*-to-*P* interaction at the bounce point causing potential ambiguity in its use for source depth analysis. Since this *sPn* phase is a long-period phenomenon, its diagnostic strength in source depth estimation is, perhaps, limited to only larger ($M_w \geq 5$) events.

The interaction of *sPmP* with *PmP* and *pPmP* phases can also be useful in constraining depths for crustal earthquakes. Figure 15 illustrates an example of this interaction observed in regional data, which is also well simulated synthetically both in the broadband data (left) and in the long-period (Press-Ewing 30–90) simulation (right). This earthquake, which occurred on March 25, 1990 (origin time: 17 h 28 m 26.8 s), was recorded at the station GARM, in the former Soviet Union. The small arrows highlight the various phases which are well constrained and correlate well with the data. The first arrow in the broadband seismograms indicates the initial *Pn* phase and the second arrow in the radial broadband seismograms (left panel) shows the arrival of *PmP*. The third arrow shows a strong depth phase which is the *sPmP*. Figure 16 illustrates the critical aspects of the *sPmP* interaction with the *PmP* phase as a function of source depth. The synthetic vertical and radial components are computed for the *PmP*, *PmP* + *pPmP* and *PmP* + *pPmP* + *sPmP* phases at a distance of 345 km and at an azimuth of 312° (panel a). They are generated using the Garm region crustal model; a source mechanism of strike = 60°, dip = 50°, and rake = −165°, and a source depth of 15 km. These same seismograms are convolved

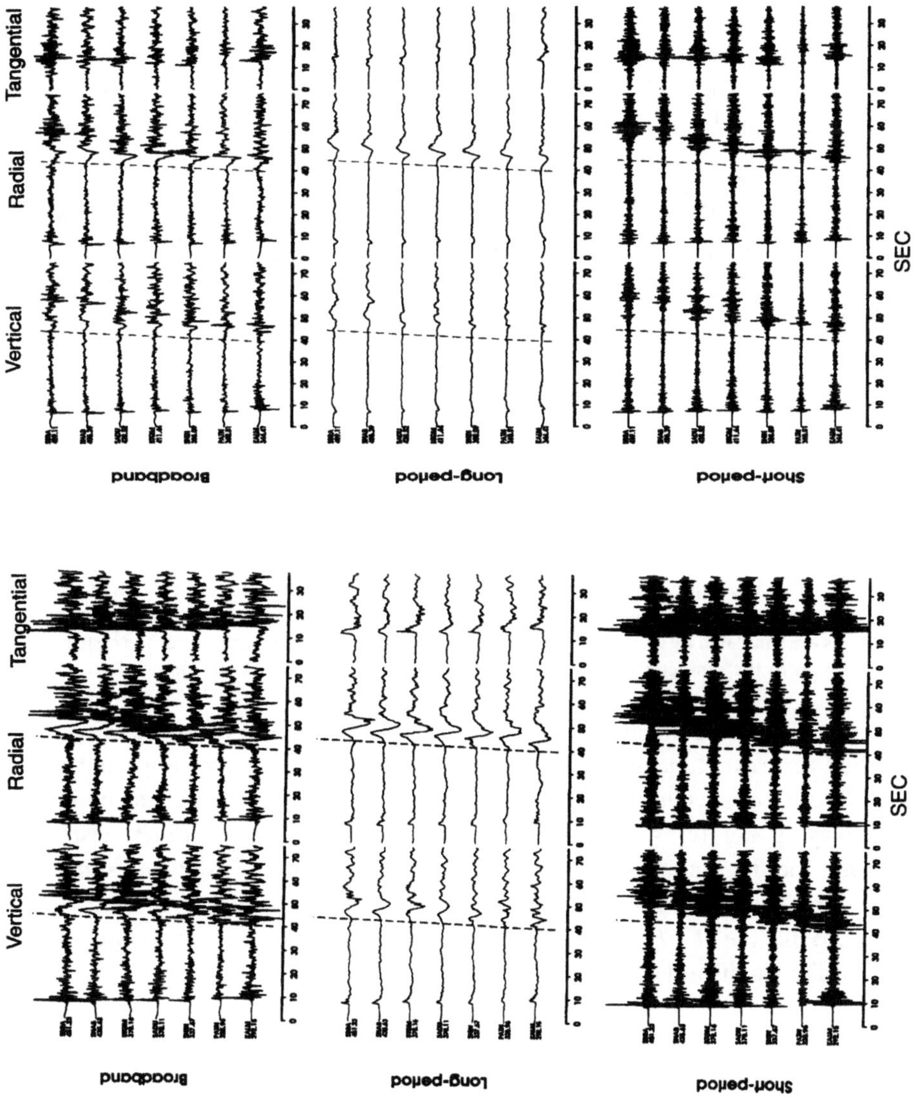

Figure 14

Seismograms from the PAKN (left) and KNET (right) networks for a Hindu-Kush event of Oct. 15, 1992 (OT: 19 h 42 m 13.8 s) which has a depth of 157 km (ISC). The top, middle and bottom panels are the broadband, long-period and short-period seismograms for a window of 30 seconds, respectively. The dashed line is the *sPn* phase, a phase that follows *S* waves to the free-surface, becomes critical and changes to *P* mode. This is a long-period feature which does not appear in short-period data, and is a diagnostic of deep events.

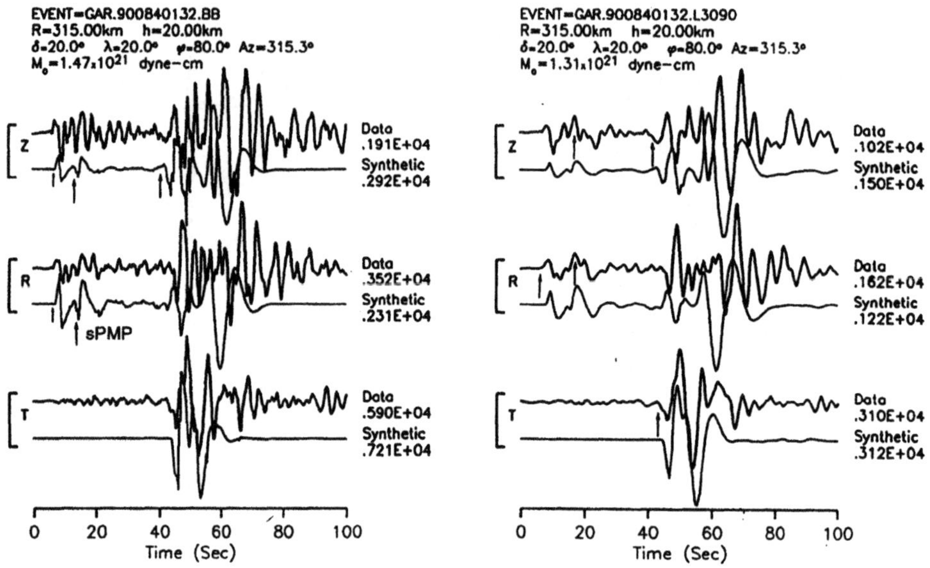

Figure 15
Modeling of regional seismograms from a Hindu-Kush event (900841728) recorded at GARM station. The figure shows the interaction of the *sPmP* and *pPmP* depth phases with *PmP*. The Press-Ewing LP3090 seismograms are simulated by convolving the recorded waveforms with the response of the instrument. For each component the synthetic seismogram is plotted immediately beneath the observed seismogram.

with a PE 30–90 instrument in panel (b). The contribution of *pPmP* relative to *PmP* (including the travel time difference from *PmP*) is small and has no apparent effect on the *PmP* seismograms, however the *sPmP* phase contributes strongly as indicated by the arrows. Panel (c) shows a profile of *PmP* + *pPmP* + *sPmP* seismograms displaying the interaction of the various phases with depth, using the same source mechanism. The uppermost seismogram is for an explosion buried at a depth of 0.64 km. While the *sPmP* waveform changes less drastically; its travel time relative to *PmP* is quite significant and is diagnostic of source depth.

Conclusions

There are many important aspects to the problem of seismic event location, especially for smaller magnitude ($M_w < 5$) events which are well recorded – that is, with good signal to noise ratio – only at nearby regional distances, for which developing an accurate regional waveguide model is very critical. With established regionalized earth models, travel-time corrections relative to a reference model can be estimated, based on accurately known event locations, and then consequently be used in the locations of other events. It is important that travel-time tables (or curves) are regionalized in order to determine accurate depths and locations which help reduce

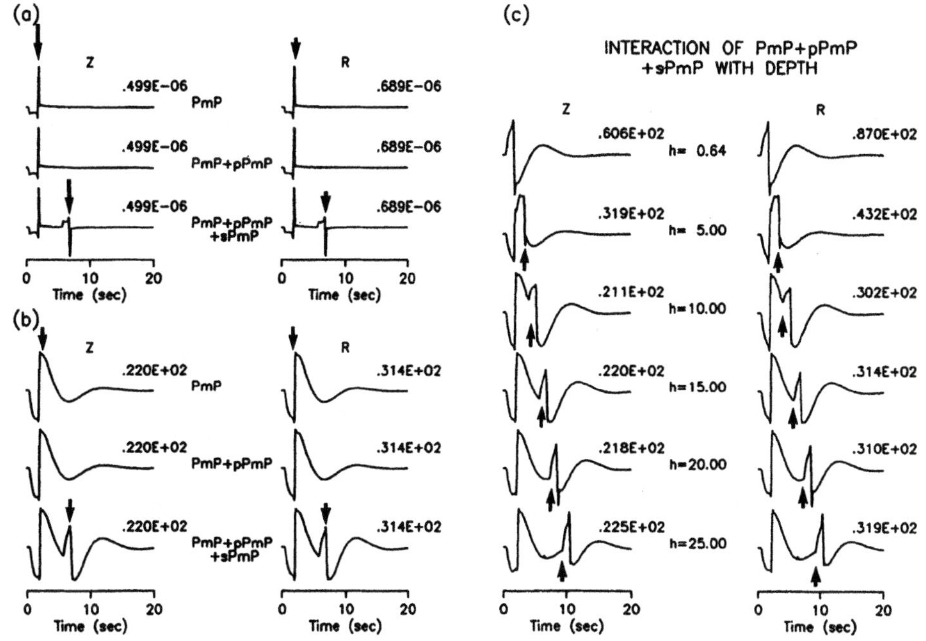

Figure 16

Ray interaction at a regional distance: (a) broadband seismograms: arrows show arrivals of the *PmP* and *sPmP* phases in the seismograms of the first and third row, respectively. Each *PmP* arrival is preceded by a long period P_n, (b) broadband seismograms after being convolved with a Streckeisen instrument, (c) ray interaction at a regional distance, especially of *sPmP* (shown by arrows on each seismogram) as a function of source depth.

the model error which is a source of location error in the teleseismic event location. In general, the most meaningful data for the regionalization are the data from large explosions with known locations and origin times. The next best data are observations from quarry blasts as their locations are also well constrained, but may have less reliable origin times. These data types provide well constrained travel-time information for the shallow focus events. However, in many regions such high quality data are not available. This paper provides a methodology which can be used to establish a set of events, generally with magnitude $M_w \geq 5$, with reliable hypocentral locations and origin times. These events have fairly well determined epicenters teleseismically and often have at least several observations at regional distances. Using the method outlined in this paper, their depths can be constrained to be consistent with both regional and teleseismic seismograms, thereby path calibrations can be used to provide the model-based travel-time estimates for various seismic phases.

In modeling the crustal waveguide of the western Mediterranean region (THIO *et al.*, 1999), we found that the path effects can be quite significant because of the mixed path contributions from the oceanic and the continental crusts. To locate small events which are expected to be recorded only by a few stations in the region, it is essential that events which are used in the regionalization of the region have accurate source parameters, especially the depths. These events require solutions which are consistent with both regional and teleseismic data. The method adopted by THIO *et al.* (1999), which uses a mixed-path regionalization scheme in conjunction with regional data, can be extended to include the teleseismic data to constrain event depths, and should be useful in regionalization of other similarly complex environments.

The key to the success of obtaining high-quality locations for the master events is in accurately establishing their depths. In general, the locations obtained by minimizing the travel-time residuals of *P* and *S* waves suffer from significant depth error. It is conventionally assumed that error in depth only trades off with the origin time. However, when events' epicenters were relocated with fixed depths, constrained by a combination of teleseismic *P*-wave modeling and regional waveform inversion, using both teleseismic and regional travel times, and a regional crustal model, the locations of seismic events can move significantly with respect to those determined from hypocentral inversions with unconstrained source depths. This was observed in central Asia (WOODS *et al.*, 1999), Pamir Hindu-Kush (SAIKIA *et al.*, 1996a; 1997) and Tibet (ZHAO and HELMBERGER, 1991a). The interaction of the Moho-reflected *PmP* phase with the depth-related, Moho-reflected *sPmP* and *pPmP* phases also provides a useful tool in determining source depth (SAIKIA and HELMBERGER, 1993; LANGSTON, 1996). This is very useful for small events which may be recorded with a strong signal-to-noise ratio only at regional distances. Even if such events are recorded at a few teleseismic stations, the depth phases are not likely to become distinctive, even at short-period. However, it will be possible to estimate the source depth for such small events by analyzing the interaction of the regional phases (SAIKIA and HELMBERGER, 1993; LANGSTON, 1996).

We also discussed a method to locate earthquakes, using an adaptive grid-search algorithm, which was applied to locate earthquakes in the Hindu-Kush region (SAIKIA *et al.*, 1996; ZHU *et al.*, 1997). A similar adaptive grid search technique has also recently been applied to test the hypothesis of locating earthquakes using a sparse network in southern California (ROMANOWICZ *et al.*, 1999). Our proposed method requires that the source depth be estimated by independent means. For smaller ($M_w < 5$) events this involves regional broadband waveform inversion, which in turn requires modeling of the surrounding regional crustal waveguide.

Acknowledgements

The authors are grateful to Donald V. Helmberger of the Caltech Seismological Laboratory for insightful comments at various phases of our investigations. We also thank the anonymous reviewers and Dr. Howard J. Patton, Los Alamos National Laboratory for their comments which enhanced this paper. This study is supported by the following agencies: Defense Special Weapons Agency (DSWA), Defense Threat Reduction Agency (DTRA) and Department of Energy (DoE); their support is gratefully acknowledged.

References

ADAMS, R. D., HUGHES, A. A., and MCGREGOR, D. M. (1982), *Analysis Procedures at the International Seismological Centre*, Phys. Earth Planet. Int. *30*, 85–93.

AKI, K., and RICHARDS, P. G., *Quantitative Seismology – Theory and Methods*, vol. I (W. H. Freeman, New York 1980).

ARVIDSSON, R., and EKSTRÖM, G. (1998), *Global CMT Analysis of Moderate Earthquakes, $M_w \geq 4.5$, Using Intermediate-period Surface Waves*, Bull. Seismol. Soc. Am. *88*, 1003–1013.

BEN-MENAHEM, A., SMITH, S. W., and TENG, T. (1965), *A Procedure for Source Studies from Spectrums of Long-period Body Waves*, Bull. Seismol. Soc. Am. *55*, 203–235.

BILLINGS, S. D., SAMBRIDGE, M. S., and KENNETT, B. L. N. (1994), *Errors in Hypocenter Location: Picking, Model, and Magnitude Dependence*, Bull. Seismol. Soc. Am. *84*, 1978–1990.

BONDAR, ISTVAN (1998), *PIDC Ground Truth Event Bulletin Technical References*, Technical Report CMR-98/21, Center for Monitoring Research, Alexandria, VA.

DI LUCCIO, F., JONES, L., ZHU, L., and HAUKSSON, E. (1998), *Comparison of Waveform Inversions and Arrival Time Techniques to Resolve Earthquake Depths and Focal Mechanisms in Southern California*, EOS, vol. *79*, no. 45, 603 pp.

DREGER, D. S., and HELMBERGER, D. V. (1993), *Determination of Source Parameters at Regional Distances with Single Station or Sparse Network Data*, J. Geophys. Res. *98*, 8107–8126.

DZIEWONSKI, A. M., CHOU, T.-A., and WOODHOUSE, J. H. (1981), *Determination of Earthquake Source Parameters from Waveform Data for Studies of Global and Regional Seismicity*, J. Geophys. Res. *86*, 2825–2852.

DZIEWONSKI, A. M., and STEIM, J. (1982), *Dispersion and Attenuation of Mantle Waves through Waveform Inversion*, Geophys. J. R. Astron. Soc. *70*, 503–527.

ENGDAHL, E. R., VAN DER HILST, R., and BULAND, R. (1998), *Global Teleseismic Earthquake Relocation with Improved Travel Times and Procedures for Depth Determination*, Bull. Seismol. Soc. Am. *88*, 722–743.

ENGDAHL, E. R., *Development of an archive of seismic ground truth events globally in support of monitoring under the CTBT*. In *Proc. 20th Ann. Seis. Res. Symp.* (eds. Fantroy, J., Heatley, D., Warren, J., Chavez, F., and Meade, C.) (DoD/NTP, Alexandria, VA, and DoE, Washington, DC 1998) pp. 11–18.

GAUR, V. K., and PRIESTLEY, K. F. (1997), *Shear-wave Velocity Structure beneath the Archean Granites around Hyderabad, Inferred from Receiver Function Analysis*, Academy Proceed. Earth. Planet. Sci. *106*, 1–8. IIS, Bangalore, India.

GEE, L. S., NEUHAUSER, D. S., DREGER, D. S., and PASYANOS, M. E. (1996), *Real-time Seismology at UC Berkeley: The Rapid Earthquake Data Integration Project*, Bull. Seismol. Soc. Am. *86*, 936–945.

GHOSE, S. M., HAMBURGER, W., and AMMON, C. J. (1998), *Source Parameters of Moderate-sized Earthquakes in the Tien Shan, Central Asia from Regional Moment Tensor Inversion*, Geophys. Res. Lett. *25*, 3181–3184.

GUPTA, I. N., ZHANG, T. R., and WAGNER, R. A. (1998), *Low-frequency Lg from NTS and Kazakh Nuclear Explosions – Observations and Interpretation*, Bull. Seismol. Soc. Am. *87*, 1115–1125.

HASKELL, N. A. (1962), *Crustal Reflection of Plane P and SV Waves*, J. Geophys. Res. *67*, 4751–4767.

HELMBERGER, D. V., and ENGEN, G. R. (1980), *Modeling the Long-period Body Waves from Shallow Earthquakes at Regional Ranges*, Bull. Seismol. Soc. Am. *70*, 1699–1714.

HELMBERGER, D. V. (1973), *Numerical Modeling of Long-period Body Waves from Seventeen to Forty Degrees*, Bull. Seismol. Soc. Am. *63*, 633–646.

HELMBERGER, D. V. (1972), *Long-period Body-wave Propagation from 4° to 13°*, Bull. Seismol. Soc. Am. *62*, 325–341.

HOLT, W. E., and WALLACE, T. C. (1989), *Crustal Thickness and Upper Mantle Velocities in the Tibetan Plateau Region from the Inversion of Regional P_{nl} Waveforms: Evidence for a Thick Upper Mantle Lid Beneath Southern Tibet*, J. Geophys. Res. *95*, 12,499–12,525.

HUDSON, J. A. (1969), *A Quantitative Evaluation of Seismic Signals at Teleseismic Distances. I: Radiation from Seismic Sources*, Geophys. J. R. Astron. Soc. *18*, 233–249.

LANGSTON, C. B. (1996), *The SsPmp Phase in Regional Wave Propagation*, Bull. Seismol. Soc. Am. *86*, 133–143.

LANGSTON, C. A., and HELMBERGER, D. V. (1975), *A Procedure for Modeling Shallow Dislocation Sources*, Geophys. J. R. Astron. Soc. *42*, 117–130.

LEFEVRE, L. V., and HELMBERGER, D. V. (1989), *Upper Mantle P-velocity Structure of the Canadian Shield*, J. Geophys. Res. *94*, 17,749–17,765.

LEVSHIN, A. L. (1985), *Effects of Lateral Inhomogeneities on Surface Wave Amplitude Measurements*, Annales Geophysicae *3*, 511–518.

MOONEY, W. D., and LI, S., *Crustal structure of China from deep seismic sounding profiles*. In *Proc. 19th Ann. Seis. Res. Symp. on Monitoring a CTBT* (eds. Shore, M., Jih, R. S., Dainty, A., and Erwin, J.) (DSWA, Alexandria, VA; AFTAC, Patrick AFB, FL; and DoE, Washington, DC 1997) pp. 104–114.

MOONEY, W. D., LASKE, G., and MASTER, T. G. (1998), *Crust 5.1: A Global Crustal Model at 5° × 5°*, J. Geophys. Res. *103*, 727–747.

NAKANISHI, I., and KANAMORI, H. (1982), *Effects of Lateral Heterogeneity and Source Process Time on the Linear Moment Tensor Inversion of Long-period Rayleigh Waves*, Bull. Seismol Soc. Am. *72*, 2063–2080.

NELSON, M. R., MCCAFFREY, R., and MOLNAR, P. (1987), *Source Parameters for 11 Earthquakes in the Tien Shan, Central Asia, Determined by P and SH Waveform Inversion*, J. Geophys. Res. *92*, 12,629–12,648.

PATTON, H. J. (1998), *Bias in the CMT Moment for Central Asian Earthquakes: Evidence from Regional Data*, J. Geophys. Res. *89*, 6929–6940.

PATTON, H. J., and WALTER, W. R. (1993), *Regional Moment: Magnitude Relations for Earthquakes and Explosions*, Geophys. Res. Lett. *20*, 277–280.

PRIESTLEY, K. F., and PATTON, H. J. (1997), *Calibration of $m_b(P_n)$, $m_b(L_g)$ Scales and Transportability of the M_0:m_b Discriminant to New Tectonic Regimes*, Bull. Seismol. Soc. Am. *87*, 1083–1099.

RITMESA, J., and LAY, T. (1993), *Rapid Source Mechanism Determination of Large $M_w > 5$ Earthquakes in the Western United States*, J. Geophys. Res. *20*, 1611–1614.

RODGER, A. J., and SCHWARTZ, S. Y. (1998), *Lithospheric Structure of the Qiangjang Terrane, Northern Tibetan Plateau, from Complete Regional Waveform Modeling, Evidence for Partial Melt*, J. Geophys. Res. *103*, 7137–7152.

RÖHM, A. H. E., TRAMPERT, J., PAULSSEN, H., and SNIEDER, R. K. (1999), *Bias in Reported Seismic Arrival Times Deduced from the ISC Bulletin*, Geophys. J. Int. *137*, 163–174.

ROMANOWICZ, B., DREGER, D., UHRHAMMER, R., and PASYANOS, M., *Earthquake locations using sparse regional broadband data*. In *Workshop on IMS Location Calibration, Tech. Doc.* (ed. Ringdal, F.) (NORSAR, Oslo 1999).

ROMNEY, C., BROOKS, B. G., MANSFIEKL, R. H., CARDER, D. S., JORDAN, J. N., and GORDON, D. W. (1962), *Travel Times and Amplitude of Principal Body Phases Recorded from Gnome*, Bull. Seismol. Soc. Am. *52*, 1057–1074.

RUSSIAN FEDERATION/UNITED STATES CALIBRATION WORKING GROUP (1998), *Evaluation of the IMS location accuracy of large chemical and nuclear explosions in Northern Eurasia and North America using regional and global P_n Travel-time Tables*. In *Proc. 20th Ann. Seis. Res. Symp.* (eds. Fantroy, J., Heatley, D., Warren, J., Chavez, F., and Meade, C.) (DoD/NTP, Alexandria, VA, and DoE, Washington, DC 1998) pp. 104–113.

RYABOY, V. (1999), *Application of 3-D crustal and upper-mantle velocity models in North America for location of regional seismograms*. In *Workshop on IMS Location Calibration, Tech. Doc.* (ed. Ringdal, F.) (NORSAR, Oslo 1999).

SAIKIA, C. K., ZHU, L., WOODS, B. B., and THIO, H. K., *Path calibration and source characterization in and around India.* In *Proc. 21st Ann. Seis. Res. Symp.: Technologies for Monitoring the CTBT*, vol. I, LA-UR-99-4700 (DoD, Alexandria, VA; and DoE, Washington, DC 1999) pp. 243–253.

SAIKIA, C. K., WOODS, B. B., THIO, H. K., SONG, X., and HELMBERGER, D. V. (1997), *Path Calibration, Source Estimation and Regional Discrimination for the Middle East and the Mediterranean*, Scientific Rept. 3, DSWA-TR-98-70 (DSWA, Special Programs Office, Alexandria, VA) 110 p.

SAIKIA, C. K., WOODS, B. B., THIO, H. K., ZHU, L., and HELMBERGER, D. V. (1996a), *Path Calibration, Source Estimation and Regional Discrimination for the Middle East: An Application to the Hindu-Kush Region*, Scientific Rept. No 1, PL-TR-96-2069 (Phillips Laboratory, HAFB, MA) 130 pp.

SAIKIA, C. K., THIO, H. K., WOODS, B. B., SONG, X., ZHU, L., and HELMBERGER, D. V. (1996b), *Path Calibration, Source Estimation and Regional Discrimination for the Middle East: An Application to the Hindu-Kush and Western Mediterranean Regions*, Scientific Rept. No. 2, PL-TR-96-2307 (Phillips Laboratory, HAFB, MA) 205 pp.

SAIKIA, C. K. (1994), *Modified Frequency-wavenumber Algorithm for Regional Seismograms using Filon's Quadrature – Modeling of L_g Waves in Eastern North America*, Geophys. J. Int. *118*, 142–158.

SAIKIA, C. K., and HELMBERGER, D. V. (1993), *Broadband Modeling of Regional Seismograms in Asia and Development of Low-magnitude Event Discrimination*, Final Rept. WCCP-R-93-04 (Phillips Laboratory, Kirtland AFB, NM) 135 pp.

SAIKIA, C. K., and BURDICK, L. J. (1991), *Fine Structure of P_{nl} Waves from Explosions*, J. Geophys. Res. *96*, 14,383–14,401.

SAIKIA, C. K., and HERRMANN, R. B. (1985), *Application of Waveform Modeling to Determine Focal Mechanisms of Four 1982 Miramichi Aftershocks*, Bull. Seismol. Soc. Am. *75*, 1021–1041.

SINGH, S. K., SANTOYO, M. A., and PACHECO, J. (1995). *Intermediate-depth Earthquakes in Central Mexico: Implications for Plate Waves*, Geophys. Res. Lett. *22*, 527–530.

SIPKIN, S. A. (1982), *Estimation of Earthquake Source Parameters by the Inversion of Waveform Data: Synthetic Seismograms*, Phys. Earth Planet. Int. *30*, 242–259.

SONG, X. J., HELMBERGER, D. V., and ZHAO, L. (1996), *Broad-band Modeling of Regional Seismograms: The Basin and Range Crustal Structure*, Geophys. J. Int. *125*, 15–29.

STEIN, S., and WIENS, D. A. (1986), *Depth Determination for Shallow Teleseismic Earthquakes: Methods and Results*, Rev. Geophys. *24*, 806–832.

TAYLOR, S. R. (1995), *Analysis of High-frequency P_g/L_g Ratios from NTS Explosions and Western U.S. Earthquakes*, Bull. Seismol. Soc. Am. *86*, 1042–1053.

TENG, T., and BEN-MENAHEM, A. (1962), *Mechanisms of Deep Earthquakes from Spectrums of Isolated Body-wave Signals. 1. The Banda Sea Earthquake of March 21, 1964*, J. Geophys. Res. *70*, 5157–5170.

THIO, H. K., SONG, X., SAIKIA, C. K., HELMBERGER, D. V., and WOODS, B. B. (1999), *Seismic Sources and Structure Estimation in the Western Mediterranean Using a Sparse Broadband Network*, J. Geophys. Res. *104*, 845–861.

THIO, H. K., and KANAMORI, H. (1995), *Moment Tensor Inversions for Local Earthquakes Using Surface Waves Recorded at TERRAscope*, Bull. Seismol. Soc. Am. *85*, 1021–1038.

VÖGFORD, K. S. (1997), *Effects of Explosion Depth and Earth Structure on the Excitation of L_g Waves: S* Revisited*, Bull. Seismol. Soc. Am. *87*, 1100–1114.

WALLACE, T. C., HELMBERGER, D. V., and MELLMAN, G. R. (1981), *A Technique for the Inversion of Regional Data in Source Parameter Studies*, J. Geophys. Res. *86*, 1679–1685.

WALTER, W. R., MAYEDA, K. M., and PATTON, H. J. (1995), *Phase and Spectral Ratio Discrimination between NTS Earthquakes and Explosions. Part I: Empirical Observations*, Bull. Seismol. Soc. Am. *85*, 1050–1067.

WOODS, B. B., SAIKIA, C. K., and THIO, H. K. (1999), *Extending regional waveform modeling to shorter periods to improve the location and identification of small seismic sources.* In *Proc. 21st Ann. Seis. Res. Symp.: Technologies for Monitoring the CTBT*, vol. I, LA-UR-99-4700 (DoD, Alexandria, VA; and DoE, Washington, DC 1999) pp. 303–311.

WOODS, B. B., SAIKIA, C. K., THIO, H. K., and PATTON, H. J. (1998), *Focal Depths and Source Parameters for Earthquakes in Northwest China*, Tech. Rept. (LANL, Los Alamos).

WOODS, B. B., and HARKRIDER, D. G. (1995), *Determining Surface Wave Magnitudes from Regional NTS Data*, Geophys. J. Int. *120*, 474–498.

Woods, B. B., Helmberger, D. V., and Kedar, S. (1993), *M_L:M_0 as a Regional Seismic Discriminant*, Bull. Seismol. Soc. Am. *83*, 1167–1183.

Zandt, G., Beck, S. L., Ruppert, S. R., Ammons, C. J., Rock, D., Minaya, E., Wallace, T. C., and Silver, P. G. (1996), *Anomalous Crust of the Bolivian Altiplano, Central Andes: Constrains from Broadband Regional Seismic Waveforms*, Geophys. Res. Lett. *23*, 1159–1162.

Zhao, L.-S., and Helmberger, D. V. (1994), *Source Estimation from Regional Broadband Seismograms*, Bull. Seismol. Soc. Am. *84*, 91–94.

Zhao, L.-S., and Helmberger, D. V. (1992), *Source Retrieval from Broadband Regional Seismograms: Hindu Kush Region*, Phys. Earth Planet. Int. *78*, 69–95.

Zhao, L.-S., and Helmberger, D. V. (1991a), *Geophysical Implications from Relocations of Tibetan Earthquakes: Hot Lithosphere*, Geophys. Res. Lett. *18*, 2205–2208.

Zhao, L.-S., and Helmberger, D. V. (1991b), *Broadband Modeling along a Shield Path. Harvard Recordings of the Saguenay Earthquake*, Geophys. J. Int. *105*, 301–312.

Zhu, L., Helmberger, D. V., Saikia, C. K., and Woods, B. B. (1997), *Regional Waveform Calibration in the Pamir-Hindu Kush Region*, J. Geophys. Res. *102*, 22,799–33,813.

Zhu, L., and Helmberger, D. V. (1996), *Advancement of Source Retrieval from Broadband Regional Seismograms*, Bull. Seismol. Soc. Am. *86*, 1634–1641.

(Received September 13, 1999, revised May 12, 2000, accepted May 31, 2000)